Environment Impact on Reproductive Health

Roberto Marci

Editor

Environment Impact on Reproductive Health

A Translational Approach

 Springer

Editor
Roberto Marci
DEPARTMENT of TRANSLATIONAL MEDICINE
University of Ferrara
Ferrara, Italy

ISBN 978-3-031-36493-8   ISBN 978-3-031-36494-5   (eBook)
https://doi.org/10.1007/978-3-031-36494-5

This work was supported by FINOX SA

This Springer imprint is published by the registered company Springer Nature Switzerland AG
The registered company address is: Gewerbestrasse 11, 6330 Cham, Switzerland

Contents

Chapter 1
Introduction to Environmental Pollutants and Human Exposure

Donatella Caserta, Flavia Costanzi, Maria Paola De Marco, Aris Besharat, and Ilary Ruscito

The new concept of health, developed in the recent years, considers the person's well-being more heterogeneously. A new model that considers the relationship between human health and the environment has strongly emerged during the last three decades. Our state of well-being is continually threatened by a series of internal and external disturbing factors, which tend to move the body away from a condition of homeostasis. The awareness of the indissoluble link between human health and the environment is increasingly widespread.

Climate change, loss of biodiversity, poor air quality, desertification, deforestation, often irreversible contamination of groundwater and the food chain, and exponential growth of the electromagnetic field (EMF) due to over-the-air communications are the direct consequences of a focused "growth" of globalized economics.

Pollution is a problem that affects organisms, especially the developing ones, such as embryos and children, in consideration of the vulnerability of their status. Prolonged exposure to minimal quantities of pollutants can progressively alter the functioning of cells, tissues, and organs, essentially interfering with deoxyribonucleic acid (DNA) expression. Unfortunately, the absolute limits of toxicity and tolerability of many pollutants are not yet known.

Today, we are detecting a rapid and progressive transformation of the molecular composition of the ecosphere and, in particular, the rapid production and diffusion of atmospheric pollutants (ultra-fine particles, heavy metals, and radiation). The World Health Organization (WHO), indeed, has recognized that environmental factors cause around 24% of diseases worldwide, and more than 33% of diseases in children under the age of 5 years are due to environmental factors [1].

D. Caserta (✉) · F. Costanzi · M. P. De Marco · A. Besharat · I. Ruscito
Gynecology Unit, Department of Medical and Surgical Sciences and Translational Medicine, Sant' Andrea University Hospital of Rome, Sapienza University of Rome, Rome, Italy
e-mail: donatella.caserta@uniroma1.it; ilary.ruscito@uniroma1.it

© The Author(s) 2023
R. Marci (ed.), *Environment Impact on Reproductive Health*,
https://doi.org/10.1007/978-3-031-36494-5_1

1

The pediatric age is much more sensitive than adults to the effect of pollutants. Children under the age of 5 years, who represent only 12% of the population, contract more than 40% of health diseases compared with adults [2]. The role of the environment has been recognized over the years in the pathophysiology of numerous pathologies. The WHO, for example, has defined obesity and diabetes as a real pandemic. Even in Italy, the phenomenon is assuming worrying proportions, particularly in children in primary schools whose obesity rate, which was 7% between 1976 and 1980 and reached 21% in 2015–2017 [3]. Childhood obesity is generally considered a systemic and multifactorial pathology, determined by several causes (excessive intake of food, sedentary lifestyle, and genetic predisposition). However, it is increasingly evident that these factors cannot alone explain the alarming phenomena, such as the recent, dramatic increase in cases and the constant anticipation of the age of onset of related diseases (in particular insulin resistance and type 2 diabetes mellitus) and the insufficient efficacy of individual therapeutic strategies. Recent studies have shown that early exposure to many pollutants can induce obesity and type 2 diabetes [4].

Numerous types of pollutants contribute to air pollution. Transportation, industrial and agricultural activities, energy production, and waste disposal plants emit thousands of tons of pollutants into the atmosphere every day, and they are the leading causes of environmental pollution. The main pollutants studied are ground-level ozone, heavy metals, polyaromatic hydrocarbons, and particulate, which constitute a serious threat to our health [5].

Air pollution affects health in various ways: The subject's health conditions, age, and duration of exposure are the main factors that affect the way through which pollutants influence our health. Air pollutants can have effects on both the respiratory tract and other organs, inducing or contributing to the onset of numerous diseases, including respiratory diseases such as asthma (especially in pediatric age), reduction in the development and functions of the apparatus respiratory tract, arteriosclerosis and cardiovascular diseases, neurodegenerative diseases, tumors, and infertility.

Human beings can be exposed to environmental contaminants through the air, water, food, and soil. Environmental contaminants can be divided into three broad categories: biological agents, chemical agents, and radiation (Fig. 1.1).

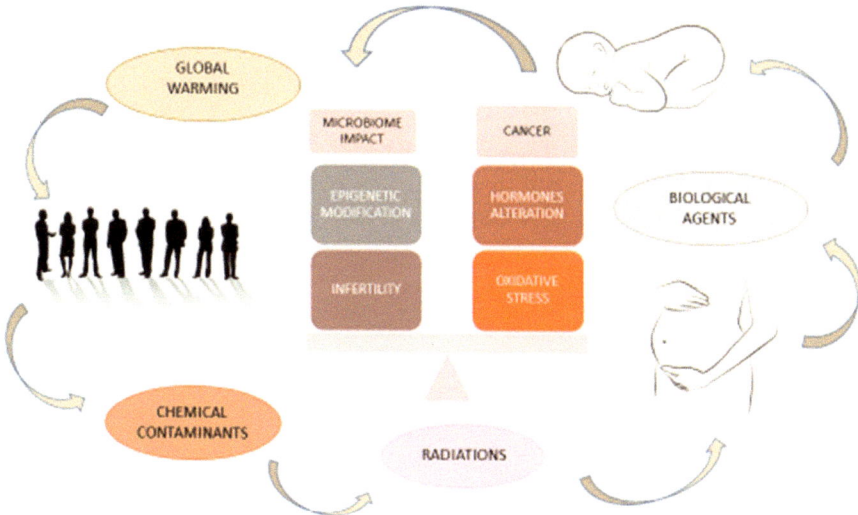

Fig. 1.1 Pollution and interaction with human health

1.1 The Biological Agents

Biological agents are living organisms such as bacteria, viruses, and fungi that are naturally present in the environment and can be responsible for gastrointestinal, allergic, and respiratory diseases. [6].

The biological agent is defined, according to the current legislation (European Directives 90/679/EEC, 93/88/EEC, and 2000/54/EC) [7], as "any microorganism, even if genetically modified, cell culture and human endoparasite, which could cause infections, allergies or poisoning".

The onset of diseases depends on many factors related to the characteristics of the single biological agent, the conditions of the subject exposed, the environmental conditions, and the methods of exposure or contact.

Although there is extensive information about the dangers of chemical and physical agents, the same cannot be said for biological agents [8].

Biological agents are infectious agents that include bacteria, rickettsiae, viruses, yeasts, molds, and single and multicellular parasites.

Each infectious agent species can have subtypes, strains, and variants that differ from the parental pathogenic potential, host specificity, transmissibility, and sensitivity to antimicrobial agents.

According to Italian Legislative Decree 81/08 Title X [9], biological agents are divided into four groups according to the risk of infection:

1. Biological agent of group 1: an agent that is unlikely to cause disease in human subjects.
2. Biological agent of group 2: an agent that can cause disease in human subjects and pose a risk to workers; it is unlikely to spread to the community; effective prophylactic or therapeutic measures are usually available.
3. Biological agent of group 3: an agent that can cause severe diseases in human subjects and constitutes a severe risk to workers; the biological agent can spread throughout the community, but effective prophylactic or therapeutic measures are usually available.
4. Biological agent of group 4: a biological agent that can cause severe diseases in human subjects and constitutes a severe risk to workers and can present a high risk of propagation in the community; there are usually no effective prophylactic or therapeutic measures available.

1.2   Radiations

Radiations are waves or particles of energy to which humans, plants, and animals are exposed. We are exposed to natural and artificial sources of high-energy ionizing radiation and low-energy nonionizing radiation, such as ultraviolet rays and electromagnetic fields.

Ionizing radiations (X-, gamma-, alpha-, beta-, and neutron rays) are electromagnetic waves or corpuscular rays with enough energy to release electrons from atoms as they pass through matter. These modifying atoms, called ions, can induce chemical reactions that cause a biological damage. At the cellular level, radiation can induce damage to DNA molecules [10]. This damage can cause cells to die or be adequately repaired by the cell's protective mechanisms. However, it can also happen that an incorrect repair is made, which still produces a viable cell. Radiation-mediated cancer is assumed to be induced by the latest mechanism.

There are two types of radiation effects on the body. The long-term effects are due to mutations produced at the cellular level. These include the onset of cancer in irradiated people and malformations in their descendants. For these effects, it is not possible to identify a threshold; the same could theoretically occur even with a shallow dose. The risk of cancer development is hypothesized to increase linearly with dose. It should be noted that children are more sensitive than adults to ionizing radiation [11].

On the other hand, the immediate effects are linked to the destruction of a large number of cells by radiation, which leads to the loss of functionality of an organ. These effects, for which there is a threshold dose (a minimum dose with which they occur), occur only at high doses. These effects include destroying the active bone marrow, intestinal mucosa, skin burns, and sterilization. [12].

Exposure to electromagnetic fields (EMFs) represented another source of health risks. For several decades, it has been known that there has been an increase in leukemia among residents near EMF [13]. In 2011, the International Agency for Research on Cancer (IARC), the European cancer research agency, definitively included cell phones and radio frequency (wireless) electromagnetic fields among the "Group 2B" carcinogens, which indicates a "possible" carcinogenic risk on humans. In addition, in this case, it is imperative to reduce the exposure of women during pregnancy and of developing subjects. In this case, too, children represent the most risk category for various reasons:

- The exposure is destined to last for decades.
- The brain is in the process of a functional organization (synapses and circuits).
- The blood–brain barrier is very permeable.
- The part of brain tissue exposed is, in proportion, much more significant than in adults.
- The bone tissue is less thick.
- The tissue has higher water content and higher cerebral concentration and therefore conducts and absorbs more energy.

1.3 Chemical Contaminants

Chemical contaminants include organic and inorganic compounds of natural and human origin. Organic compounds contain carbon, usually combined with hydrogen and other elements such as fluorides, chlorides, bromides, iodides, nitrogen, sulfur, and phosphorus. Examples of organic compounds are pesticides, polychlorinated biphenyls (PCBs), polycyclic aromatic hydrocarbons (PAHs), and trihalomethanes (THMs). Inorganic compounds include air pollutants, such as ozone, nitrogen oxides, and sulfur dioxide; metals; lead; and fluorides. Chemicals occur naturally in our environment due to weathering and erosion and are also released by human activities, such as agriculture, industry, power generation, transport, and the use and disposal of consumer products. Exposure to high levels of chemical contaminants can result in a variety of health effects, including allergies; skin and eye irritation; heart, respiratory, reproductive, kidney, or neurological problems; and cancer [14].

In particular, ozone is an odorless and a colorless gas present both in the earth's upper atmosphere (stratosphere) and at the ground level (troposphere). The ozone present at high altitudes constitutes a protective band from solar radiation. In the lower layers of the atmosphere (tropospheric ozone), on the other hand, it behaves as a pollutant, constituting a severe problem for public health. Ground-level ozone is a secondary pollutant produced by the reaction of oxygen with nitrogen dioxide (NO_2) and the contribution of volatile organic compounds in intense solar radiation and high temperatures [15].

Particulate matter is a mixture of solid and liquid particles suspended in the air that reaches its maximum concentration in winter. It includes particles of various sizes into which dust, earth, materials from roads, pollen, molds, spores, bacteria, viruses, and thousands of chemicals can converge and causes respiratory, cardiovascular, and neurodegenerative diseases. The primary sources of particulate matter are vehicular traffic, industrial activities, and heating systems. The most dangerous particulate fraction is the particulate resulting from the product of thermochemical reactions in foundries, cement factories, steel mills, waste incinerators, diesel engines, and other combustion processes. Thanks to its submicroscopic dimensions, the particulate passes through the alveoli and penetrates the arteries, the brain, and the cell nuclei, opening the way to many chronic degenerative, inflammatory, and cancerous diseases [16].

1.3.1 Endocrine Disruptors

Since the 1990s, an increasingly growing scientific interest has been placed in studying endocrine disruptors (EDs). In 2016, the UN Environment Programme commissioned the International Panel on Chemical Pollution (IPCP) to develop three reports relating to the various EDs and their mechanisms of action [17–19]. In June 2018, the European Food Safety Authority (EFSA) and the European Chemicals Agency (ECHA) published a guide to identifying substances with the characteristics of EDs [20]. There are numerous substances considered probable EDs. Currently, 107 substances have passed the complete evaluation process to be identified as EDs (Table 1.1) as the European Union (EU) regulated under the Plant Protection Products Regulation (PPPR), of the Biocides Products Regulation (BPR) or Registration, Evaluation, Authorisation and Restriction (REACH) (the list of candidates and authorizations). EDs include a wide range of chemicals that can alter the hormonal balance of living organisms, including humans. EDs interact with the standard biochemical signals released by the glands of our body, which are responsible for regulating extremely delicate functions: immune, endocrine, metabolic, reproductive, and neuropsychic. The pathologies induced by frequent exposure to minimal doses of EDs are thyroid and neurodevelopmental disorders, abortion, infertility, genital and reproductive anomalies, endometriosis, obesity and type 2 diabetes, tumors, and immune-mediated diseases [21]. According to the Istituto Superiore Di Sanità, "an ED is an exogenous substance, or a mixture, which alters the functionality of the endocrine system, causing adverse effects on the health of an organism, or of its progeny or (under) population".

EDs act subtly, even at minimal doses, especially in crucial stages of development, such as intrauterine life [22–26] or childhood [27]. Exposure to EDs can also alter gametes and implantation mechanisms [28]. Harmful effects have been found in various female pathologies [29] and neoplasms such as endometrial cancer [30, 31].

The impact of EDs on the environment can be considered for their ubiquitous presence, in some cases, their persistence and their potential effects on living beings.

The primary sources of EDs' environmental risk are behaviors that do not comply with current legislation, industrial processing and disposal processes, and incorrect

Table 1.1 Substances identified as endocrine disruptors at EU level (data latest update 2022/4: https://edlists.org/)

Name and abbreviation	Health effects	Environmental effects
(±)-1,7,7-trimethyl-3-[(4-methylphenyl)methylene]bicyclo[2.2.1] heptan-2-(4-MBC)	X	X
(1R,3E,4S)-1,7,7-trimethyl-3-(4- methylbenzylidene) bicyclo[2.2.1]heptan-2-one (4-MBC)	X	
(1R,3Z,4S)-1,7,7-trimethyl-3-(4- methylbenzylidene) bicyclo[2.2.1]heptan-2-one (4-MBC)	X	
(1R,4S)-1,7,7-trimethyl-3-(4-methylbenzylidene)bicyclo[2.2.1] heptan-2-one (4-MBC)	X	
(1S,3E,4R)-1,7,7-trimethyl-3-(4- methylbenzylidene) bicyclo[2.2.1]heptan-2-one (4-MBC)	X	
(1S,3Z,4R)-1,7,7-trimethyl-3-(4-methylbenzylidene) bicyclo[2.2.1]heptan-2-one (4-MBC)	X	
(3E)-1,7,7-trimethyl-3-(4-methylbenzylidene)bicyclo[2.2.1] heptan-2-one (4-MBC)	X	
1,7,7-trimethyl-3-[(4-methylphenyl)methylene]bicyclo[2.2.1] heptan-2-one; 3-BC		X
14-(nonylphenoxy)-3,6,9,12-tetraoxatetradecan-1-ol		X
17-(4-nonylphenoxy)-3,6,9,12,15-pentaoxaheptadecan-1-ol		X
2-(4-nonylphenoxy)ethanol		X
2-[2-(4-nonylphenoxy)ethoxy]ethanol		X
2-[2-[2-[2-(4-nonylphenoxy)ethoxy]ethoxy]ethoxy]ethanol		X
2-[4-(1,1,3,3-tetramethylbutyl)phenoxy]ethanol		X
2-[4-(3,6-dimethylheptan-3-yl)phenoxy]ethanol		X
2-{2-[4-(2,4,4-trimethylpentan-2-yl)phenoxy]ethoxy}ethanol		X
2-{2-[4-(3,6-dimethylheptan-3-yl)phenoxy]ethoxy}ethanol		X
20-(4-nonylphenoxy)-3,6,9,12,15,18-hexaoxaicosan-1-ol		X
20-[4-(1,1,3,3-tetramethylbutyl) phenoxy]-3,6,9,12,15,18-hexaoxaicosan-1-ol		X
23-(nonylphenoxy)-3,6,9,12,15,18,21-heptaoxatricosan-1-ol		X
26-(4-Nonylphenoxy)-3,6,9,12,15,18,21,24- octaoxahexacosan -1-ol		X
26-(nonylphenoxy)-3,6,9,12,15,18,21,24-octaoxahexacosan-1-ol		X
3,6,9,12-Tetraoxatetradecan-1-ol, 14-(4-nonylphenoxy)-		X
4-(1-ethyl-1-methylhexyl)phenol		X
4-(1-Ethyl-1,3-dimethylpentyl)phenol		X
4-(1-Ethyl-1,4-dimethylpentyl)phenol		X
4-(1,1,3,3-tetramethylbutyl)phenol		X
4-(1,1,5-Trimethylhexyl)phenol		X
4-(2-methylhexan-2-yl)phenol		X
4-(2,2-dimethylpentan-3-yl)phenol		X
4-(2,3-dimethylpentan-2-yl)phenol		X
4-(2,3,3-trimethylbutan-2-yl)phenol		X

(continued)

Table 1.1 (continued)

Name and abbreviation	Health effects	Environmental effects
4-(2,4-dimethylpentan-2-yl)phenol		X
4-(2,4-dimethylpentan-3-yl)phenol		X
4-(3-ethylheptan-2-yl)phenol		X
4-(3-ethylpentan-3-yl)phenol		X
4-(3-ethylpentyl)phenol		X
4-(3-methylhexan-2-yl)phenol		
4-(3-methylhexan-3-yl)phenol		X
4-(3-methylhexyl)phenol		X
4-(3,3-dimethylpentan-2-yl)phenol		X
4-(4-methylhexan-2-yl)phenol		X
4-(4-methylhexyl)phenol		X
4-(4,4-dimethylpentan-2-yl)phenol		X
4-(5-methylhexan-2-yl)phenol		X
4-(5-methylhexan-3-yl)phenol		X
4-(5-methylhexyl)phenol		X
4-(heptan-2-yl)phenol		X
4-(heptan-3-yl)phenol		X
4-(heptan-4-yl)phenol		X
4-heptylphenol		X
4-isododecylphenol	X	X
4-Nonylphenol, branched, ethoxylated		X
4-Nonylphenol, branched, ethoxylated 1–2.5 moles ethoxylated		X
4-Nonylphenol, ethoxylated 1–2.5 moles ethoxylated		X
4-t-Nonylphenol-diethoxylate		X
4-tert-butylphenol		X
4,4′-(1-methylpropylidene)bisphenol: Bisphenol B	X	X
4,4′-isopropylidenediphenol; Bisphenol A	X	X
Benzyl butyl phthalate (BBP)	X	
Bis(2-ethylhexyl) phthalate (DEHP)	X	X
Butyl 4-hydroxybenzoate; Butylparaben	X	
Cholecalciferol	X	X
Dibutyl phthalate (DBP)	X	
Dicyclohexyl phthalate (DCHP)	X	
Diisobutyl phthalate (DIBP)	X	
Formaldehyde, reaction products with branched and linear heptylphenol, carbon disulfide and hydrazine		X
Formaldehyde, reaction products with phenol heptyl derivs. and 1,3,4-thiadiazolidine-2,5-dithione		X
Isononylphenol		X
Isononylphenol, ethoxylated		X
Mancozeb	X	X

Table 1.1 (continued)

Name and abbreviation	Health effects	Environmental effects
Nonylphenol		X
Nonylphenol, branched, ethoxylated		X
Nonylphenol, branched, ethoxylated 1–2.5 moles ethoxylated		X
Nonylphenol, ethoxylated		X
Nonylphenol, ethoxylated (10-EO)		X
Nonylphenol, ethoxylated (15-EO)		X
Nonylphenol, ethoxylated (6,5-EO)		X
Nonylphenol, ethoxylated (8-EO)		X
Nonylphenol, ethoxylated (EO = 10)		X
Nonylphenol, ethoxylated (EO = 4)		X
Nonylphenol, ethoxylated (polymer)		X
Nonylphenolpolyglycolether		X
p-(1-methyloctyl)phenol		X
p-(1,1-dimethylheptyl)phenol		X
p-(1,1-dimethylpropyl)phenol		X
p-isononylphenol		X
p-nonylphenol		X
Phenol, (tetrapropenyl) derivatives	X	X
Phenol, 4-(1-ethyl-1,2-dimethylpropyl)-		X
Phenol, 4-dodecyl, branched	X	X
Phenol, 4-isododecyl-	X	X
Phenol, 4-nonyl-, branched		X
Phenol, 4-nonyl-, phosphite (3:1)		X
Phenol, 4-tert-heptyl-		X
Phenol, dodecyl-, branched	X	X
Phenol, heptyl derivs		X
Phenol, nonyl-, branched		X
Phenol, p-isononyl-, phosphite (3:1)		X
Phenol, p-sec-nonyl-, phosphite		X
Phenol, tetrapropylene-	X	X
Poly (oxy-1,2-ethanediyl), alpha-(nonylphenyl)-omega-hydroxy-, branched		X
Poly(oxy-1,2-ethanediyl), a-(nonylphenyl)-w-hydroxy-		X
Poly(oxy-1,2-ethanediyl),α-[(1,1,3,3-tetramethylbutyl)phenyl]-ω-hydroxy-		X
Polyethylene glycol p-(1,1,3,3-tetramethylbutyl)phenyl ether		X
Tris (4-nonylphenol, branch) phosphorous acid ester		X
Tris(nonylphenyl) phosphite		

disposal of products containing plastics, glues, and paints. EDs characterized by high environmental persistence have a greater accumulation capacity in organisms.

The pollutants that can interfere with the function of sex hormones are of particular importance for all organisms, especially for their effects on the conservation of species and the maintenance of biodiversity. Transfer from one organism to another occurs through the food chain, increasing concentrations along the food chain. The presence of EDs in the environment is assessed through environmental monitoring using water samples, soil and sediments, and sentinel animals (indicator organisms). By comparing the data obtained, the state of environmental quality and the effects on organisms are determined. The PREVIENI project, for example, is an integrated study on the risk assessment of contaminants (perfluorooctane sulfonate [PFOS] and perfluorooctanoic acid [PFOA], di-2-ethylhexyl phthalate (DEHP) and its active metabolite mono-2-ethylhexyl phthalate (MEHP), and bisphenol A [BPA]) in ecosystems and the human population promoted by the Ministry of the Environment and the Protection of the Territory and the Sea and with the support of the World Wildlife Fund (WWF), Italy. EDs and the related biomarkers measured by the PREVIENI in sentinel organisms (four animal species representative of different habitats: earthworm, barbel, trout, and coot) represent values that can be associated with a state of exhibition background, contributing to the evaluation of reference values ([32]). The second research program of the study is based on the case–control approach for the assessment of fertility. The results show how factors associated with lifestyle and environmental exposure to EDs represent independent risk factors for human reproductive health [33].

Due to their complex nature, exposure to EDs can result in numerous clinical phenotypes: Understanding the mechanisms of action represents a constantly evolving field of research. The heterogeneity of the compounds helps to hinder the identification of a common mechanism of action. Suppose some substances exhibit some similar characteristics, such as molecular weight or the presence of specific highly reactive groups. In such case, no common characteristic can be identified among all the EDs, and therefore, the generalization, or even less the prediction, of the exact mechanism of action is impossible [34]. However, it is possible to divide the entire group of EDs into three broad categories:

- Hormone agonists, whose intake involves, directly or indirectly, a phenomenon of receptor activation (hyperstimulation); phytoestrogens and thyroid-stimulating substances are into this category.
- Hormone antagonists can interact with hormone receptors preventing, directly or indirectly, their physiological activation (inhibition); for example, substances with antiestrogenic and antiandrogenic actions belong to this group.
- -Metabolic modifiers can interfere with the physiological endogenous hormone secretion or other stages of the regular action of hormones, including their transport in the blood, intracellular preprocessing or postprocessing and, therefore, their degradation and elimination. Substances that stimulate hepatic metabolism or can chelate circulating hormones can be included in this category.

Despite significant advances in this field, research in the world of EDs has not so far led to irrefutable scientific evidence. The main criticalities encountered in the study of EDs are remarkably heterogeneous. Many substances have a short half-life, others have very low molecular weights, and others still act through metabolites; therefore, identifying these substances is not simple and requires adequate instrumentation and considerable clinical background. In many cases, the intake of

EDs does not have immediate effects. Exposure before puberty or during intrauterine life can, for example, lead to significant effects on fertility after many years [24, 26, 35]. The foundations of many adult pathologies could be traced back to the exposure to EDs during life in utero. Epigenetic effects, as in the case of the modulation of DNA expression by methylation, can be transmitted to generation, sometimes without giving apparent clinical manifestations in indirectly exposed subjects [36]. Most studies on EDs aim to identify the effects and mechanisms of action of individual substances. However, given the heterogeneous nature of EDs, it is safe to assume that their distribution is ubiquitous and that simultaneous exposure to multiple substances is anything but theoretical. The possible interactions between different substances are largely ignored: The mechanisms of action of several EDs may be additive or even synergistic, leading to more striking manifestations in the face of less exposure to the individual components. In addition, long-term exposure to low doses of EDs often makes it challenging to identify a causal link between the agents and the clinic.

   Another problem is related to animal models that are commonly used to evaluate the effects of some substances in vivo; however, it is difficult to provide an adequate estimate of how much the results can be extrapolated to the human being, if only for the temporal dynamics. Moreover, studies aimed at identifying the effects of individual substances are hardly representative of reality. Furthermore, some substances may be inert if studied individually and biologically active following the presence of other agents or within complex biological matrices such as blood. Furthermore, it should be considered that a similar exposure to a mixture of EDs can result in clinical manifestations of a different entity in male or female subjects. Despite all these limitations, animal models still represent the most reliable study method for EDs.

1.4   Global Warming

Man exerts an increasing influence on the Earth's climate and temperature through fossil fuels, deforestation, and cattle breeding.

   These activities add vast amounts of greenhouse gases to those naturally present in the atmosphere, fueling the greenhouse effect and global warming.

   The leading cause of climate change is the greenhouse effect. Some gases in the Earth's atmosphere act like glass in a greenhouse: They capture the sun's heat, preventing it from returning to space and causing global warming.

   Many of these gases occur naturally, but human activity increases the concentrations of some of them in the atmosphere, in particular carbon dioxide (CO_2), methane, nitric oxide, and fluorinated gases.

   The CO_2 produced by human activities is the main element of global warming. In 2020, its concentration in the atmosphere was 48% above the preindustrial level (before 1750).

Other greenhouse gases are emitted by human activity in smaller quantities. Nitric oxide, like CO_2, is a long-lived greenhouse gas that accumulates in the atmosphere for decades and even centuries. Methane is a more potent greenhouse gas than CO_2 but has a shorter atmospheric life.

Natural causes, such as changes in solar radiation or volcanic activity, are estimated to have contributed less than 0.1 °C to total warming between 1890 and 2010.

The period 2011–2020 was the hottest decade, with an average global temperature of 1.1 °C above the preindustrial levels in 2019. Human-induced global warming is currently increasing at a rate of 0.2 °C.

A 2 °C increase over the preindustrial temperature is associated with severe impacts on the natural environment and human health and well-being, including a much higher risk of dangerous and catastrophic changes in the global environment [37].

For this reason, the international community has recognized the need to keep warming well below 2 °C and to continue efforts to limit it to 1.5 °C.

References

1. Valent F, Little D, Bertollini R, et al. Burden of disease attributable to selected environmental factors and injury among children and adolescents in Europe. Lancet. 2004;363:2032–9. https://doi.org/10.1016/S0140-6736(04)16452-0.
2. Reinhardt U, Cheng T. The world health report 2000–health systems: improving performance. Bull World Health Organ. 2000;78(8):1064.
3. Breda J, McColl K, Buoncristiano M, et al. Methodology and implementation of the WHO European childhood obesity surveillance initiative (COSI). Obes Rev. 2021;22:e13215. https://doi.org/10.1111/obr.13215.
4. Neel BA, Sargis RM. The paradox of progress: environmental disruption of metabolism and the diabetes epidemic. Diabetes. 2011;60:1838–48. https://doi.org/10.2337/db11-0153.
5. Sexton K, Selevan SG, Wagener DK, et al. Estimating human exposures to environmental pollutants: availability and utility of existing databases. Arch Environ Health. 1992;47:398–407. https://doi.org/10.1080/00039896.1992.9938381.
6. Szulc J. Biological Agents. In: Respiratory protection against hazardous biological agents. Boca Raton: CRC Press; 2020.
7. European Parliament And Council. Gazzetta ufficiale delle Comunità europee. 2000. https://eurlex.europa.eu/legalcontent/IT/TXT/PDF/?uri=CELEX:32000L0054&from=NL
8. INTERNATIONAL AGENCY FOR RESEARCH ON CANCER. Biological agents volume 100 B. a review of human carcinogens. IARC Monogr Eval Carcinog Risks Hum. 2012;100:1–441.
9. Ministry of labor and social policies. Testo unico sulla salute e sicurezza sul lavoro. 2016. https://www.lavoro.gov.it/documenti-e-norme/studi-e-statistiche/Documents/Testo%20Unico%20sulla%20Salute%20e%20Sicurezza%20sul%20Lavoro/Testo-Unico-81-08-Edizione-Giugno%202016.pdf
10. Chaturvedi A, Jain V. Effect of ionizing radiation on human health. Int J Plant Environ. 2019;5:200–5. https://doi.org/10.18811/ijpen.v5i03.8.
11. Ozdemir F, Kargi A. Electromagnetic waves and human health. In: Electromagnetic waves. Rijeka: IntechOpen; 2011.

12. Martinez-López W, Hande MP. Health effects of exposure to ionizing radiation. In: Advanced security and safeguarding in the nuclear power industry. New York: Academic Press; 2020. p. 81–97.
13. Liu X, Xiangiang Y, Shujun Z, et al. The effects of electromagnetic fields on human health: recent advances and future. J Bionic Eng. 2021;18:210–37.
14. Thompson LA, Darwish WS. Environmental chemical contaminants in food: review of a global problem. J Toxicol. 2019;2019:2345283. https://doi.org/10.1155/2019/2345283.
15. Norval M, Lucas RM, Cullen AP. The human health effects of ozone depletion and interactions with climate change. Photochem Photobiol Sci. 2011;10:199–225. https://doi.org/10.1039/c0pp90044c.
16. Kim KH, Kabir E, Kabir S. A review on the human health impact of airborne particulate matter. Environ Int. 2015;74:136–43. https://doi.org/10.1016/j.envint.2014.10.005.
17. The International Panel on Chemical Pollution (IPCP). Overview report I: Worldwide initiatives to identify endocrine disrupting chemicals (EDCs) and potential EDCs. United nations (UN) environment programme. 2017a.
18. The International Panel on Chemical Pollution (IPCP). Overview Report II: An overview of current scientific knowledge on the life cycles, environmental exposures, and environmental effects of select endocrine disrupting chemicals (EDCs) and potential EDCs. United nations (UN) environment programme. 2017b.
19. The International Panel on Chemical Pollution (IPCP). Overview report III: Existing national, regional, and global regulatory frameworks addressing Endocrine Disrupting Chemicals (EDCs). United nations (UN) environment programme. 2017c.
20. ECHA (European Chemicals Agency), EFSA (European Food Safety Authority), et al. Guidance for the identification of endocrine disruptors in the context of regulations (EU) no 528/2012 and (EC) no 1107/2009. EFSA J. 2018;16:5311. https://doi.org/10.2903/j.efsa.2018.5311.
21. Kahn LG, Philippat C, Nakayama SF, et al. Endocrine-disrupting chemicals: implications for human health. Lancet Diabet Endocrinol. 2020;8:703–18. https://doi.org/10.1016/S2213-8587(20)30129-7.
22. Mallozzi M, Bordi G, Garo C, et al. The effect of maternal exposure to endocrine disrupting chemicals on fetal and neonatal development: a review on the major concerns. Birth Defects Res C Embryo Today. 2016;108:224–42. https://doi.org/10.1002/bdrc.21137.
23. Caserta D, Pegoraro S, Mallozzi M, et al. Maternal exposure to endocrine disruptors and placental transmission: a pilot study. Gynecol Endocrinol. 2018;34:1001–4. https://doi.org/10.1080/09513590.2018.1473362.
24. Caserta D, Graziano A, Lo Monte G, et al. Heavy metals and placental fetal-maternal barrier: a mini-review on the major concerns. Eur Rev Med Pharmacol Sci. 2013a;17:2198–206.
25. Caserta D, Di Segni N, Mallozzi M, et al. Bisphenol a and the female reproductive tract: an overview of recent laboratory evidence and epidemiological studies. Reprod Biol Endocrinol. 2014;12:37. https://doi.org/10.1186/1477-7827-12-37.
26. Caserta D, Bordi G, Ciardo F, et al. The influence of endocrine disruptors in a selected population of infertile women. Gynecol Endocrinol. 2013b;29:444–7. https://doi.org/10.3109/09513590.2012.758702.
27. Ghassabian A, Vandenberg L, Kannan K, et al. Endocrine-disrupting chemicals and child health. Annu Rev Pharmacol Toxicol. 2022;62:573–94. https://doi.org/10.1146/annurev-pharmtox-021921-093352.
28. Caserta D, Costanzi F, De Marco MP, et al. Effects of endocrine-disrupting chemicals on endometrial receptivity and embryo implantation: a systematic review of 34 mouse model studies. Int J Environ Res Public Health. 2021;18:6840. https://doi.org/10.3390/ijerph18136840.
29. Caserta D, Maranghi L, Mantovani A, et al. Impact of endocrine disruptor chemicals in gynaecology. Hum Reprod Update. 2008;14:59–72. https://doi.org/10.1093/humupd/dmm025.
30. Mallozzi M, Leone C, Manurita F, et al. Endocrine disrupting chemicals and endometrial cancer: an overview of recent laboratory evidence and epidemiological studies. Int J Environ Res Public Health. 2017;14:334. https://doi.org/10.3390/ijerph14030334.

31. Caserta D, De Marco MP, Besharat AR, et al. Endocrine disruptors and endometrial cancer: molecular mechanisms of action and clinical implications, a systematic review. Int J Mol Sci. 2022;23:2956. https://doi.org/10.3390/ijms23062956.
32. Guerranti C, Perra G, Alessi E, et al. Biomonitoring of chemicals in biota of two wetland protected areas exposed to different levels of environmental impact: results of the "PREVIENI" project. Environ Monit Assess. 2017;189:456. https://doi.org/10.1007/s10661-017-6165.
33. La Rocca C, Alessi E, Bergamasco B, et al. Exposure and effective dose biomarkers for perfluorooctane sulfonic acid (PFOS) and perfluorooctanoic acid (PFOA) in infertile subjects: preliminary results of the PREVIENI project. Int J Hyg Environ Health. 2012;215:206–11. https://doi.org/10.1016/j.ijheh.2011.10.016.
34. Combarnous Y, Nguyen TMD. Comparative overview of the mechanisms of action of hormones and endocrine disruptor compounds. Toxics. 2019;7:5. https://doi.org/10.3390/toxics7010005.
35. Caserta D, Mantovani A, Marci R, et al. Environment and women's reproductive health. Hum Reprod Update. 2011;17:418–33. https://doi.org/10.1093/humupd/dmq061.
36. Alavian-Ghavanini A, Rüegg J. Understanding epigenetic effects of endocrine disrupting chemicals: from mechanisms to novel test methods. Basic Clin Pharmacol Toxicol. 2018;122:38–45. https://doi.org/10.1111/bcpt.12878.
37. Rossati A. Global warming and its health impact. Int J Occup Environ Med. 2017;8:7–20. https://doi.org/10.15171/ijoem.2017.963.

Chapter 2
Cellular Mechanisms of Endocrine Disruption

Roberta Rizzo, Daria Bortolotti, Sabrina Rizzo, and Giovanna Schiuma

2.1 Endocrine Disruptors: Impact on Health

With industrialization, the production of chemicals and their introduction into the environment have increased massively. These new agents included many chemical classes and comprise an integral part of the world economy and commerce [1]. Nevertheless, several of the chemicals used today are called endocrine-disrupting chemicals (EDCs).

A chemical agent is classified as an endocrine-disrupting chemical when it can interfere with the synthesis, metabolism, and action of endogenous hormones.

These substances have been even more tested systematically for endocrine-disrupting effects in organisms as the production of large amounts of synthetic industrial and biomedical chemicals, as well as unwanted pollutants, poses destructive consequences to our ecosystem and imposes negative health effects to wildlife and humans [2, 3].

Several compounds have been identified as potential EDCs due to their actions of estrogenic and androgenic gene regulation, altering the synthesis, metabolism, and action of endogenous hormones interfering with different human mechanisms of male and female reproductive systems. The main chemical and natural compounds that are currently classified and approved as EDCs that deserve a special attention are summarized in Table 2.1.

R. Rizzo (✉) · D. Bortolotti · S. Rizzo · G. Schiuma
Department of Chemical, Pharmaceutical and Agricultural Sciences, University of Ferrara, Ferrara, Italy
e-mail: rbr@unife.it; daria.bortolotti@unife.it; sabrina.rizzo@unife.it; giovanna.schiuma@unife.it

© The Author(s) 2023
R. Marci (ed.), *Environment Impact on Reproductive Health*,
https://doi.org/10.1007/978-3-031-36494-5_2

Table 2.1 Type of emerging EDCs, classification, and effects on human health

Endocrine-disrupting chemical	Class	General effect on human health	Binding receptor
Atrazine	Estrogenic, a herbicide of the triazine class	Causes increased estrogen production and androgen inhibition; disruption of the hypothalamic control of luteinizing hormone and prolactin levels; adrenal glands damage and reduction of steroid hormone metabolism [4]; low fetal weight and heart, urinary, and limb defects [5]; incidence of gastroschisis [6] and reproductive problems [7].	Acts as an agonist for estrogen receptors (ERs)
Bisphenol A (BPA)	Estrogenic, used for the production of epoxy resin and polycarbonate plastics	Estrogenic activity causes endocrine disruption [8] associated with female fertility problems as polycystic ovary syndrome and endometriosis, decrease in oocytes' number and decrease in antral follicle counts [9–11]. BPA is associated with increased allergic sensitization [12]	Binds several receptors, including estrogen receptors (ERs), aryl hydrocarbon receptor (AhR), and others (such as peroxisome proliferator-activated receptors [PPARs])
Chlorambucil	A drug for chemotherapy, an alkylating agent	Causes infertility or leads to birth defects. In men, causes damage to sperm and significantly reduces sperm count (oligospermia) [13, 14]. It leads to increases in deletions and other mutations in germ cells [15, 16]	Binds several receptors including estrogen receptors (ERs) and androgen receptors (ARs)
Decitabine (5-aza-CdR)	Estrogenic, a drug for chemotherapy	Affects embryo survival and sperm morphology, decreasing sperm motility and capacity [17–19]	Activates estrogen receptor alpha (ERα receptor)

Table 2.1 (continued)

Endocrine-disrupting chemical	Class	General effect on human health	Binding receptor
Diethylstilbestrol (DES)	Estrogenic, a drug	Induces susceptibility to tumor in testis and reproductive tract tissues [20], in particular allowing to uterus epithelial tumors and to vaginal and cervical cancer. It induces reproductive tract abnormalities [21]	Acts as an agonist for the estrogen receptors (ERs)
Diethylhexylphthalate (DEHP)	An androgen antagonist, phthalates (used as plasticizers)	It affects reproductive functions. Prenatal phthalate exposure has been shown to associate with imparired reproductive functions in adolescent males [22]. It induces lower plasma thyroxine levels and decreased uptake of iodine in thyroid follicular cells [23]. DEHP increases cell proliferation, decreases apoptosis, and causes oxidative damage [24]	Acts as an antagonist for androgen receptors (ARs)
Genistein (GE)	Estrogenic/ antiestrogenic, phytoestrogen (plant hormones)	Increases ERs expression [25, 26]	Full agonist of estrogen receptor beta (ERβ) and partial agonist of estrogen receptor alpha (ERα)
Hexachlorocyclohexane (HCH or lindane)	Pesticide	Increases insulin and estradiol blood serum concentrations and decreases thyroxine concentrations. It leads to reduction in estrous cycles and luteal progesterone concentrations [27, 28]	Competitive binding to androgen receptors (ARs), estrogen receptors (ERs), and progesterone receptors (PRs)

(continued)

Table 2.1 (continued)

Endocrine-disrupting chemical	Class	General effect on human health	Binding receptor
Melphalan	A drug, an alkylating agent	Increase deletions in germ cells, increases mutations in both spermatogonial and postspermatogonial germ cells [29]	Binds several receptors, including estrogen receptors (ERs) and androgen receptors (ARs)
Methoxychlor (MTX)	Estrogenic, a pesticide	Adversely affects both male and female reproductive systems: aberrant folliculogenesis [30]; in men it induces defects in spermatogenic capacity, an increase in adult sperm cell apoptosis, and abnormal testis development [31, 32]	Binds estrogen receptors (ERs)
p-Nonylphenol	Estrogenic, derived from nonylphenol ethoxylates (used as industrial surfactants)	Alters spermatogenesis impairing testicular mass and sperm count [33]	Bind estrogen receptors (ERs)
Phthalates	Antiandrogenic, environmental xenoestrogen (to produce plastic)	Damages deoxyribonucleic acid (DNA) in sperm, causing inhibition of male fertility, semen quality, sperm motility, and semen volume [34]	Acts as an antagonist for androgen receptors (ARs)
Tetrachlorodibenzodioxin (TCDD)	Estrogenic, a dioxin	Causes hepatotoxicity, peripheral and central neurotoxicity; increases atherosclerosis, hypertension, diabetes, and tumor promotion. It induces neural system damage, including neuropsychological impairment [35]	Binds aryl hydrocarbon receptor (AhR)

Table 2.1 (continued)

Endocrine-disrupting chemical	Class	General effect on human health	Binding receptor
Vinclozolin (VCZ)	Antiandrogenic, a fungicide	Mimics male hormones, like testosterone, and binds androgen receptors compromising male fertility: not only in the first generation that was exposed in utero, but in males born for three generations and beyond [36]. It induces transgenerational defects in spermatogenic capacity, an increase in adult sperm cell apoptosis, abnormal testis development [31, 37] and predisposition to tumors, and prostate disease. These transgenerational effects correlate with epigenetic changes, specifically an alteration in DNA methylation in the male germ line [38]. It interferes with steroid hormone metabolism [39–41], leading to kidney diseases, immune abnormalities [42], and anxiety behavior [43]	Acts as a competitive antagonist for the androgen receptors (ARs)

Epidemiological studies suggest an association between the increasing exposure to chemicals and the development of some of the main ailments of the industrialized world (e.g., disturbances in reproduction, hormone-related cancers, and metabolic disorders like obesity and type 2 diabetes) [44]. In particular, some of these chemicals act as endocrine disruptors as they disturb endogenous hormone signaling pathways.

The consequences of endocrine disruption in human health are discussed. In particular, the possible involvement of EDCs' exposure on the development of metabolic diseases and their possible responsibility in health disorders are the basic reasons for understanding the mechanisms of action of EDCs. This research field is, however, the subject of scientific and public controversy due to the lack of knowledge about the possible molecular mechanisms underlying endocrine disruption and disease development [45].

A recent review reports that about 40% of human death (62 million per year) is attributed to the effect of exposure to chemical pollutants [46]. Moreover, a study from the US Center for Disease Control (CDC) reported that in the bodies of Americans of all ages have been found over 116 extraneous chemicals [47] and in the cord blood of American infants have been detected over 358 industrial chemicals and pesticides [48].

In some cases, it has been suggested that precise diseases (cancer, neurological disorders, allergies, and reproductive disorders) may be associated with exposure to chemical agents, like EDCs [49]. Focusing on the reproductive disorders, according to the World Health Organization (WHO)'s data, about 80 million people worldwide were estimated to be affected by infertility [50] because of the interference with external agents.

According to these evidences, these compounds raise serious concerns about their potential health impact. These chemicals are prevalently synthetic molecules from an industrial origin [51, 52], but also include some natural molecules [53, 54]. In the latter case, the presence of these compounds ubiquitously in the environment makes really difficult to avoid exposure to them.

The endocrine system of vertebrates is an intricate web of stimulatory and inhibitory hormone signals that control basic functions of body such as metabolism, growth, digestion, and cardiovascular function, as well as more specialized processes such as behavior, sexual differentiation (during embryogenesis), sexual maturation (during puberty), and adult reproduction [55]. EDCs affect human health by disturbing normal endocrine activity through interaction with different receptors involved in key metabolic interaction strategies. This detrimental effect is due to the ability of EDCs to interfere with or mimic endogenous hormones and other signaling molecules of the endocrine system [56, 57].

Due to the huge impact of EDCs on hormonal system, these substances can cause different clinical conditions, including infertility, alterations in sperm quality, abnormalities in sex organs and growth, endometriosis, early puberty [58, 59], altered nervous system function, neurological and learning disabilities [60, 61], immune function [62], certain cancers [63, 64], respiratory problems, metabolic issues, diabetes, obesity, cardiovascular problems [44], thyroid function alterations [65], immune diseases [62], and more (Table 2.2). This aspect highlights the importance of the evaluation of EDC exposure on a developing organism in order to prevent the development of diseases or dysfunctions later in life [74].

Table 2.2 Main effects of EDCs in different human systems

System	Dysfunctions and diseases
Reproductive system	Infertility [66], sexual organ abnormalities, alteration of organs maturation, reduction of male and female fertility, alteration of steroidogenesis, and oligospermia [57–59]
Metabolic system	Obesity, atherosclerosis [67], type 2 diabetes [81], liver abnormalities, high blood pressure [68, 69], decreased adipocyte differentiation , dyslipidemia [70], cardiovascular disease [71], metabolic syndrome (MetS), insulin resistance [44], thyroid function alterations [65], and cancer [64, 72]
Nervous system	Neurological disorders, altered nervous system functions, and learning disabilities [61]
Immune system	Allergies, respiratory problems, and immune system defects [62, 73]

2.1.1 Effects of EDCs on Metabolism

Due to the ability of EDCs to interfere with hormone signaling and metabolism, exposure to these substances could lead to the development of metabolic diseases.

In particular, these chemicals cause consequences on the metabolic system by interacting with the hormone receptors (HRs) of the nuclear receptor (NR) family [75] that represents a family of structurally related transcription factors that are involved in various essential biological functions (e.g., fetal development, homeostasis, reproduction, metabolism, and response to xenobiotic substances).

Some EDCs can bind directly to these receptors either as agonists or antagonists, thus enhancing or inhibiting the effect of hormones [44]. In this way, EDCs interfere with the process of hormone biosynthesis, transport to the target tissue, levels of hormone-binding proteins, and hormone catabolism and deregulate hormone availability [76, 77]. One of the main alterations involved in the development of metabolic diseases after EDC exposure is represented by the onset of metabolic syndrome (MetS), due to alteration in fat metabolism and glucose uptake because of the engagement of NRs by EDCs [78] (Table 2.2) that lead to obesity [68, 69] (Table 2.2). For this reason, EDCs involved in MetS onset are called "obesogens," and it was assumed that already in utero and onward exposure may play a role in the development of obesity and related diseases during life [78–80] (Table 2.2).

In addition to EDCs–NRs interaction, the involvement of other receptors in the modulation of the metabolic system has been reported, for example, the aryl hydrocarbon receptor (AhR) [81, 82], estrogen receptors (ERs), androgen receptors (ARs), retinoid X receptor (RXR), and peroxisome proliferator-activated receptors (PPARs) [83].

According to the data reported in the literature, there are specific molecules that interact with the aforementioned receptors, causing a negative effect on human metabolism: For example, dioxin exposure has been reported to increase the risk for type 2 diabetes [84] (Table 2.2), while persistent organic pollutants (POPs), a group of AhR ligands, have been found to be associated with both diabetes and MetS in an epidemiological study of human serum samples [70, 85]. Other authors conclude that low-dose exposure to polychlorinated biphenyls-77 (PCB-77), which could be accumulated in the adipose tissue, may contribute to the development of obesity and atherosclerosis [67] (Table 2.2). In particular, it is known that estrogen and its receptors play an important role in adipogenesis, and exogenous estrogens have an impact on adipose metabolism; for example, octylphenol, a chemical widely used as a surfactant and frequently found in wastewater, decreases adipocyte differentiation [86] (Table 2.2).

Again, bisphenol A (BPA), a monomer with strong estrogenic properties, largely used in a lining of food and beverage containers, medical tubing, and dental fillings, has been associated with cardiovascular disease, liver abnormalities, and diabetes [71] (Table 2.2).

2.1.2   EDCs and Cancer

As mentioned earlier, EDCs are substances that interfere with the endocrine system and therefore with hormones, which are involved in the evolution of cancer.

Exposure to EDCs, in particular to estrogen- or androgen-mimicking EDCs, can promote tumor formation, especially prostate and breast cancers, the latter in particular during the prenatal period, whereas the exposure to some EDCs may affect mammary gland development and increase breast cancer risk later in life [72].

Moreover, EDC exposure could also interfere with hormonal cancer therapy (Table 2.2).

Regard the exposure to EDCs in the prenatal period, there are some indications confirming that exposure to endocrine-disrupting substances in utero can confer an increased risk of cancer in humans (Table 2.2). A clear example is the diethylstilbestrol, reported to be associated with increased incidence of clear-cell carcinoma of the vagina when the exposure occurred during fetal life [87].

Moreover, perinatal exposure to low doses of BPA was reported to result in altered mammary gland morphogenesis and carcinoma in situ [88]. Similar to what is reported referring to the mammary gland, the fetal development of the prostate is affected by the exposure to BPA, which increases the adult prostate size, promoting prostate cancer development [89] (Table 2.2).

One of the main mechanisms responsible for tumorigenesis due to EDC exposure involves epigenetic changes on chromatin and DNA [90, 91]. These kinds of modifications, even if are not affecting DNA sequence, are stable over rounds of cell division and could also be heritable. Epigenetic modifications, such as chromatin methylation or histone modifications, lead to alteration in gene expression that may have repercussions at a phenotypical level, inducing syndromes or tumors, during both prenatal period and adult life.

2.1.3   Effects of EDCs on Reproduction, Growth, and Development

The impairment of the endocrine system regulation may raise abnormal function and development of the reproductive systems [59] (Table 2.2).

In particular, early-life exposures to high level of EDCs have been linked to developmental abnormalities and may increase the risk for a variety of diseases later in life [92]. In fact, EDC exposures during fetal development and childhood can cause long-lasting health effects, as during this developmental period, hormones regulate both formation and maturation of organs [57].

In particular, these hormones are subjected to hypothalamic control during fetal and early postnatal life and are crucial to enable a correct sexual differentiation and a successful reproduction in adulthood [93, 94]. Vilahur et al. [95] showed that prenatal exposure to endocrine-disrupting chemicals causes changes in DNA methylation that were manifested as the latent development of male infertility, reproductive cancers, and other dysfunctions (Table 2.2).

Some classes of EDCs (dichlorodiphenyltrichloroethanes [DDTs], BPA, phthalates, PCBs, and others) can mimic or block the effects of male and female sex hormones, exhibiting a diverse effect on the reproductive and development sphere in men and women, leading to various hormonal changes [59]. In the case of women, EDCs are implicated in the development of some gynecologic pathologies and fertility problems: de Cock et al. (1994) found that reduced fecundability ratio and longer time-to-pregnancy are associated with the application of pesticides fecundability [96] (Table 2.2). Moreover, a positive association between EDCs, especially estrogen-like BPA [97], and gynecological problems highlighting recurrent miscarriages was observed.

On the other hand, men are known to be more susceptible to steroidogenesis, the process for steroid hormone production. Several evidences supported the role of EDCs in interfering with steroidogenesis (particularly through interaction with NRs), modulating the release of endogenous steroid hormones that may cause subsequent reproductive dysfunction [58] (Table 2.2).

The effects of EDCs on fetal testis seem to be more striking as the disruption of steroidogenesis at this early developmental stage can also affect the proliferation of germ cells and Sertoli cells [98–100] (Table 2.2), which supports the maximum number of sperms that can be produced in adulthood [101, 102].

2.2 Targets of EDCs: Genomic and Nongenomic Modulation

Insecticides, plasticizers, and detergents can be classified as potential EDCs, and the exposure to these chemicals in the environment in the food chain, or occupational exposures, may affect human health inducing developmental effects and birth defects [103].

EDCs act on human health through various mechanisms, which can be classified as genomic and nongenomic (Fig. 2.1). EDCs can affect every possible cellular hormonal pathway: For example, some of these compounds can bind directly to hormone receptors either as agonists or antagonists, thus enhancing or inhibiting the effect of a hormone regulating gene expression [45]. Despite this, nongenomic modulation was observed, referring to epigenetic modifications, as a system of interaction on human health [104] (Fig. 2.1).

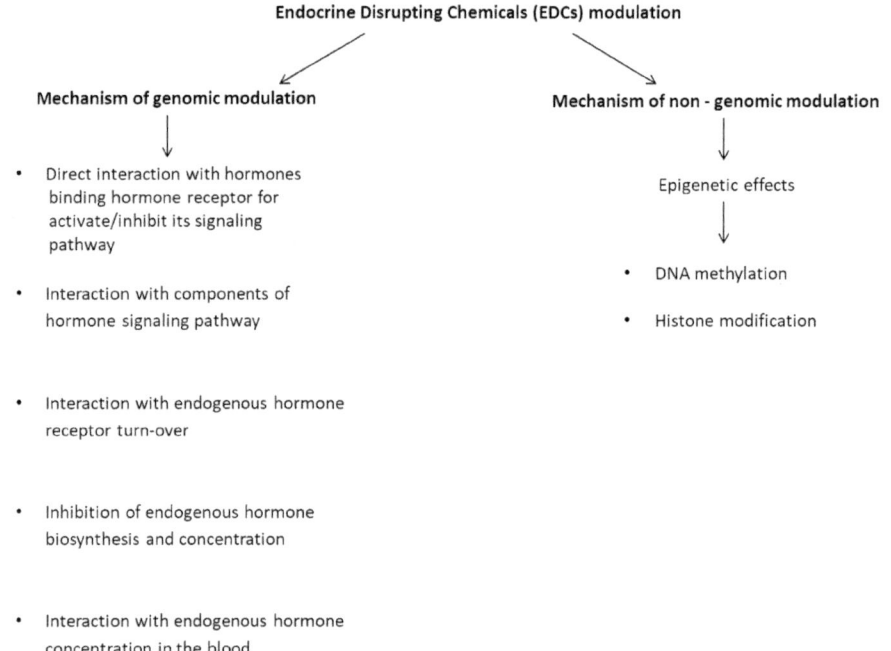

Fig. 2.1 Schematic representation of the types of modulation of the EDCs, genomic and nongenomic

2.2.1 EDCs and Genomic Modulation

A wide range of EDCs exert their effects using a hormone-type mimicking mechanism. These synthetic molecules can be present in rather high quantities and therefore can compete with endogenous hormones despite their lower affinity toward hormone receptors (HRs).

EDCs exhibit structures that are similar to hormones [105]; thus, they can interfere with their binding site as a agonist, activating the receptor, or as an antagonist, inhibiting HRs (Fig. 2.1). This mechanism is defined as "genomic" due to the fact that hormone receptors, like androgen receptors (ARs), estrogen receptors (ERs), and aryl hydrocarbon receptors (AhRs), are nuclear receptors (NRs) that have a direct effect into DNA as a transcription factor [75]. In the specific case of estrogen receptors, p-nonylphenol and bisphenol A act as agonists for ERs, and like estradiol [103], vinclozolin [106], DDT [107], atrazine, and lindane [108] are AR antagonists and have antiandrogenic effects, reducing the expression of ARs [56].

In addition to this mechanism, EDCs are able to interact directly with hormonal regulation by acting on the components of the hormonal signaling pathway (Fig. 2.1). The interaction between EDCs and these components involves a modification of the hormone biosynthesis, positively or negatively, or their degradation [45]. Indeed, xenobiotics and many EDCs have an indirectly effect on hormone regulation through the activation/inhibition of receptors that induced the expression of enzymes involved in activation, conjugation, and elimination of endogenous hormone [56] (Fig. 2.1). The most striking example is in the interference of these compounds with enzyme cytochrome P450, modifying the hormone synthesis as in the case of steroid synthesis [109]. Despite this, also many other metabolic enzymes involved in the synthesis, elimination, and conversion of steroid hormones, such as testosterone to 17β-estradiol (E2) and progesterone to testosterone [110], could be affected by EDCs.

EDCs could also disturb the balance of circulating and local tissue concentration of hormones disturbing the normal functions of the endocrine system. In particular, steroid hormones are hydrophobic and require to be bound by blood proteins to be transported. This characteristic is shared also by EDCs, which have this characteristic of hydrophobicity [111]. In this mechanism, EDCs do not compete with hormones at the receptor level, but at the level of their circulating binding proteins; in fact, these endocrine disruptors are susceptible to compete with hydrophobic hormone-binding transport proteins [112]; otherwise, other EDCs can affect the biosynthesis or degradation of hormone-binding transport protein [113] (Fig. 2.1).

Finally, a further mechanism responsible for EDCs alteration with the endocrine system is represented by the inhibition of receptor expression, modifying endogenous hormone receptor turnover [114] (Fig. 2.1).

Each single mechanism of interaction does not exclude the other; EDCs represent a category of compounds with various and complex activities in the endogenous system.

2.2.2   EDCs and Nongenomic Modulation

The conjugation of the term "epigenetics" [115] gives importance to the effects of the environment on the human gene apparatus. Epigenetics relates the changes that occur in gene expression and phenotype with the inheritance of these, not altering the DNA sequences and identifying external agents from the environment as the cause. The environment can be perturbed by EDCs, and endocrine disruptors affect the epigenetic processes such as DNA methylation [116, 117] and histone modification [118]; this can be considered the nongenomic modulation by EDCs.

Nongenomic modulation by EDCs has importance during development, particularly in three windows of exposure when the epigenome is susceptible to reprogramming by EDCs during gamete maturation [119], implantation of fetus [120], and differentiation of pluripotent cells [121].

The reprogram of the epigenome caused by exposure to EDCs is supported by several evidences; in fact, chemical disruptors such as DES [122], BPA, genistein [123], and vinclozolin [124] all have been shown to alter patterns of DNA methylation. Moreover, Newbold [125] and Tang [126] explored DNA methylation changes induced by developmental exposure to EDCs, specifically to DES.

Finally, Anway et al. [31] demonstrated the transgenerational effects of vinclozolin on spermatogenesis and fertility in males after gestational exposure to these EDCs. This supports the fact that EDCs induce transgenerational inheritance by assuming transmission via germ line alterations [127].

2.3   Hormone Mimicry and Disruption by EDCs: Modulation of Hormone Activity

The name "EDCs" itself refers to the ability of a specific group of environmental endocrine-active chemicals to interfere with the endocrine system, as well as other systems (mainly the reproductive system), disturbing endogenous hormone signaling pathways [44]. As already mentioned, EDCs can influence and modulate the hormonal system, by acting as endogenous hormones and other signaling molecules of the endocrine system, or even mimicking them. This particular property represents a great potential risk for human health, and therefore, the increasing release of contaminants in the environment should be a concern of a global interest.

The principal molecular mechanism of action exhibited by EDCs is represented by their interaction with hormone receptors (HRs), influencing hormonal activity directly (Table 2.3). Receptors have evolved to be protected against binding with endogenous molecules other than hormones; however, the growing environmental contamination of synthetic toxicants having the shape and size of the actual hormone could not have been encountered before during receptor evolution. For this reason, even though EDCs have lower receptor affinity compared with physiological hormones, because of their abundance in the environment, these chemicals can compete with endogenous hormones [105].

Table 2.3 Molecular mechanisms of action employed by EDCs for endocrine disruption and modulation of hormonal activity

EDCs' mechanism of action	Consequence on hormonal system
Binding to hormone receptor	Activation/inhibition of its signaling pathway
Interaction with components of hormone signaling pathway (downstream of receptor)	Activation/inhibition of the signaling pathway
Influencing endogenous hormone biosynthesis/degradation	Increase/decrease of hormone concentration
Binding to circulating hormone-binding protein	Decreasing hormone transport and concentration in blood
Influencing hormone-binding protein synthesis/degradation	Increase/decrease of hormone-binding proteins, reflecting in hormone concentration in blood
Influencing hormone receptor turnover	Increase/decrease of hormone receptors
Epigenetic modification	Alteration of hormone signaling pathway downstream of receptor and aberrant receptor turnover

Endocrine disruptors are notably able to bind to the family of nuclear receptors (NRs), including the estrogen receptors (ERs) and androgen receptors (ARs) [44], but they can also be associated with some membrane receptors. Nevertheless, thyroid hormone receptors, retinoid X receptor (RXR), and peroxisome proliferator-activated receptors (PPARs) have been recently identified as additional binding targets too [44]. This is the case of estradiol and bisphenol A that can bind both the nuclear receptors ERα and ERβ and a transmembrane receptor called G protein-coupled estrogen receptor 1 (GPER) (GPR30) [128, 129].

Importantly, depending on the binding of EDCs with NRs or membrane receptors, it is possible to observe different kinds of effects. In fact, in the case of NR engagement by EDCs, we observe a modulation of gene expression that exerts a long-term effect on target cells' phenotype [5, 130] due to their transcriptional factor function, while in the case of binding with membrane receptors the result is a short-term and more acute effect [5, 131].

On the other hand, behaving like agonists, EDCs can also hinder endogenous hormones, by occupying HRs and antagonizing the proper ligand–hormone interaction (Table 2.3) [132]. Many of these EDCs showing antiestrogenic, antiandrogenic, antiprogesteronic, and anti-ER activities have been detected in wastewater [133].

The antagonism exerted by EDCs could be explained by the ability of some chemical compounds to block receptor conformation in their inactive state, resulting in the inhibition of their signaling pathways [45]. For example, polychlorinated biphenyls (PCBs) can prevent the association between triiodothyronine (T3) and thyroid hormone receptor (THR), with the consequent dissociation of the transcriptionally active THR/retinoid X receptor heterodimer complex from the thyroid response element (TRE) [113].

Besides the hormonal activity exhibited by EDCs via receptor binding, toxicants can influence hormonal system activating other signaling pathways, that is, interacting with components of hormone signaling pathways downstream of receptor activation (Table 2.3) [45]. Since this phenomenon does not involve any binding with hormone receptors, such EDCs may present different structures from endogenous hormones [45]. An example is represented by fluoxetine (FLX), a selective serotonin reuptake inhibitor (SSRI) active substance present in antidepressant drugs, which has the potential to alter many intracellular signaling pathways in different cellular types, without HR association [134–136]. In addition, some bisphenols can interact with Ras small G proteins (e.g., K-Ras4B) and activate the Ras signaling cascade, causing the increase of pERK and pAKT [137]. Finally, the herbicide atrazine can inhibit cAMP-specific phosphodiesterase 4 (PDE4) [130], leading to cAMP intracellular accumulation, and tolylfluanid is able to reduce insulin receptor substrate-1 (IRS-1) levels, downstream from the insulin receptor, in human adipocytes [131].

Another way by which exogenous molecules exert endocrine disruption is by directly affecting the endogenous hormone biosynthesis or degradation (Table 2.3). Again, in this specific case, there is no interaction with hormone receptors; therefore, chemicals can exhibit different structures than physiological hormones [45].

Evidences of altered hormone biosynthesis have been shown after exposition to different EDCs; for example, low dose of BPA inhibits adiponectin secretion in vitro in human adipocytes [114, 138, 139], 4-nonyphenol (4-NP) inhibits the synthesis of testosterone by Leydig cells following the stimulation by human chorionic gonadotropin [140], or triclosan stimulates vascular endothelial growth factor (VEGF) secretion by human prostate cancer cells [141]. Concerning EDCs' influence on hormone degradation, polybrominated diphenyl ethers (PBDEs) have been described to potentially increase thyroxine (T4) elimination, lowering its concentration level in blood [113], while parabens inhibit estrogen degradation [142], increasing the hormone concentration in blood.

Since the majority of the hormones are hydrophobic, like steroids and thyroid ones, they necessarily must be transported in association with specific binding proteins through the bloodstream. Thus, since EDCs present structural similarity to hormones, they are hydrophobic and can bind the same hormone-binding transport proteins, competing with endogenous hormones (Table 2.3) [143–145]. The result of the association between binding proteins and toxicants is the decrease in hormone transport that decreased their concentration in blood. Examples of transport proteins that are usually subjected to be bound by chemicals instead of hormones include steroid hormone-binding protein (SHBG) or α-fetoprotein (AFP) [112, 146].

Furthermore, EDCs can modulate endogenous free active hormone concentration not only by taking their place in binding transport protein, but also via the direct modification of these protein levels in the bloodstream, affecting their biosynthesis or degradation (Table 2.3) [45]. In fact, a lower availability of transport proteins means at the same time a lower concentration of hormones. EDCs responsible for this mechanism are binding-independent, so they can exhibit different structures from those of endogenous hormones. The advantage of such mechanism of action is that chemicals mainly target liver because this is the classic degrading organ, and at the same time, liver is the place where binding transport proteins are synthetized and degraded [45]. PBDEs, for example, may downregulate transport protein transthyretin (TTR) level [113] and consequently lower T4 amount in blood.

Among the several mechanisms affecting hormonal system to mention, there is also the ability of EDCs to regulate HR turnover through the stimulation or inhibition of their expression (Table 2.3). The absence of the binding with hormone receptors in this specific mechanism allows such chemicals to exhibit structural differences from hormones [45]. BPA, for example, has been shown to induce leptin receptor expression in ovarian cancer cells in vitro [31], cadmium increased estrogen receptor beta (ERβ) and Cyp19a1 enzymes in endothelial human umbilical vein endothelial cells (HUVECs) in vitro, and a dose-dependent decrease of androgen receptor (AR) expression levels was observed after 24 h of exposure [38]. Conversely, the inhibition of receptors has been shown after the administration of a low oral dose of BPA to rats, which decreased estrogen receptor expression in their hypothalamic cells [39]. The inhibition of androgen receptors by BPA has also been observed in vivo [29] and in vitro cells of patients with breast or prostate cancer [40].

Nowadays, it is well known that EDCs may also influence human epigenome by acting through various mechanisms, such as DNA methylation and histone code alteration [1]. Epigenetic modifications consist of changes in gene expression and resulting phenotype, without any alteration in nucleotide sequence of DNA. The alteration of gene expression can act on differential physiological pathways and can affect normal hormonal activity, for example, by alternating hormone signaling pathway downstream of receptor or provoking aberrant receptor turnover, therefore constituting another nontraditional mechanism of action of EDCs on hormonal system (Table 2.3). Waalkes et al. [147] demonstrated that inorganic arsenic exposure in utero induced a significantly decreased promoter methylation of ER in the liver, resulting in increased ER expression, which also correlated with the increased incidence of hepatocellular carcinoma in exposed animals. The soybean isoflavone genistein (GE) has been shown to prevent breast cancer and to induce epigenetic reactivation of estrogen receptors [148]. Recent studies have suggested that GE, besides its ability to enhance the anticancer capacity of the estrogen antagonist tamoxifen (TAM) in ERα-positive breast cancer cells, can also reactivate ERα expression in ERα-negative breast cancer cells. This positive influence on estrogen receptor expression was enhanced when combined with a histone deacetylase (HDAC) inhibitor, named trichostatin A (TSA). GE treatment also resensitized ERα-dependent cellular responses to the activator 17β-estradiol (E2) and the antagonist TAM. The following research revealed that GE can remodel chromatin structure and consequently reactivate ERα gene expression.

Although the spectrum of chemical compounds to which we may be exposed is broad and can influence human health by multiple mechanisms of action, EDCs can be classified according to their endocrine effect in the main classes of estrogenic, antiestrogenic, androgenic, and antiandrogenic EDCs. The name of the class they belong to suggests the type of endogenous hormonal pathway that is disrupted by those specific chemicals.

Among the endocrine-disrupting chemicals, environmental estrogens were the first to cause concern. Lately, it was discovered that also other hormonal systems are susceptible to disruption, like androgen signaling pathway. At last, thyroid hormone was identified as a target for endocrine disruption too [149], as well as peroxisome proliferator-activated receptor (PPAR)/retinoid X receptor (RXR) system [150, 151].

2.4 Mechanism of Action of EDCs and Estrogens

17-beta-Estradiol (E2) is a key hormone involved in many biological processes in humans, like development and maintenance of the female reproductive tract, brain, bone, and cardiovascular system. Moreover, E2 is a key component in male development too [152]. In female adulthood, estrogen takes part in metabolism and coordinates the morphological alterations occurring during physiological menstrual cycle and pregnancy, together with differentiation and proliferation of target tissues [152].

The classical and conventional estrogenic function is notably mediated by the interaction of E2 with specific estrogen receptors ERα and ERβ, which are steroid receptors belonging to the superfamily of nuclear receptors, in particular type I NRs [152, 153]. ERα and ERβ are tissue-specific; therefore, they are thought to possess distinct physiological roles [154]. ERα is expressed in breast, uterus, pituitary, testis, and kidney, while ERβ is found in cardiovascular system, prostate, hypothalamus, gastrointestinal tract, ovary, kidney, lungs [155–159], and breast [160–162]. The tissue distribution pattern of ERβ suggests its role both in male and female reproduction and development and in central neuroendocrine regulation. Even if most tissues show a predominance of either ERα or ERβ, many others coexpress both [163]. To further support the different regulatory roles that distinguish the two types of ERs, ERα contains dissimilar domains compared with ERβ, suggesting a potential for different ligand specificity that results in different effects [103].

As already mentioned before, all nuclear receptors directly participate in the cellular response to hormone, affecting gene expression by acting as transcription factors. Ligand-activated ER generally induces the increase of target gene expression in target tissues, even if in some other cases it can decrease specific gene transcription [164]. However, the mechanism of negative regulation is still less characterized.

Thus, estrogens mainly work through the binding to ERs in order to transactivate the expression of estrogen-responsive genes. The latter contain an estrogen-responsive element (ERE) in their promoters/enhancers, which represents a recognition sequence to which ER binds. Before activation by association with E2 ligand, ERs exhibit inactivation due to the binding with the heat shock protein-90 (HSP90) [152] and display a diffuse nuclear localization [163]. After ER binding with the E2, ER dimerizes and becomes an active transcription factor acting on the expression of several important genes such as progesterone receptor (another nuclear hormone receptors), vascular endothelial growth factor, c-Fos/c-Jun (proto-oncogenes), and cyclin D1 (cell cycle regulators) [152], stimulating proliferation or differentiation at the cellular level. Several coactivators are also recruited to the complex and, together with ER, participate in remodeling of chromatin structure in order to provide access for transcription machinery to the target gene promoter and allow the transcription to begin [165].

Finally, an additional form of nuclear regulation has been discovered via ERs named "composite regulation," which is based on protein–protein interaction with other transcription factor, requiring the association between activated steroid receptor and members of the activator protein 1 (AP-1) complex, Fos and Jun [103, 166–170], resulting in either gene activation or repression [166]. Nevertheless, it is not still clear how negative, positive, and composite regulations participate together in the complex cellular functions related to the action of estrogens, like differentiation, organ development, and growth.

Among the EDCs, a huge variety of compounds have been identified as estrogenic chemicals, also called "xenoestrogens" [171, 172]. EDCs with the ability to interfere with estrogenic signaling are, for example, genistein, octylphenol, nonylphenol, bisphenol A (BPA), and diethylstilbestrol (DES), and many of these xenoestrogens are ligands for ERs and therefore can act either as agonists (estrogenic) or as antagonists (antiestrogenic) toward endogenous estrogens.

Several EDCs (e.g., DES, BPA, methoxychlor [173], and genistein) show ER-mediated effects on gene expression, competing with E2 for binding to the same ERs. However, xenoestrogens exhibit a different affinity for the two types of ERs: While DES and methoxychlor have a higher affinity for ERα, genistein and BPA mostly bind ERβ [174]. Interestingly, recent studies have identified a natural compound proposed to be an ERβ-specific ligand [175], which may play a role in preventing progression to prostate cancer by activating prostatic ERβ [176]. Differential ERα- or ERβ-mediated effects may partially account for parallel differential effects of EDCs on different target tissues. This is confirmed in the reproductive tract, where the deleterious effects of EDCs are xenoestrogen-specific [177]. Thus, xenoestrogens can disrupt the endocrine system, both showing differential effects on ERα or ERβ or via differential binding affinity.

Similar to the physiological estrogenic pathway, the association between xenoestrogens and ERs leads to the expression of estrogen-responsive genes (Fig. 2.2), as observed following DES, 7-methylbenz[a]anthracene-3,9-diol (MBA), coumestrol, and genistein (GE) [178] exposure. On the contrary, other xenoestrogens can act as hormonal antagonists (Fig. 2.2) employing multiple mechanisms, for example, by preventing the binding of ER to DNA (e.g., BPA) [179] or inhibiting the binding of ER coactivators [174] to avoid transactivation of gene expression.

However, in addition to the classical ER binding, a possible parallel pathway of estrogenic regulation may involve the existence and binding of alternative "estrogenic receptors," that is, orphan receptors called ER-related receptors (ERRs), -1 and -2 (Fig. 2.2), which share a similar sequence to ER [180]. Several researchers have focused on the nature (constitutive or liganded) of their transcriptional activities. Moreover, it has been showed that ERRs can interfere with estrogen signaling in various ways, either positively or negatively [181]. Therefore, the identification of possible modulators (positive or negative) of ERR activities could be highly useful in understanding some estrogen-related pathologies.

Besides mimicking and antagonizing endogenous estrogens, EDCs may also mediate estrogenic biological effects by inducing enzymes that accelerate the metabolism of estradiol (Fig. 2.2) [103]. Dioxins, for example, have provided

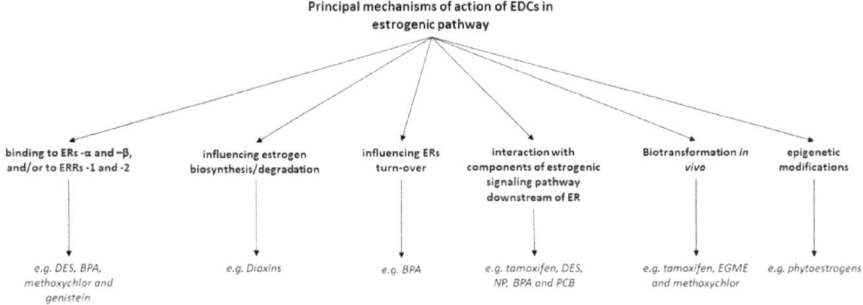

Fig. 2.2 Summary of the main mechanisms of action of EDCs involved in estrogenic pathway

additional evidence for their antiestrogenic effects inducing an increased metabolism of E2 [182, 183].

An additional endocrine-disrupting effect of EDCs on estrogenic responses consists in their ability to increase or decrease the amount of available ERs (Fig. 2.2) [184–186]. Many members of NR superfamily are degraded by the ubiquitin-proteasome pathway in a ligand-dependent manner, in order to prevent cells from overstimulation by endogenous hormones or other activating signals [187]. Consequently, EDCs might act on proteasome-mediated degradation of nuclear receptors, altering physiological estrogenic pathway. Masuyama et al. [188] compared the effects of BPA and estradiol treatments on ER-mediated transcription, to try to explain the observations relating to differential effects of BPA treatment on ER levels [189]. In the presence of estradiol, both ERα and ERβ interacted directly with suppressor for Gal 1 (SUG1) of the proteasome. In contrast, BPA activated ER-mediated transcription, without enhancing the interaction between ERβ and SUG1. In the presence of BPA, ubiquitination and degradation of ERβ were also slower than those in the presence of estradiol or phthalic acid, suggesting that BPA may affect the ERβ-mediated transcription of target genes by inhibiting ERβ degradation [190].

Lastly, an attractive hypothesis contemplates the conversion of endocrine chemicals from inactive to active ER-binding estrogens, due to the action of a biotransformation happening in vivo (Fig. 2.2) [103]. A phenomenon of biotransformation has been reported for tamoxifen [191], after the isolation of its metabolite 4-OH-tamoxifen from animals and humans treated with the compound. 4-OH-tamoxifen induces antiestrogenic effects, and it has an even higher affinity for ERs compared with its parent compound. Additional EDCs that are metabolically converted into active agents include ethylene glycol monomethyl ether (EGME) [192] and methoxychlor [193].

Endocrine compounds can also exert apparent estrogenic growth effects by interacting with different cellular factors that could occur downstream from ER of the estrogenic regulatory cascade in target cells (Fig. 2.2) [103]. Tamoxifen, for example, affects calmodulin regulation without ER mediation and directly inhibits protein kinase C through non-ER mechanisms [194, 195]. Other xenoestrogens may rapidly modify other signaling pathways, such as DES, nonylphenol (NP), BPA, and polychlorinated biphenyl (PCB), which are able to alter phosphorylation state of proteins belonging to the large family of mitogen-activated protein kinases (MAPK) in mussel hemocytes [196, 197]. Finally, phytoestrogens possess a variety of nonhormonal properties; for example, dietary phytoestrogens are capable of inhibiting tyrosine kinase activity, which are involved in various growth factor signaling pathways implicated in control of cell growth and differentiation [163]. Anyway, some phytoestrogens, like isoflavones in soy and resveratrol in grapes, have been identified as active agents responsible for benefits to human health exerting hormonal mimicry or antagonism through endocrine pathways or endocrine disruption. Recently, the Food and Drug Administration recognized that the Asian diet, high in soy consumption [198], can be associated with lowered cholesterol and reduction in cardiac risk [199], and the same was sustained by the "French Paradox" that associates red wine consumption to decreased cardiac risk [200]. New scientific data would link the dietary assumption of phytoestrogens with the epigenetic mechanism of histone acetylation (Fig. 2.2) through the phenomenon of gene superinduction. The term "superinduction" means the increased expression of nuclear receptor-activated genes to higher levels than those observed with the established ligand, for example, estradiol for ERs [103]. These observations could explain how soy phytoestrogens and grape may act as nontraditional molecular mechanism, leading to health benefits and anticancer effects. In fact, some of these beneficial effects are now associated with a particular family of histone deacetylases (HDACs), including the human SirT1. Again, the action of fiber assumption in lowering colon cancer incidence has been explained by the presence of butyrate and through the effects on histone acetylation status [201, 202]. Lately, several investigations have asserted that the soy isoflavone phytoestrogens, genistein and daidzein, can be linked to resveratrol, butyrate, and histone acetylation state. Both the isoflavone phytoestrogens act as superinducers of estrogen signaling pathways [203]. This evidence could be the proof that histone acetylation status can also be affected by these compounds. Furthermore, similar superinduction properties are also seen with grape phytoestrogen, resveratrol [103, 204], which has been linked to the effects on HDAC Sir2/SirT1 and, thus, also histone acetylation status [205, 206].

2.5 Mechanism of Action of EDCs and Androgens

Testosterone constitutes the key hormone of androgen hormonal pathway in human, and male testis already produces it around gestational day 65 [207], in order to guarantee the proper establishment of sexual behaviors, male reproductive tract development, and masculinization of other organs. In fact, androgens mediate a wide range of biological responses, such as testicular and accessory sex gland development and function, pubertal sexual maturation, maintenance of spermatogenesis and maturation of sperm, male gonadotropin regulation through feedback loops, and various male secondary characteristics like bone mass, musculature, fat distribution, and hair patterning [208]. Moreover, testosterone becomes critical for brain development too, thanks to its aromatization in 17β-estradiol (E2) by the action of the aromatase CYP19 [209].

Similar to estrogens, both testosterone and its metabolite dihydrotestosterone (DHT) can bind type I NRs, called androgen receptors (ARs). ARs, as well as all the members included in the family of nuclear receptors, are a class of ligand-activated proteins that can enter the nucleus functioning as transcription factors and regulating specific gene expression. ARs can be found in multiple organs, such as hypothalamus [210], pituitary, kidney, prostate, and adrenals (and ovary) [211, 212].

During the perinatal period of programming of the endocrine axis, the hormonal feedback from the gonads to the hypothalamus and pituitary gland represents an event of extreme importance and sensitivity toward endogenous and exogenous stimuli [152]. While female sexual differentiation is considered as a default developmental pathway since it is independent of estrogens and androgens, male sexual differentiation is driven by the fetal testes and it is entirely androgen-dependent [213]. For this reason, male sexual differentiation is highly susceptible to androgen disruptors that could affect developmental programming and reproductive tract maturation [214]. In fact, the eventual lack of testosterone in male fetus due to antiandrogenic exposure, to a genetic mutation in AR or to a blocked metabolism of the hormone, can induce the development of phenotypic female with testes.

The exposure to EDCs during male reproductive tract development may alter testosterone–AR association or the endogenous hormone metabolism, resulting in permanent reprogramming of male reproductive tract and its hormonal communication with the entire hypothalamic–pituitary–gonadal (HPG) axis. In adulthood, HPG axis is already established, and antiandrogen compounds can cause aberration in sperm production and libido in males [215, 216].

The mechanisms of action exploited by endocrine compounds to disrupt androgen pathway are multiple. Among these, we can mention the influence on receptor turnover by decreasing AR levels (Fig. 2.3), the alteration of luteinizing hormone (LH) stimulation (Fig. 2.3), the interference with androgen synthesis, metabolism and clearance (Fig. 2.3), and the alteration of proper folding of the ligand-binding domain (LBD) in ARs after EDC binding [107, 214, 217–221]. Aberrant LBD misfolding means AR inactivity, due to its inability to recruit coactivators and to initiate transcription.

Although there are many sites of action for chemicals to interfere with androgen signaling, endocrine chemicals are classified into two main categories: those that interfere with androgen biosynthesis or metabolism (non-receptor-mediated disruptors) (Fig. 2.3) and those that interact with ARs to interfere with the ligand-dependent transcriptional function (receptor-mediated disruptors) (Fig. 2.3). Despite these two principal classes, some chemicals such as PCBs [219], DES [222],

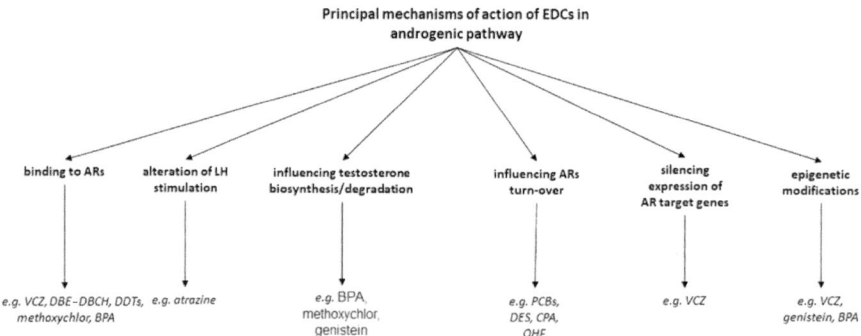

Fig. 2.3 Summary of the main mechanisms of action of EDCs involved in the androgenic pathway

cyproterone acetate (CPA), and hydroxyflutamide (OHF) [217] affect AR activity through the reduction of its expression and level. In addition, another group of EDCs disrupts androgen pathway by inhibiting AR ligand, binding, dimerization, and DNA binding or by silencing expression of AR target genes affecting down-stream cellular response.

Androgen receptor–mediated disruptors can be further divided into agonists and antagonists. Agonists bind to androgen receptors and activate a response mimicking the action of endogenous androgens; on the contrary, antagonists block AR transactivation.

A pilot study by Araki et al. [223] reported that some industrial or environmental chemicals show AR agonist activity, and in particular, the compound 1,2-dibromo-4-(1,2-dibromoethyl) cyclohexane (DBE-DBCH) was recently identified as the first potent environmental activator of the human AR [224, 225]. DBE-DBCH can exist in four diastereoisomeric forms: α and β that can be converted into γ and δ at specific conditions [226]. Several analyses showed that diastereomers γ and δ are more potent activators of human AR than α and β, but all the DBE-DBCH diastereomers induced the expression of the downstream target prostate-specific antigen (PSA) in vitro [227].

Contrariwise, EDCs with antiandrogenic action are, for example, dichlorodiphenyltrichloroethanes (DDTs), whose isomers [228] and metabolite [41] were shown to reduce the association between DHT and AR in vivo and to inhibit DHT-induced transcriptional activation in vitro [107]. In addition to AR antagonistic effects of DDT, high concentrations of its metabolite have been shown to function as inhibitors of 5α-reductase that converts testosterone to DHT [229], providing a clear example of how chemical compounds can affect androgen signaling at multiple sites of action. Fetal and neonatal DDT exposure in male produced demasculinizing effects with a high incidence of epididymal and testicular lesions [107, 230] and reduced prostate growth and inflammation [231]. Methoxychlor has a similar structure to DDT and, beyond its well-known estrogenic activity, also shows affinity to the AR at comparable or even higher levels than DDTs [41]. In addition to methoxychlor, BPA is first believed to act in estrogen signaling pathway; however, many scientific evidences have shown its association with AR [41] and its antagonist activity [232]. Lastly, vinclozolin also exerts an endocrine-disrupting potential as an AR antagonist through its primary metabolites [233]. Its mechanism of action consists of inhibiting AR transactivation and androgen-dependent gene expression. In vivo administration of vinclozolin at different doses, routes, and periods (gestation, lactation, puberty, and adulthood) is closely related to a different kind of effect on the male reproductive tract [214]. In addition, vinclozolin exposure during sex determination in developing male germ cells (fetal days 8–14) may lead to transgenerational effect. This phenomenon is based on epigenetic modification on male germ cells that consist of perturbations in DNA methylation patterns, underlining the presence of a relationship between this antiandrogenic compound and epigenetics [38].

2.6 Conclusions

Today, several compounds are classified as endocrine disruptor compounds (EDCs), intended as chemical agents that can interfere with the synthesis, metabolism, and action of endogenous hormones [56]. The EDCs are present in the environment ubiquitously and include both natural molecules [53, 54] and synthetic molecules [51, 52]. Several of these compounds used today have been tested systematically for endocrine-disrupting effects in organisms, as demonstrated by epidemiological studies that suggest an association between the exposure to chemicals and the development of some of the main ailments (e.g., metabolic disorders like obesity and type 2 diabetes) [44] (Table 2.2). This action is due to the ability of EDCs to interfere, with different strategies, with endogenous hormones and other signaling molecules of the endocrine system [56, 57].

The effect of these compounds on the endocrine system contributes to the emergence of several problems in the metabolism and systems of the human body. The first system to be negatively affected by these compounds is the metabolic system, through the interaction of EDCs with the hormone receptors (HRs) of the nuclear receptor (NR) family [75], mainly estrogen receptors (ERs) and androgen receptors (ARs) [103]. The direct receptor interaction of EDCs as agonists or antagonists enhances or inhibits the hormones' action [44], leading to the development of conditions such as the metabolic syndrome (MetS) [78, 79, 125] (Table 2.2). Another negative manifestation of the effect of EDCs on health is the development of cancer [72], caused by both the induction due to these substances of epigenetic changes [54] and the interference of these with the endocrine system and hormones, which are involved in the evolution of cancer (Table 2.2).

EDCs, in addition to the evolution of cancer and problems related to metabolism, affect most of all reproduction, growth, and development. In particular, early-life exposures to high level of EDCs have been associated with developmental abnormalities and may increase the risk for a variety of diseases later in life [23]. In particular, some classes of EDCs can mimic or block the effects of male and female sex hormones, reducing fecundability and alternating reproductive development in men and women [59] (Table 2.2).

Due to the importance and the impact on human health of EDCs, even more studies have begun to focus on potentially harmful compounds.

As illustrated in Table 2.3, endocrine-active compounds can exert their disrupting potential toward hormonal signaling through a huge variety of mechanisms acting at different levels.

According to their endocrine effect, EDCs can be classified as estrogenic, antiestrogenic, androgenic, and antiandrogenic based on their effects on the hormone system.

Of course, the main classification of EDCs is represented by division between estrogenic and androgenic compounds, intended as products that bind estrogen receptors (ERs) and androgen receptors (ARs) with an activating or inhibiting function (Table 2.1). The EDCs–hormone receptor interaction causes various effects, especially on the reproductive system [56]. Within this classification of EDCs, there are several compounds that are included in the classes of pesticides [49], phytoestrogens [26], plastics or associated chemicals (such as phthalates) [23], and drugs (especially anticancer) [234] (Table 2.1). These molecules have effects on human health mainly through two mechanisms: genomic and nongenomic (Fig. 2.1). The genomic modulation induces the regulation of hormone gene expression [45] due to the fact that hormone receptors bound by EDCs are nuclear receptors (NRs) and have a direct effect on DNA as transcription factors [75]. Despite this, EDCs also show nongenomic modulation, referring to epigenetic modifications [127] (Fig. 2.1).

EDCs have the ability to modulate endocrine system and hormonal activity by hormone mimicry and disruption as the type of genomic modulation mechanism [44]. In fact, EDCs may act as endogenous hormones and other signaling molecules of the endocrine system or even mimic them.

EDCs that compete with endogenous hormones as agonists bind HRs inducing their activation and the initiation of the hormonal signaling pathway [105]. On the contrary, EDCs can also act as antagonists of endogenous hormones (Table 2.3), by binding the same hormone receptors but occupying receptor binding site and antagonizing the proper ligand–hormone interaction [132].

Besides the hormonal activity exhibited by EDCs via receptor binding, toxicants can influence hormonal system activating other signaling pathways, that is, interacting with components of hormone signaling pathways downstream of receptor activation (Table 2.3) [45]. A parallel way to modulate hormonal system consists of altering endogenous hormone biosynthesis or degradation (Table 2.3). Again, EDCs can affect hormonal activity by regulating HR turnover through the stimulation or inhibition of their expression (Table 2.3) [45]. Finally, some EDCs exert endocrine disruption on hormonal system via epigenetic modifications as nongenomic modulation (Table 2.3), via DNA methylation and histone code alteration [1] that occurs during the development, when the epigenome is more susceptible to reprogramming by EDCs [119, 120].

This chapter introduced the general characteristics of endocrine-disrupting chemical compounds and summarized their main molecular mechanisms of action, which involve both genomic and nongenomic modulations. Endocrine disruption is associated with inappropriate regulation of hormone activity, underlining the complexity of physiological hormonal signaling and suggesting a large number of potential targets for EDC disruption. Increased studies concerning EDCs' action in hormone effects, signaling, and transcriptional regulation could provide a better understanding of the danger and their potential consequences of endocrine-active substances in human health.

References

1. Guerrero-Bosagna C, Valladares L. Endocrine disruptors, epigenetically induced changes, and transgenerational transmission of characters and epigenetic states. In: Endocrine-disrupting chemicals. Cham: Springer; 2007. p. 175–89.
2. Jablonka E, et al. The genome in context: biologists and philosophers on epigenetics. BioEssays. 2002;24(4):392–4.
3. Singal R, Ginder GD. DNA methylation. Blood. 1999;93(12):4059–70.
4. Cooper RL, et al. Atrazine disrupts the hypothalamic control of pituitary-ovarian function. Toxicol Sci. 2000;53(2):297–307.
5. MH Yu, H.T., M Tsunoda, Environmental toxicology: biological and health effects of pollutants. 2011.
6. Waller SA, et al. Agricultural-related chemical exposures, season of conception, and risk of gastroschisis in Washington state. Am J Obstet Gynecol. 2010;202(3):241 e1–6.
7. Victor-Costa AB, et al. Changes in testicular morphology and steroidogenesis in adult rats exposed to atrazine. Reprod Toxicol. 2010;29(3):323–31.
8. Maffini MV, et al. Endocrine disruptors and reproductive health: the case of bisphenol-a. Mol Cell Endocrinol. 2006;254–255:179–86.
9. Kandaraki E, et al. Endocrine disruptors and polycystic ovary syndrome (PCOS): elevated serum levels of bisphenol A in women with PCOS. J Clin Endocrinol Metab. 2011;96(3):E480–4.
10. Caserta D, et al. Bisphenol A and the female reproductive tract: an overview of recent laboratory evidence and epidemiological studies. Reprod Biol Endocrinol. 2014;12(1):37.
11. Souter I, et al. The association of bisphenol-a urinary concentrations with antral follicle counts and other measures of ovarian reserve in women undergoing infertility treatments. Reprod Toxicol. 2013;42:224–31.
12. Bauer SM, et al. The effects of maternal exposure to bisphenol a on allergic lung inflammation into adulthood. Toxicol Sci. 2012;130(1):82–93.
13. Connor TH, et al. Reproductive health risks associated with occupational exposures to antineoplastic drugs in health care settings: a review of the evidence. J Occup Environ Med. 2014;56(9):901–10.
14. Richter P, Calamera JC, Morgenfeld MC, Kierszenbaum AL, Lavieri JC, Mancini RE. Effect of chlorambucil on spermatogenesis in the human with malignant lymphoma. Cancer. 1970;25(5):1026–30.
15. Russell LB, et al. Chlorambucil effectively induces deletion mutations in mouse germ cells. Proc Natl Acad Sci U S A. 1989;86(10):3704–8.
16. Russell LB, Hunsicker PR, Shelby MD. Melphalan, a second chemical for which specific-locus mutation induction in the mouse is maximum in early spermatids. Mutat Res Lett. 1992;282(3):151–8.
17. Ning Y, et al. 5-Aza-2′-deoxycytidine inhibited PDGF-induced rat airway smooth muscle cell phenotypic switching. Arch Toxicol. 2013;87(5):871–81.
18. Stenzig J, et al. DNA methylation in an engineered heart tissue model of cardiac hypertrophy: common signatures and effects of DNA methylation inhibitors. Basic Res Cardiol. 2016;111(1):9.
19. Klaver R, et al. Direct but no transgenerational effects of decitabine and vorinostat on male fertility. PLoS One. 2015;10(2):e0117839.
20. Newbold RR, et al. Proliferative lesions and reproductive tract tumors in male descendants of mice exposed developmentally to diethylstilbestrol. Carcinogenesis. 2000;21(7):1355–63.
21. Li S, et al. Environmental exposure, DNA methylation, and gene regulation: lessons from diethylstilbesterol-induced cancers. Ann N Y Acad Sci. 2003;983(1):161–9.
22. Axelsson J, Rylander L, Rignell-Hydbom A, Lindh CH, Jönsson BA, Giwercman A. Prenatal phthalate exposure and reproductive function in young men. Environ Res. 2015;138:264–70.

23. Kim SH, Park MJ. Phthalate exposure and childhood obesity. Ann Pediatr Endocrinol Metab. 2014;19(2):69–75.
24. Rusyn I, Corton JC. Mechanistic considerations for human relevance of cancer hazard of di(2-ethylhexyl) phthalate. Mutat Res. 2012;750(2):141–58.
25. Jung T, et al. Effects of the protein phosphorylation inhibitor genistein on maturation of pig oocytes in vitro. J Reprod Fertil. 1993;98(2):529–35.
26. Helferich WG, Andrade JE, Hoagland MS. Phytoestrogens and breast cancer: a complex story. Inflammopharmacology. 2008;16(5):219–26.
27. Rawlings NC, Cook SJ, Waldbillig D. Effects of the pesticides carbofuran, chlorpyrifos, dimethoate, lindane, triallate, trifluralin, 2,4-D, and pentachlorophenol on the metabolic endocrine and reproductive endocrine system in ewes. J Toxicol Environ Health A. 1998;54(1):21–36.
28. Beard AP, Rawlings NC. Thyroid function and effects on reproduction in ewes exposed to the organochlorine pesticides lindane or pentachlorophenol (PCP) from conception. J Toxicol Environ Health A. 1999;58(8):509–30.
29. Witt KL, Bishop JB. Mutagenicity of anticancer drugs in mammalian germ cells. Mutat Res-Fund Mol Mech Mutagen. 1996;355(1–2):209–34.
30. Sifakis S, et al. Human exposure to endocrine disrupting chemicals: effects on the male and female reproductive systems. Environ Toxicol Pharmacol. 2017;51:56–70.
31. Anway MD, et al. Epigenetic transgenerational actions of endocrine disruptors and male fertility. Science. 2005;308(5727):1466–9.
32. Cupp AS, et al. Effect of transient embryonic in vivo exposure to the endocrine disruptor methoxychlor on embryonic and postnatal testis development. J Androl. 2003;24(5):736–45.
33. Jager C, Bornman MS, Horst G, I. The effect of p-nonylphenol, an environmental toxicant with oestrogenic properties, on fertility potential in adult male rats. Andrologia. 2009;31(2):99–106.
34. Zamkowska D, et al. Environmental exposure to non-persistent endocrine disrupting chemicals and semen quality: an overview of the current epidemiological evidence. Int J Occup Med Environ Health. 2018;31(4):377–414.
35. Miyagawa S, Sato T, Iguchi T. Nonylphenol. Handbook of hormones comparative endocrinology for basic and clinical research. Academic Press. 2016; Subchapter 101A5: 73–574.
36. Gilbert SF, Epel D. Ecological developmental biology: integrating epigenetics, Medicine, and evolution. Sunderland, MA: Sinauer Associates Inc; 2009.
37. Brieno-Enriquez MA, et al. Exposure to endocrine disruptor induces transgenerational epigenetic deregulation of microRNAs in primordial germ cells. PLoS One. 2015;10(4):e0124296.
38. Guerrero-Bosagna C, et al. Epigenetic transgenerational actions of vinclozolin on promoter regions of the sperm epigenome. PLoS One. 2010;5(9):e13100.
39. Andersen HR, et al. Effects of currently used pesticides in assays for estrogenicity, androgenicity, and aromatase activity in vitro. Toxicol Appl Pharmacol. 2002;179(1):1–12.
40. Mikamo E, et al. Endocrine disruptors induce cytochrome P450 by affecting transcriptional regulation via pregnane X receptor. Toxicol Appl Pharmacol. 2003;193(1):66–72.
41. Fang H, et al. Study of 202 natural, synthetic, and environmental chemicals for binding to the androgen receptor. Chem Res Toxicol. 2003;16(10):1338–58.
42. Anway MD, Leathers C, Skinner MK. Endocrine disruptor vinclozolin induced epigenetic transgenerational adult-onset disease. Endocrinology. 2006;147(12):5515–23.
43. Skinner MK, et al. Transgenerational epigenetic programming of the brain transcriptome and anxiety behavior. PLoS One. 2008;3(11):e3745.
44. Swedenborg E, et al. Endocrine disruptive chemicals: mechanisms of action and involvement in metabolic disorders. J Mol Endocrinol. 2009;43(1):1–10.
45. Combarnous Y, Nguyen TMD. Comparative overview of the mechanisms of action of hormones and endocrine disruptor compounds. Toxics. 2019;7(1):5.

46. Pimentel D, et al. Ecology of increasing diseases: population growth and environmental degradation. Hum Ecol Interdisc J. 2007;35(6):653–68.
47. National Center for Environmental Health (U.S.). Division of Laboratory Sciences.; National Health and Nutrition Examination Survey (U.S.). Third national report on human exposure to environmental chemicals. 2005:1–467.
48. Environmental Working Group. Pollution in people: Cord blood contaminants in minority newborns. 2009;6(2011):3–60.
49. Mnif W, et al. Effect of endocrine disruptor pesticides: a review. Int J Environ Res Public Health. 2011;8(6):2265–303.
50. Jarow JP, et al. Best practice policies for male infertility. J Urol. 2002;167(5):2138–44.
51. Westerhoff P, et al. Fate of endocrine-disruptor, pharmaceutical, and personal care product chemicals during simulated drinking water treatment processes. Environ Sci Technol. 2005;39(17):6649–63.
52. Zoeller RTJM. Environmental chemicals as thyroid hormone analogues: new studies indicate that thyroid hormone receptors are targets of industrial chemicals? Mol Cell Endocrinol. 2005;242(1–2):10–5.
53. Lecomte S, et al. Phytochemicals targeting estrogen receptors: Beneficial rather than adverse effects? Int J Mol Sci. 2017;18(7):1381.
54. Wynne-Edwards KE. Evolutionary biology of plant defenses against herbivory and their predictive implications for endocrine disruptor susceptibility in vertebrates. Environ Health Perspect. 2001;109(5):443–8.
55. Lopez-Rodriguez D, et al. Cellular and molecular features of EDC exposure: consequences for the GnRH network. Nat Rev Endocrinol. 2021;17:1–14.
56. Greathouse KL, Walker CL. Environmental impacts on reproductive health and fertility. mechanisms of endocrine disruption. Cambridge University Press. 2010:72.
57. Gore AC. Developmental programming and endocrine disruptor effects on reproductive neuroendocrine systems. Front Neuroendocrinol. 2008;29(3):358–74.
58. Yeung BH, et al. Endocrine disrupting chemicals: multiple effects on testicular signaling and spermatogenesis. Spermatogenesis. 2011;1(3):231–9.
59. Caserta D, et al. Impact of endocrine disruptor chemicals in gynaecology. Hum Reprod Update. 2008;14(1):59–72.
60. Kajta M, Wojtowicz AK. Impact of endocrine-disrupting chemicals on neural development and the onset of neurological disorders. Pharmacol Rep. 2013;65(6):1632–9.
61. Kajta M, Wójtowicz AK. Impact of endocrine-disrupting chemicals on neural development and the onset of neurological disorders. Pharmacol Rep. 2013;65(6):1632–9.
62. Segovia-Mendoza M, et al. How microplastic components influence the immune system and impact on children health: focus on cancer. Birth Defects Res. 2020;112(17):1341–61.
63. Brisken C. Endocrine disruptors and breast cancer. Chimia Int J Chem. 2008;62(5):406–9.
64. Soto AM, Sonnenschein C. Environmental causes of cancer: endocrine disruptors as carcinogens. Nat Rev Endocrinol. 2010;6(7):364–71.
65. Moriyama K, et al. Thyroid hormone action is disrupted by bisphenol a as an antagonist. J Clin Endocrinol Metab. 2002;87(11):5185–90.
66. Diamanti-Kandarakis E, et al. The impact of endocrine disruptors on endocrine targets. Horm Metab Res. 2010;42(8):543–52.
67. Arsenescu V, et al. Polychlorinated biphenyl-77 induces adipocyte differentiation and proinflammatory adipokines and promotes obesity and atherosclerosis. Environ Health Perspect. 2008;116(6):761–8.
68. Després J-P, et al. Abdominal obesity and the metabolic syndrome: contribution to global cardiometabolic risk. Arterioscler Thromb Vasc Biol. 2008;28(6):1039–49.
69. Phillips LK, Prins JB. The link between abdominal obesity and the metabolic syndrome. Curr Hypertens Rep. 2008;10(2):156–64.
70. Lee D-H, et al. A strong dose-response relation between serum concentrations of persistent organic pollutants and diabetes: results from the National Health and Examination Survey 1999–2002. Diabetes Care. 2006;29(7):1638–44.

71. Lang IA, et al. Association of urinary bisphenol a concentration with medical disorders and laboratory abnormalities in adults. JAMA. 2008;300(11):1303–10.
72. Brisken C. Endocrine disruptors and breast cancer. Chimia. 2008;62(5):406–9.
73. Anway MD, Skinner MK. Epigenetic programming of the germ line: effects of endocrine disruptors on the development of transgenerational disease. Reprod BioMed Online. 2008;16(1):23–5.
74. Dickerson SM, Gore AC. Estrogenic environmental endocrine-disrupting chemical effects on reproductive neuroendocrine function and dysfunction across the life cycle. Rev Endocr Metab Disord. 2007;8(2):143–59.
75. Gronemeyer H, Gustafsson JA, Laudet V. Principles for modulation of the nuclear receptor superfamily. Nat Rev Drug Discov. 2004;3(11):950–64.
76. Baker ME, Medlock KL, Sheehan DM. Flavonoids inhibit estrogen binding to rat alpha-fetoprotein. Proc Soc Exp Biol Med. 1998;217(3):317–21.
77. Boas M, et al. Environmental chemicals and thyroid function. Eur J Endocrinol. 2006;154(5):599–611.
78. Baillie-Hamilton PF. Chemical toxins: a hypothesis to explain the global obesity epidemic. J Altern Complement Med. 2002;8(2):185–92.
79. Newbold RR, et al. Developmental exposure to endocrine disruptors and the obesity epidemic. Reprod Toxicol. 2007;23(3):290–6.
80. Newbold RR, et al. Effects of endocrine disruptors on obesity. Int J Androl. 2008;31(2):201–8.
81. McMillan BJ, Bradfield CA. The aryl hydrocarbon receptor sans xenobiotics: endogenous function in genetic model systems. Mol Pharmacol. 2007;72(3):487–98.
82. Beischlag TV, et al. The aryl hydrocarbon receptor complex and the control of gene expression. Crit Rev Eukaryot Gene Expr. 2008;18(3):207–50.
83. Gore AC. Endocrine-disrupting chemicals: from basic research to clinical practice. Totowa, NJ: Humana Press; 2007.
84. Warner M, et al. Diabetes, metabolic syndrome, and obesity in relation to serum dioxin concentrations: the Seveso women's health study. Environ Health Perspect. 2013;121(8):906–11.
85. Wang CX, et al. Exposure to persistent organic pollutants as potential risk factors for developing diabetes. Sci Chin-Chem. 2010;53(5):980–94.
86. Bonefeld-Jørgensen EC, et al. Endocrine-disrupting potential of bisphenol A, bisphenol A dimethacrylate, 4-n-nonylphenol, and 4-n-octylphenol in vitro: new data and a brief review. Environ Health Perspect. 2007;115(Suppl 1):69–76.
87. Baird DD, Newbold R. Prenatal diethylstilbestrol (DES) exposure is associated with uterine leiomyoma development. Reprod Toxicol. 2005;20(1):81–4.
88. Mustieles Miralles V. Maternal and paternal preconception exposure to phenols and preterm birth. Environ Int. 2020;137:105523.
89. Taylor JA, et al. Interactive effects of perinatal BPA or DES and adult testosterone and estradiol exposure on adult urethral obstruction and bladder, kidney, and prostate pathology in male mice. Int J Mol Sci. 2020;21(11):3902.
90. Anway MD, Skinner MK. Epigenetic transgenerational actions of endocrine disruptors. Endocrinology. 2006;147(6 Suppl):S43–9.
91. Chang HS, et al. Transgenerational epigenetic imprinting of the male germline by endocrine disruptor exposure during gonadal sex determination. Endocrinology. 2006;147(12):5524–41.
92. Watkins DJ, et al. Impact of phthalate and BPA exposure during in utero windows of susceptibility on reproductive hormones and sexual maturation in peripubertal males. Environ Health. 2017;16(1):69.
93. Dumesic DA, Abbott DH, Padmanabhan V. Polycystic ovary syndrome and its developmental origins. Rev Endocr Metab Disord. 2007;8(2):127–41.
94. Padula AM. The freemartin syndrome: an update. Anim Reprod Sci. 2005;87(1–2):93–109.
95. Vilahur N, et al. Prenatal exposure to mixtures of xenoestrogens and repetitive element DNA methylation changes in human placenta. Environ Int. 2014;71:81–7.

96. de Cock J, et al. Time to pregnancy and occupational exposure to pesticides in fruit growers in The Netherlands. Occup Environ Med. 1994;51(10):693–9.
97. Sugiura-Ogasawara M. Reply to: 'limitations of a case–control study on bisphenol a (BPA) serum levels and recurrent miscarriage'. Hum Reprod. 2006;21(2):566–7.
98. Scott HM, Mason JI, Sharpe RM. Steroidogenesis in the fetal testis and its susceptibility to disruption by exogenous compounds. Endocr Rev. 2009;30(7):883–925.
99. Sharpe RM. Pathways of endocrine disruption during male sexual differentiation and masculinization. Best Pract Res Clin Endocrinol Metab. 2006;20(1):91–110.
100. Sharpe RM. Environmental/lifestyle effects on spermatogenesis. Philos Trans R Soc Lond Ser B Biol Sci. 2010;365(1546):1697–712.
101. Orth JM, Gunsalus GL, Lamperti AA. Evidence from Sertoli cell-depleted rats indicates that spermatid number in adults depends on numbers of Sertoli cells produced during perinatal development. Endocrinology. 1988;122(3):787–94.
102. Orth JM, et al. Gonocyte-Sertoli cell interactions during development of the neonatal rodent testis. Curr Top Dev Biol. 2000;50(50):103–24.
103. Adler SR. Cellular mechanisms of endocrine disruption. In: Endocrine-disrupting chemicals. Cham: Springer; 2007. p. 135–74.
104. Martini M, Corces VG, Rissman EF. Mini-review: epigenetic mechanisms that promote transgenerational actions of endocrine disrupting chemicals: applications to behavioral neuroendocrinology. Horm Behav. 2020;119:104677.
105. Montes-Grajales D, Olivero-Verbel J. EDCs DataBank: 3D-structure database of endocrine disrupting chemicals. Toxicology. 2015;327:87–94.
106. Wong C, et al. Androgen receptor antagonist versus agonist activities of the fungicide vinclozolin relative to hydroxyflutamide. J Biol Chem. 1995;270(34):19998–20003.
107. Kelce WR, et al. Persistent DDT metabolite p,p'-DDE is a potent androgen receptor antagonist. Nature. 1995;375(6532):581–5.
108. Daxenberger A. Pollutants with androgen-disrupting potency. Eur J Lipid Sci Technol. 2002;104(2):124–30.
109. Klaassen CD. Casarett and Doull's toxicology: the basic science of poisons. Ann Intern Med. 1992;117(5)
110. Norman AW, Henry HL. Hormones. Amsterdam: Elsevier; 1997.
111. Wan Q, et al. Research progress on the relationship between sex hormone-binding globulin and male reproductive system diseases. Andrologia. 2021;53:e13893.
112. Sheikh IA, et al. Endocrine disruption: computational perspectives on human sex hormone-binding globulin and phthalate plasticizers. PLoS One. 2016;11(3):e0151444.
113. Boas M, Feldt-Rasmussen U, Main KM. Thyroid effects of endocrine disrupting chemicals. Mol Cell Endocrinol. 2012;355(2):240–8.
114. Qiu LL, et al. Decreased androgen receptor expression may contribute to spermatogenesis failure in rats exposed to low concentration of bisphenol a. Toxicol Lett. 2013;219(2):116–24.
115. Waddington CH. The epigenotype. 1942. Int J Epidemiol. 2012;41(1):10–3.
116. Holliday R, Pugh JE. DNA modification mechanisms and gene activity during development. Science. 1975;187(4173):226–32.
117. Riggs AD. X inactivation, differentiation, and DNA methylation. Cytogenet Cell Genet. 1975;14(1):9–25.
118. Quivy V, et al. Gene activation and gene silencing: a subtle equilibrium. Cloning Stem Cells. 2004;6(2):140–9.
119. Schaefer CB, et al. Epigenetic decisions in mammalian germ cells. Science. 2007;316(5823):398–9.
120. Santos F, et al. Dynamic reprogramming of DNA methylation in the early mouse embryo. Dev Biol. 2002;241(1):172–82.
121. Sanz LA, et al. A mono-allelic bivalent chromatin domain controls tissue-specific imprinting at Grb10. EMBO J. 2008;27(19):2523–32.

122. Alworth LC, et al. Uterine responsiveness to estradiol and DNA methylation are altered by fetal exposure to diethylstilbestrol and methoxychlor in CD-1 mice: effects of low versus high doses. Toxicol Appl Pharmacol. 2002;183(1):10–22.
123. Dolinoy DC, Huang D, Jirtle RL. Maternal nutrient supplementation counteracts bisphenol A-induced DNA hypomethylation in early development. Proc Natl Acad Sci U S A. 2007;104(32):13056–61.
124. Anway MD, Rekow SS, Skinner MK. Transgenerational epigenetic programming of the embryonic testis transcriptome. Genomics. 2008;91(1):30–40.
125. Newbold RR, et al. Developmental exposure to diethylstilbestrol alters uterine gene expression that may be associated with uterine neoplasia later in life. Mol Carcinog. 2007;46(9):783–96.
126. Tang WY, et al. Persistent hypomethylation in the promoter of nucleosomal binding protein 1 (Nsbp1) correlates with overexpression of Nsbp1 in mouse uteri neonatally exposed to diethylstilbestrol or genistein. Endocrinology. 2008;149(12):5922–31.
127. Champagne FA. Epigenetic mechanisms and the transgenerational effects of maternal care. Front Neuroendocrinol. 2008;29(3):386–97.
128. Thomas P, Dong J. Binding and activation of the seven-transmembrane estrogen receptor GPR30 by environmental estrogens: a potential novel mechanism of endocrine disruption. J Steroid Biochem Mol Biol. 2006;102(1–5):175–9.
129. Hiroi H, et al. Differential interactions of bisphenol a and 17beta-estradiol with estrogen receptor alpha (ERalpha) and ERbeta. Endocr J. 1999;46(6):773–8.
130. Kucka M, et al. Atrazine acts as an endocrine disrupter by inhibiting cAMP-specific phosphodiesterase-4. Toxicol Appl Pharmacol. 2012;265(1):19–26.
131. Sargis RM, et al. The novel endocrine disruptor tolylfluanid impairs insulin signaling in primary rodent and human adipocytes through a reduction in insulin receptor substrate-1 levels. Biochim Biophys Acta. 2012;1822(6):952–60.
132. Tabb MM, Blumberg B. New modes of action for endocrine-disrupting chemicals. Mol Endocrinol. 2006;20(3):475–82.
133. Rao K, et al. In vitro agonistic and antagonistic endocrine disrupting effects of organic extracts from waste water of different treatment processes. Front Environ Sci Eng. 2013;8(1):69–78.
134. Stepulak A, et al. Fluoxetine inhibits the extracellular signal regulated kinase pathway and suppresses growth of cancer cells. Cancer Biol Ther. 2008;7(10):1685–93.
135. Liu XL, et al. Fluoxetine regulates mTOR signalling in a region-dependent manner in depression-like mice. Sci Rep. 2015;5(1):16024.
136. Ofek K, et al. Fluoxetine induces vasodilatation of cerebral arterioles by co-modulating NO/ muscarinic signalling. J Cell Mol Med. 2012;16(11):2736–44.
137. Schopel M, et al. Allosteric activation of GDP-bound Ras isoforms by Bisphenol derivative plasticisers. Int J Mol Sci. 2018;19(4):1133.
138. Ptak A, Gregoraszczuk EL. Bisphenol a induces leptin receptor expression, creating more binding sites for leptin, and activates the JAK/stat, MAPK/ERK and PI3K/Akt signalling pathways in human ovarian cancer cell. Toxicol Lett. 2012;210(3):332–7.
139. Hugo ER, et al. Bisphenol A at environmentally relevant doses inhibits adiponectin release from human adipose tissue explants and adipocytes. Environ Health Perspect. 2008;116(12):1642–7.
140. Jambor T, et al. In vitro effect of 4-nonylphenol on human chorionic gonadotropin (hCG) stimulated hormone secretion, cell viability and reactive oxygen species generation in mice Leydig cells. Environ Pollut. 2017;222:219–25.
141. Derouiche S, et al. Activation of TRPA1 channel by antibacterial agent Triclosan induces VEGF secretion in human prostate cancer stromal cells. Cancer Prev Res (Phila). 2017;10(3):177–87.
142. Engeli RT, et al. Interference of paraben compounds with estrogen metabolism by inhibition of 17beta-Hydroxysteroid dehydrogenases. Int J Mol Sci. 2017;18(9):2007.

143. Ishihara A, Sawatsubashi S, Yamauchi K. Endocrine disrupting chemicals: interference of thyroid hormone binding to transthyretins and to thyroid hormone receptors. Mol Cell Endocrinol. 2003;199(1–2):105–17.
144. Déchaud H, et al. Xenoestrogen interaction with human sex hormone-binding globulin (hSHBG)1. Steroids. 1999;64(5):328–34.
145. Yamauchi K, Ishihara A. Thyroid system-disrupting chemicals: interference with thyroid hormone binding to plasma proteins and the cellular thyroid hormone signaling pathway. Rev Environ Health. 2006;21(4):229–51.
146. Hong H, et al. Human sex hormone-binding globulin binding affinities of 125 structurally diverse chemicals and comparison with their binding to androgen receptor, estrogen receptor, and alpha-fetoprotein. Toxicol Sci. 2015;143(2):333–48.
147. Waalkes MP, et al. Estrogen signaling in livers of male mice with hepatocellular carcinoma induced by exposure to arsenic in utero. J Natl Cancer Inst. 2004;96(6):466–74.
148. Li Y, et al. Epigenetic reactivation of estrogen receptor-alpha (ERalpha) by genistein enhances hormonal therapy sensitivity in ERalpha-negative breast cancer. Mol Cancer. 2013;12(1):9.
149. Schmutzler C, et al. Thyroid hormone biosynthesis is a sensitive target for the action of endocrine disrupting chemicals (EDC). Exp Clin Endocrinol Diabetes. 2006;114(S 1):OR8_44.
150. Grun F, Blumberg B. Environmental obesogens: organotins and endocrine disruption via nuclear receptor signaling. Endocrinology. 2006;147(6 Suppl):S50–5.
151. Kanayama T, et al. Organotin compounds promote adipocyte differentiation as agonists of the peroxisome proliferator-activated receptor gamma/retinoid X receptor pathway. Mol Pharmacol. 2005;67(3):766–74.
152. Woodruff TJ, et al. Environmental impacts on reproductive health and fertility. Cambridge University Press; 2010.
153. Beato M. Gene regulation by steroid hormones. In: Gene Expression. Springer; 1993. p. 43–75.
154. Gustafsson JA. Estrogen receptor beta--a new dimension in estrogen mechanism of action. J Endocrinol. 1999;163(3):379–83.
155. Couse JF, et al. Tissue distribution and quantitative analysis of estrogen receptor-alpha (ERalpha) and estrogen receptor-beta (ERbeta) messenger ribonucleic acid in the wild-type and ERalpha-knockout mouse. Endocrinology. 1997;138(11):4613–21.
156. Mitchner NA, Garlick C, Ben-Jonathan N. Cellular distribution and gene regulation of estrogen receptors alpha and beta in the rat pituitary gland. Endocrinology. 1998;139(9):3976–83.
157. Sar M, Welsch F. Differential expression of estrogen receptor-beta and estrogen receptor-alpha in the rat ovary. Endocrinology. 1999;140(2):963–71.
158. Shughrue PJ, Komm B, Merchenthaler I. The distribution of estrogen receptor-beta mRNA in the rat hypothalamus. Steroids. 1996;61(12):678–81.
159. Shughrue PJ, et al. Comparative distribution of estrogen receptor-alpha (ER-alpha) and beta (ER-beta) mRNA in the rat pituitary, gonad, and reproductive tract. Steroids. 1998;63(10):498–504.
160. Dotzlaw H, et al. Expression of estrogen receptor-beta in human breast tumors. J Clin Endocrinol Metab. 1997;82(7):2371–4.
161. Ferguson AT, Lapidus RG, Davidson NE. The regulation of estrogen receptor expression and function in human breast cancer. In: Biological and hormonal therapies of cancer. Cham: Springer; 1998. p. 255–78.
162. Vladusic EA, et al. Expression of estrogen receptor beta messenger RNA variant in breast cancer. Cancer Res. 1998;58(2):210–4.
163. Yoon K, et al. Estrogenic endocrine-disrupting chemicals: molecular mechanisms of actions on putative human diseases. J Toxicol Environ Health B Crit Rev. 2014;17(3):127–74.
164. Beato M. Transcriptional control by nuclear receptors. FASEB J. 1991;5(7):2044–51.
165. Klinge CM. Estrogen receptor interaction with co-activators and co-repressors. Steroids. 2000;65(5):227–51.

166. Diamond MI, et al. Transcription factor interactions: selectors of positive or negative regulation from a single DNA element. Science. 1990;249(4974):1266–72.
167. Tzukerman M, Zhang XK, Pfahl M. Inhibition of estrogen receptor activity by the tumor promoter 12-O-tetradeconylphorbol-13-acetate: a molecular analysis. Mol Endocrinol. 1991;5(12):1983–92.
168. Jonat C, et al. Antitumor promotion and antiinflammation: down-modulation of AP-1 (Fos/Jun) activity by glucocorticoid hormone. Cell. 1990;62(6):1189–204.
169. Lucibello FC, et al. Mutual transrepression of Fos and the glucocorticoid receptor: involvement of a functional domain in Fos which is absent in FosB. EMBO J. 1990;9(9):2827–34.
170. Yang-Yen H-F, et al. Transcriptional interference between c-Jun and the glucocorticoid receptor: mutual inhibition of DNA binding due to direct protein-protein interaction. Cell. 1990;62(6):1205–15.
171. Vondracek J, Kozubik A, Machala M. Modulation of estrogen receptor-dependent reporter construct activation and G0/G1-S-phase transition by polycyclic aromatic hydrocarbons in human breast carcinoma MCF-7 cells. Toxicol Sci. 2002;70(2):193–201.
172. Coldham NG, et al. Evaluation of a recombinant yeast cell estrogen screening assay. Environ Health Perspect. 1997;105(7):734–42.
173. Eroschenko VP, Rourke AW, Sims WF. Estradiol or methoxychlor stimulates estrogen receptor (ER) expression in uteri. Reprod Toxicol. 1996;10(4):265–71.
174. Routledge EJ, et al. Differential effects of xenoestrogens on coactivator recruitment by estrogen receptor (ER) alpha and ERbeta. J Biol Chem. 2000;275(46):35986–93.
175. Pak TR, et al. The androgen metabolite, 5α-Androstane-3β, 17β-diol, is a potent modulator of estrogen receptor-β1-mediated gene transcription in neuronal cells. Endocrinology. 2005;146(1):147–55.
176. Imamov O, Lopatkin NA, Gustafsson JA. Estrogen receptor beta in prostate cancer. N Engl J Med. 2004;351(26):2773–4.
177. Welshons WV, et al. Low-dose bioactivity of xenoestrogens in animals: fetal exposure to low doses of methoxychlor and other xenoestrogens increases adult prostate size in mice. Toxicol Ind Health. 1999;15(1–2):12–25.
178. Oostenbrink C, van Gunsteren WF. Free energies of ligand binding for structurally diverse compounds. Proc Natl Acad Sci U S A. 2005;102(19):6750–4.
179. Gould JC, et al. Bisphenol a interacts with the estrogen receptor alpha in a distinct manner from estradiol. Mol Cell Endocrinol. 1998;142(1–2):203–14.
180. Evans RM. The steroid and thyroid hormone receptor superfamily. Science. 1988;240(4854):889–95.
181. Horard B, Vanacker JM. Estrogen receptor-related receptors: orphan receptors desperately seeking a ligand. J Mol Endocrinol. 2003;31(3):349–57.
182. Spink DC, et al. Stimulation of 17 beta-estradiol metabolism in MCF-7 cells by bromochloro- and chloromethyl-substituted dibenzo-p-dioxins and dibenzofurans: correlations with antiestrogenic activity. J Toxicol Environ Health. 1994;41(4):451–66.
183. Spink DC, et al. 17β-estradiol hydroxylation catalyzed by human cytochrome P450 1A1: a comparison of the activities induced by 2,3,7,8-tetrachlorodibenzo-p-dioxin in MCF-7 cells with those from heterologous expression of the cDNA. Arch Biochem Biophys. 1992;293(2):342–8.
184. La Rosa P, et al. Xenoestrogens alter estrogen receptor (ER) alpha intracellular levels. PLoS One. 2014;9(2):e88961.
185. Sheeler CQ, Dudley MW, Khan SA. Environmental estrogens induce transcriptionally active estrogen receptor dimers in yeast: activity potentiated by the coactivator RIP140. Environ Health Perspect. 2000;108(2):97–103.
186. Khurana S, Ranmal S, Ben-Jonathan N. Exposure of newborn male and female rats to environmental estrogens: delayed and sustained hyperprolactinemia and alterations in estrogen receptor expression. Endocrinology. 2000;141(12):4512–7.
187. Dennis AP, Haq RU, Nawaz Z. Importance of the regulation of nuclear receptor degradation. Front Biosci. 2001;6:D954–9.

188. Masuyama H, et al. Endocrine disrupting chemicals, phthalic acid and nonylphenol, activate Pregnane X receptor-mediated transcription. Mol Endocrinol. 2000;14(3):421–8.
189. Inoshita H, Masuyama H, Hiramatsu Y. The different effects of endocrine-disrupting chemicals on estrogen receptor-mediated transcription through interaction with coactivator TRAP220 in uterine tissue. J Mol Endocrinol. 2003;31(3):551–61.
190. Masuyama H, Hiramatsu Y. Involvement of suppressor for Gal 1 in the ubiquitin/proteasome-mediated degradation of estrogen receptors. J Biol Chem. 2004;279(13):12020–6.
191. Jordan VC, et al. A monohydroxylated metabolite of tamoxifen with potent antioestrogenic activity. J Endocrinol. 1977;75(2):305–16.
192. Johanson G. Toxicity review of ethylene glycol monomethyl ether and its acetate ester. Crit Rev Toxicol. 2000;30(3):307–45.
193. Cummings AM. Methoxychlor as a model for environmental estrogens. Crit Rev Toxicol. 1997;27(4):367–79.
194. O'Brian CA, et al. Inhibition of protein kinase C by tamoxifen. Cancer Res. 1985;45(6):2462–5.
195. Issandou M, et al. Opposite effects of tamoxifen on in vitro protein kinase C activity and endogenous protein phosphorylation in intact MCF-7 cells. Cancer Res. 1990;50(18):5845–50.
196. Canesi L, et al. Effects of PCB congeners on the immune function of Mytilus hemocytes: alterations of tyrosine kinase-mediated cell signaling. Aquat Toxicol. 2003;63(3):293–306.
197. Canesi L, et al. Environmental estrogens can affect the function of mussel hemocytes through rapid modulation of kinase pathways. Gen Comp Endocrinol. 2004;138(1):58–69.
198. Rose DP, Boyar AP, Wynder EL. International comparisons of mortality rates for cancer of the breast, ovary, prostate, and colon, and per capita food consumption. Cancer. 1986;58(11):2363–71.
199. Food and A Drug. Food labeling health claims; soy protein and coronary heart disease. Fed Regist. 1999;64:57699–733.
200. Wu JM, et al. Mechanism of cardioprotection by resveratrol, a phenolic antioxidant present in red wine (review). Int J Mol Med. 2001;8(1):3–17.
201. Marlett JA, et al. Position of the American dietetic association: health implications of dietary fiber. J Am Diet Assoc. 2002;102(7):993–1000.
202. Hinnebusch BF, et al. The effects of short-chain fatty acids on human colon cancer cell phenotype are associated with histone hyperacetylation. J Nutr. 2002;132(5):1012–7.
203. Villar-Garea A, Esteller M. Histone deacetylase inhibitors: understanding a new wave of anticancer agents. Int J Cancer. 2004;112(2):171–8.
204. Gehm BD, et al. Resveratrol, a polyphenolic compound found in grapes and wine, is an agonist for the estrogen receptor. Proc Natl Acad Sci U S A. 1997;94(25):14138–43.
205. Howitz KT, et al. Small molecule activators of sirtuins extend Saccharomyces cerevisiae lifespan. Nature. 2003;425(6954):191–6.
206. Wood JG, et al. Sirtuin activators mimic caloric restriction and delay ageing in metazoans. Nature. 2004;430(7000):686–9.
207. Kelce WR, Wilson EM. Antiandrogenic effects of environmental endocrine disruptors. In: Endocrine disruptors—part I. cham: Springer; 2001. p. 39–61.
208. Matsumoto T, et al. Androgen receptor functions in male and female physiology. J Steroid Biochem Mol Biol. 2008;109(3–5):236–41.
209. Holloway CC, Clayton DF. Estrogen synthesis in the male brain triggers development of the avian song control pathway in vitro. Nat Neurosci. 2001;4(2):170–5.
210. Beyer C, Green SJ, Hutchison JB. Androgens influence sexual differentiation of embryonic mouse hypothalamic aromatase neurons in vitro. Endocrinology. 1994;135(3):1220–6.
211. Burek M, et al. Tissue-specific distribution of the androgen receptor (AR) in the porcine fetus. Acta Histochem. 2007;109(5):358–65.
212. Davison SL, Davis SR. Androgens in women. J Steroid Biochem Mol Biol. 2003;85(2–5):363–6.
213. Nef S, Parada LF. Hormones in male sexual development. Genes Dev. 2000;14(24):3075–86.
214. Luccio-Camelo DC, Prins GS. Disruption of androgen receptor signaling in males by environmental chemicals. J Steroid Biochem Mol Biol. 2011;127(1–2):74–82.

215. Chapin RE, et al. Endocrine modulation of reproduction. Fundam Appl Toxicol. 1996;29(1):1–17.
216. Welsch F. How can chemical compounds alter human fertility? Eur J Obstet Gynecol Reprod Biol. 2003;106(1):88–91.
217. List HJ, et al. Effects of antiandrogens on chromatin remodeling and transcription of the integrated mouse mammary tumor virus promoter. Exp Cell Res. 2000;260(1):160–5.
218. Massaad C, et al. How can chemical compounds alter human fertility? Eur J Obstet Gynecol Reprod Biol. 2002;100(2):127–37.
219. Portigal CL, et al. Polychlorinated biphenyls interfere with androgen-induced transcriptional activation and hormone binding. Toxicol Appl Pharmacol. 2002;179(3):185–94.
220. Bonefeld-Jorgensen EC, et al. Endocrine-disrupting potential of bisphenol A, bisphenol A dimethacrylate, 4-n-nonylphenol, and 4-n-octylphenol in vitro: new data and a brief review. Environ Health Perspect. 2007;115(Suppl 1):69–76.
221. Ostby J, et al. The fungicide procymidone alters sexual differentiation in the male rat by acting as an androgen-receptor antagonist in vivo and in vitro. Toxicol Ind Health. 1999;15(1–2):80–93.
222. McKinnell C, et al. Suppression of androgen action and the induction of gross abnormalities of the reproductive tract in male rats treated neonatally with diethylstilbestrol. J Androl. 2001;22(2):323–38.
223. Araki N, et al. Screening for androgen receptor activities in 253 industrial chemicals by in vitro reporter gene assays using AR-EcoScreen cells. Toxicol in Vitro. 2005;19(6):831–42.
224. Larsson A, et al. Identification of the brominated flame retardant 1,2-dibromo-4-(1,2-dibromoethyl)cyclohexane as an androgen agonist. J Med Chem. 2006;49(25):7366–72.
225. Nyholm JR, et al. Maternal transfer of brominated flame retardants in zebrafish (Danio rerio). Chemosphere. 2008;73(2):203–8.
226. Arsenault G, et al. Structure characterization and thermal stabilities of the isomers of the brominated flame retardant 1,2-dibromo-4-(1,2-dibromoethyl)cyclohexane. Chemosphere. 2008;72(8):1163–70.
227. Khalaf H, et al. Diastereomers of the brominated flame retardant 1,2-dibromo-4-(1,2 dibromoethyl)cyclohexane induce androgen receptor activation in the hepg2 hepatocellular carcinoma cell line and the lncap prostate cancer cell line. Environ Health Perspect. 2009;117(12):1853–9.
228. Danzo BJ. Environmental xenobiotics may disrupt normal endocrine function by interfering with the binding of physiological ligands to steroid receptors and binding proteins. Environ Health Perspect. 1997;105(3):294–301.
229. Lo S, et al. Effects of various pesticides on human 5alpha-reductase activity in prostate and LNCaP cells. Toxicol in Vitro. 2007;21(3):502–8.
230. Wolf C Jr, et al. Administration of potentially antiandrogenic pesticides (procymidone, linuron, iprodione, chlozolinate, p,p'-DDE, and ketoconazole) and toxic substances (dibutyl- and diethylhexyl phthalate, PCB 169, and ethane dimethane sulphonate) during sexual differentiation produces diverse profiles of reproductive malformations in the male rat. Toxicol Ind Health. 1999;15(1–2):94–118.
231. You L, Brenneman KA, Heck H. In utero exposure to antiandrogens alters the responsiveness of the prostate to p,p'-DDE in adult rats and may induce prostatic inflammation. Toxicol Appl Pharmacol. 1999;161(3):258–66.
232. Xu LC, et al. Evaluation of androgen receptor transcriptional activities of bisphenol A, octylphenol and nonylphenol in vitro. Toxicology. 2005;216(2–3):197–203.
233. Kelce WR, et al. Environmental hormone disruptors: evidence that vinclozolin developmental toxicity is mediated by antiandrogenic metabolites. Toxicol Appl Pharmacol. 1994;126(2):276–85.
234. Mittendorf R. Teratogen update: carcinogenesis and teratogenesis associated with exposure to diethylstilbestrol (DES) in utero. Teratology. 1995;51(6):435–45.

Chapter 3
Endocrine Disruptors, Epigenetic Changes, and Transgenerational Transmission

Roberta Rizzo, Daria Bortolotti, Sabrina Rizzo, and Giovanna Schiuma

3.1 Effects of Epigenetics on Human Reproduction: An Introduction

Recent discoveries in the field of molecular biology are focused on phenomena like chromatin condensation, histone (H) modification, and deoxyribonucleic acid (DNA) methylation, as well as the action of small non-coding ribonucleic acid (RNA), which together belong to the branch of epigenetics. The term "epigenetics" was coined in 1940 by Conrad Waddington [1] who described it as "the branch of biology which studies the causal interactions between genes and their product which bring phenotypes into being." In fact, epigenetics includes all those mechanisms that are able to regulate DNA expression without modifying nucleotide sequence.

Among the main epigenetic mechanisms mentioned earlier, DNA methylation is the most widely known and most studied modification (Table 3.1). The process of DNA methylation constitutes a postreplicative modification, in which a methyl group is added covalently to a DNA residue [10]. The methylation occurs at the carbon 5 of the cytosine ring in $5'$-$3'$-oriented CG dinucleotides (named as CpGs), and it is catalyzed by the action of DNA methyltransferases (DNMTs) [11]. Furthermore, recent evidences have shown that also RNA factors, such as small RNAs (small interfering RNA [siRNA] and microRNA [miRNA]), have the ability to direct DNA methylation through a mechanism called RNA-directed DNA methylation (RdDM), performed by double-stranded RNA (dsRNA), which may be produced after the transcription of inverted repeats [12].

R. Rizzo (✉) · D. Bortolotti · S. Rizzo · G. Schiuma
Department of Chemical, Pharmaceutical and Agricultural Sciences, University of Ferrara, Ferrara, Italy
e-mail: rbr@unife.it; daria.bortolotti@unife.it; sabrina.rizzo@unife.it; giovanna.schiuma@unife.it

© The Author(s) 2023
R. Marci (ed.), *Environment Impact on Reproductive Health*,
https://doi.org/10.1007/978-3-031-36494-5_3

Table 3.1 Main epigenetic mechanisms, molecular targets, and effects on gene expression

Epigenetic modification	Molecular target	Effect on gene expression
DNA methylation	Cytosine residue of DNA	Gene silencing
Small noncoding RNAs (miRNA and siRNA)	Messenger RNA (mRNA)	Gene silencing [2, 3]
Histone acetylation	Lysine of histone H2A, H2B, H3, and H4	Gene activation [4]
Histone methylation	Lysine and arginine of histone proteins	Gene activation/ silencing [5]
Histone phosphorylation	Serine and threonine of histone proteins	Gene activation/ silencing [6]
Histone sumoylation	Lysine of histone proteins	Gene silencing [7]
Histone ubiquitination	Lysine of histone proteins	Gene activation [8]
Histone ADP ribosylation	Lysine of histone proteins	Gene activation [9]

Furthermore, also the action of small noncoding RNA, transcribed from noncoding DNA, was identified as another epigenetic process involved in chromosome remodeling and transcriptional or posttranscriptional regulation, by influencing RNA stability and gene expression (Table 3.1) [13–16].

Besides the mentioned mechanisms, chromatin condensation and histone modification are also key processes involved in epigenetic modification, particularly acting on a chromosome structure. In fact, in eukaryotic organisms, genome is compacted by basic proteins named histones, which allow the organization of DNA into chromatin [17] that is susceptible to modification depending on specific stimuli such as transcriptional repressors, functional RNA, or other accessory factors [18]. For these reasons, epigenetic regulation of chromatin, and the consequent variation in gene expression, may be environmentally dependent. In particular, histones are susceptible to a large variety of posttranslational modifications such as phosphorylation, acetylation, methylation, ubiquitination, sumoylation, adenosine triphosphate (ADP) ribosylation, glycosylation, biotinylation, and carbonylation [17] that are involved in chromatin state alteration and consequently act on gene expression (Table 3.1). The combination of the different histone modifications mentioned above constitutes the histone code [19].

The data available in the literature suggest that epigenetic mechanisms, involving molecular regulators such as histone variant, histone posttranslational modifications, nucleosome positioning chromosome looping, DNA structural variations, and RNA-mediated regulation [20–25], are closely related to chromatin state and therefore affect normal gene expression, as shown in Table 3.1. The molecular explanation of the influence of these mechanisms on gene activation or silencing is represented exactly by their ability to modulate chromatin conformation, which can be condensed with the consequent inhibition of polymerase accessibility for gene expression that causes lack of gene transcription and translation, leading to gene silencing. Anyway, DNA methylation, noncoding RNAs, and histone modification, with their consequences on chromatin state, are deeply interlinked to each other and represent a more integrated epigenetic system rather than disconnected events.

Epigenetic modifications, because of their effects on gene expression, play a central role in the regulation of gene expression during embryo development, from gametogenesis (oogenesis and spermatogenesis) to organogenesis, acting on chromatin state through DNA methylation and histone alterations and influencing later development [26].

During oogenesis and in particular during critical stages of oocyte growth and meiotic maturation, chromatin modifications control different key processes including gene expression, the establishment of maternal-specific DNA methylation marks, and chromosome stability [27]. The presence of a correct epigenome is responsible for proper chromosome segregation, for silencing of repetitive elements and potentially dangerous transposons and for meiotic centromere stability. The mammalian oocyte needs epigenetics, as dynamic chromatin alterations, to gain meiotic and developmental potential. Chromatin conformation is modified by chromatin remodeling proteins and histone modifications, which regulate heterochromatin formation and centromere function in the female germ line [28]. Differential methylation patterns are established during gametogenesis, both oogenesis and spermatogenesis, guaranteeing allele-specific parental identity by the process of genome imprinting [29]. The expression of these imprinted genes depends on regulatory sequences called imprinted control regions (ICRs) that are differentially methylated in the germ line [30]. Maternal methylation pattern at specific loci during oogenesis begins to be established during a critical period of postnatal oocyte growth starting from Day 5 of postnatal development, coincident with the transitional stage from primary to secondary follicles [31–33].

Epigenetics also regulates the biological process of spermatogenesis, in which the expression of several genes in the testes is controlled by modifications like DNA methylation, histone modifications, and chromatin remodeling. Testicular DNA has a unique pattern of methylation, which is eight times hypomethylated compared with somatic tissues. In particular, Sertoli cells had low levels of methylation in euchromatin and high levels in juxtacentromeric regions [34]. The characteristic methylation pattern of testis germ cells is established before meiosis, when these different demethylation/methylation processes act [35]. Testicular germ cells have distinct methylation patterns that depend on genomic sequence and usually occur on nonrepetitive genomic regions that are methylated de novo [35, 36]. Although the methylation status of certain genes may be changed during the different stages of spermatogenesis, it may or may not correspond to the gene's expression pattern [37]. Besides methylation, some variants of histones, called H2AL1, H2AL2, and H2BL1, have been identified in mature spermatozoa, and these probably take part in reprogramming pericentric heterochromatic regions during spermatogenesis [38]. In addition, the testes-specific linker histone variant H1T is normally exchanged for H1 during spermatogenesis [39], but H1T can also be replaced by the linker histone named HILS1, which may influence chromatin state and promote condensing spermatids [39, 40]. Thus, epigenetics is responsible for proper regulation of spermatogenesis to ensure physiologic sperm function and embryonic development; in fact, epigenetic aberrations are linked with male infertility [36].

As well as ensuring the correct gametogenesis in both males and females, epigenetic mechanisms also take part in cellular specialization during embryo development [41], because this process consists of changing patterns of gene expression that allow the specification of the cells of the early embryo from totipotency (the potential to form any kind of cells of the body, the extraembryonic membranes, and the placenta) to a discrete cell population. The lineage-specific pattern of gene expression is based on modifications of chromatin structure and function, which do not involve any change in nucleotide sequence of DNA, at the same way of every other epigenetic modification. Now, it is well known that differentiation of pluripotent cells is related to the methylation of the promoter of the essential pluripotency transcription factor, Oct4 [42].

Several variations in gene expression related to the chromatin state (condensed or uncondensed) observed during embryogenesis have been studied. In fact, considering embryo development, chromatin state is crucial in determining whether some genomic regions can be transiently condensed and accordingly silenced or, on the contrary, expressed when uncondensed [43], providing a fine mechanism of control of gene expression that guarantees the correct development of tissues and organs [41, 44].

Epigenetic mechanisms are active players in many physiological processes of human life, and for this reason, they are strongly involved in reproduction and some related genetic diseases. The embryo inherits two copies of the same gene, one from the mother and the other from the father, but each one of these can be silenced by epigenetic modifications causing many illnesses and disorders. This phenomenon, that is, the "turning off" of one parent's gene during gametogenesis (oogenesis and spermatogenesis), is defined as genetic imprinting.

The alteration of the genetic imprinting due to epigenetic disorders leads to the arise of reproductive genetic diseases, including Angelman syndrome and Prader-Willi syndrome (PWS) [45]. These two syndromes are characterized by an aberrant chromosome silencing that causes the loss of expression of specific maternal or paternal genes: In Angelman syndrome, a small deletion in chromosome 15 [46] affects the gene UBE3A, leading to the expression of the maternal gene and the silencing of paternal one determining nervous system impairment, while Prader-Willi syndrome (PWS) is a genetic disorder characterized by extreme feeding problems including hyperphagia or insatiable appetite and obsession with food, as well as decreased muscle tone in affected children due to epigenetic repression of PWS genes on maternal chromosome [47].

During the entire embryo development, there are two major rounds of epigenetic reprogramming [48, 49] that include a global loss of DNA methylation in cells of the early embryo. A phase of demethylation acts on paternally inherited genome in the first cell cycle after fertilization, while the maternally inherited one loose methylation progressively, because of the failure of the maintenance of methylation pattern together with replication after each cell division. The result is the hypomethylation of the entire embryo by the blastocyst stage [49] that is followed instead by a process of methylation during implantation [50]. Some earlier studies proposed that the formation of 5′-hydroxymethylcytosine, derived from the oxidation of 5-methylcytosine in the paternal pronucleus in fertilized oocyte, could represent a possible intermediate for global DNA demethylation of the paternal genome [51]. Moreover, it has been found that this modification is stable in both paternally and maternally derived genomes, and therefore, it may provide its own unique form of epigenetic information [52], even because 5′-hydroxymethylcystosine is quite a stable modification of the genome [53].

However, exposure of the embryo to a range of stresses during the period of its physiological epigenetic reprogramming can result in abnormal developmental outcomes. Moreover, external stimuli increase their own detrimental potential when act in the early stages of organism development. In fact, in these phases, the embryo is more sensitive to these stimuli because the alteration resulting from the stresses could be transmitted through the germ line and be hidden until much later in life [54, 55]. Therefore, any environmental stress during the early stages of development produces more systemic consequences than the same exposure in adulthood, which on the contrary has more local and limited consequences [56].

Recently, growing evidences suggest that the environment in which the embryo, fetus, and neonate develop seems to be involved in alteration of physiological epigenetic program. Nowadays, we are surrounded by thousands of compounds, chemicals and not, which daily interact with us and inevitably have effects on our state of health. Among all the different toxicants we are dealing with, there is a group of interest which is known as endocrine-disrupting chemicals (EDCs) that have the ability to act on human epigenome, affecting mostly reproduction [57]. In addition, the environmental effects on epigenetic settings during germ cell formation have the potential for the transgenerational inheritance of some of these induced modifications, and for these reasons, the epigenetic effect of EDCs on human reproduction should be taken into account.

3.2 Association Between EDCs and Epigenetics on Human Fertility

The study of epigenetic regulation during reproduction and development focused on the molecular mechanisms of gene expression related to developmental biology [58], in order to define the environmental mechanisms involved in alterations of gene expression patterns without affecting DNA sequence [59]. Environmental factors have a significant impact on biology, particularly referring to toxic compounds [57]. These environmental toxicants can modulate biological systems and influence physiology, even promoting disease states. Chemical compounds can be found in pharmaceutical drugs, personal care products, food additives, and food containers. These products could interfere with human endocrine systems and have the ability to induce diseases such as prostate and breast cancers or metabolic diseases [60–62] or have effects on the human reproductive, thyroid, cardiovascular, and neuroendocrinology systems. However, the effect of external toxicants on human health mainly consists of epigenetic alteration of genome without any modification to DNA sequence, thanks to the fact that DNA developed a general resistance against external attacks in order to maintain genome stability during evolution.

Among the thousands of contaminants released in the environment with a clear impact on human health, there is a group of endocrine-active substances mentioned before, the endocrine-disrupting chemicals (EDCs), which can interact directly or indirectly with the endocrine system and subsequently result in an effect on the endocrine system, organs, and tissues and may affect reproductive function. The term EDCs itself, coined at the Wingspread Conference in Wisconsin (in the USA) in 1991 [63], already refers to exogenous chemical entities or mixtures of compounds that are capable of interfering with or mimicking endogenous hormones and other signaling molecules of the endocrine system. Among these molecules that are able to interact with EDCs, the family of nuclear receptors (NRs) that includes orphan receptors (whose ligand is not known [64, 65]) such as steroid and xenobiotic receptors (SXRs) can recognize and bind many classes of EDCs [56], inducing endocrine responses and accordingly underlining the association between nuclear receptors and endocrine disruption coming from the external environment [66]. Orphan NRs family includes the estrogen receptor-related receptors (ERRs) [67] that share target genes, coregulators, and promoters [68, 69] with estrogen receptors (ERs), but contrast the classic ER-mediated estrogen-responsive signal [70, 71]. In fact, also EERs, similarly to SXRs, have the ability to bind EDCs [56, 72]. In addition, some EDCs have been reported to probably bind other crucial receptors involved in the hormonal signaling, like aryl hydrocarbon receptor (AhR) [73] and thyroid hormone receptor [74].

This area of investigation has grown in the last years introducing as EDCs a variety of substances [63], including xenoestrogen [75], environmental hormones [76, 77], hormonally actives agents [78], and environmental agents [79]. All these chemicals are categorized in classes and are integral part of the world economy and commerce. The global, social, and economic importance of EDCs is confirmed by the attention paid to them by various international organizations, such as the United States Environmental Protection Agency (USEPA) and the Organization for Economic Cooperation and Development (OECD) that have set up a task force to identify, prioritize, and validate test methods for the detection of endocrine disruptors [56, 80].

EDCs are categorized to different classes. For example, they can be first classified according to their endocrine effect (Fig. 3.1) that could be related to antiandrogenic, androgenic, estrogenic, or aryl hydrocarbon receptor agonists; inhibitors of steroid hormone synthesis; antithyroid substances; and retinoid agonists. In addition, endocrine disruptors can be classified based on their usage in agriculture and daily life: For example, pesticides (dichlorodiphenyltrichloroethane [DDT] and methoxychlor [MTX]), fungicides (vinclozolin), herbicides (atrazine), industrial chemicals (polychlorinated biphenyls [PCBs] and dioxins), plastics (phthalates,

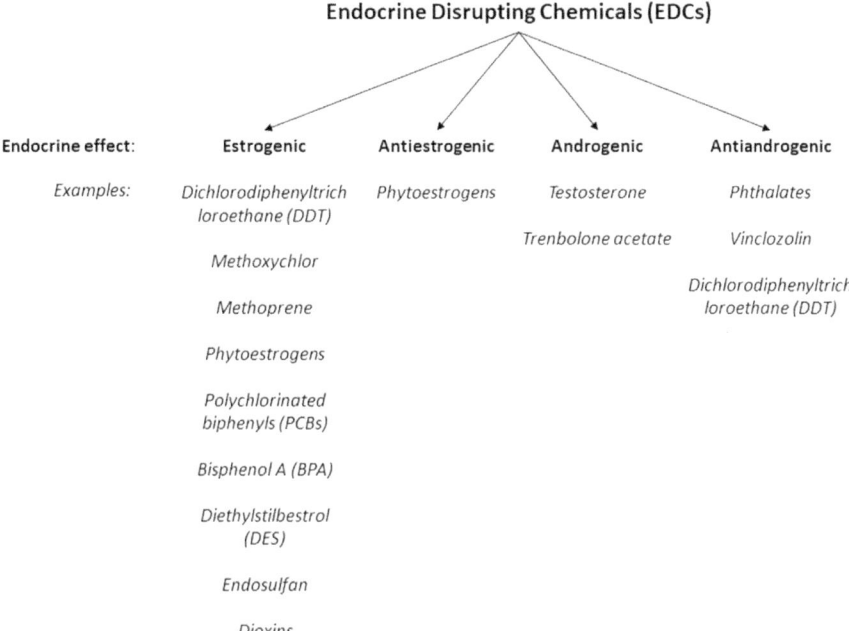

Fig. 3.1 Schematic representation of EDC classification based on their endocrine mechanisms

bisphenol A [BPA], and alkylphenols), and plant hormones (phytoestrogens). Lastly, some pharmaceuticals, personal care products, and nutraceuticals are also known as endocrine disruptors [81].

Considering pesticides, an increasing number of these substances have been recognized with androgen antagonist activity (antiandrogen), like dichlorodiphenyltrichloroethane (DDT) with its metabolites or other insecticides [82, 83]. Moreover, also linuron, another compound classified as an herbicide with a toxic effect specifically on human fertility, has been shown to compete with ligand for binding with the androgen receptor, resulting in the alteration of androgen-dependent gene expression [84, 85].

In addition to those synthetic EDCs already described, endocrine function can also be disrupted by chemicals originated in living organisms. Chemical compounds named phytochemicals or phytoestrogens are produced by plants and act as endogenous signals within the plant or are secreted for communications with other organisms, for example, to inhibit predatory herbivores [86]. Interestingly, phytoestrogens are isoflavones capable of binding to estrogen receptors alpha and beta (ERα and ERβ) and acting as weak agonists of estradiol [87, 88], partially exhibiting the estrogenic activity of the actual hormone [89, 90].

During the past decades, a particular attention has been given to the harmful effects of EDCs in the reproductive system as it has been reported that compounds with endocrine-disrupting mechanism of action can seriously affect human reproduction with a negative influence on human fertility [91–94]. In fact, several studies have demonstrated a considerable decrease in fertility biomarkers, notably sperm counts, in human populations that have been exposed to EDCs [95–98]. For example, bisphenol A (BPA) has been shown to affect fertility in mouse model, [99] and studies on its effects and mechanisms are continuing. Fertility has also been shown to be affected by medical prescriptions and drugs, like diethylstilbestrol (DES), a synthetic drug whose effects were only detected in the offspring 20 years after its administration to pregnant women [100]. The synthetic estrogen DES, in fact, was inappropriately prescribed to pregnant women between 1940 and 1970 to prevent miscarriage, premature labor, and complications of pregnancy, but it has been identified as the trigger of a rare vaginal clear-cell adenocarcinoma [101]. Again, some of these drugs can also alter gonad quality and reduce subsequent fertility and effectiveness of reproduction [102].

Besides the direct effect of ECDs on the endocrine system, these compounds may also exert their harmful effects by inducing epigenetic changes in the genome, particularly when they act during critical periods of the ontogeny of the organism exposed [103, 104]. In this specific case, epigenetic modifications due to EDCs (Table 3.2) are capable of inducing adult onset diseases than can also be transmitted through multiple generations by the germ line [114]. Transgenerational epigenetic inheritance has been proposed to be mediated by DNA methylation, histone modifications, and specific miRNA expression [115, 116]. In practice, prenatal exposure to EDCs may affect human fertility altering primordial germ line differentiation and development, inducing transgenerational epigenetic disorders. However, in the early stage of development of mammals, uterus and placenta represent barriers against which external factors are strongly buffered in their concentration, but despite these important forms of protection during pregnancy, in some cases, EDCs can cross placental and brain barriers, interfering with normal embryo development and organ functions [63].

EDCs can reach the fetus through two principal ways: The first is via oviductal and uterine endometrial secretions [117] that together contribute to constitute the environment within the uterus where the embryogenesis happens. However,

Table 3.2 Epigenetic mechanisms induced by endocrine-disrupting chemicals on the human reproductive system and some of their effects

Epigenetic mechanism of EDCs	Adverse phenotypes
DNA methylation	• Decreased sperm quality and subfertility, decreased testes weight, and decreased testosterone and estradiol levels [105]; • Primary follicle depletion and inhibition of oocyte maturation [106–108]; • Abnormal testis development and increased spermatogenic cell apoptosis with decreased sperm concentration [109]; • Male infertility [109, 110]
Histone modifications	• Decreased sperm quality and testes testosterone [111]
Noncoding RNAs (miRNAs)	• Increased cell death and changes in cell function [112]; • Increased progesterone production [113]

maternal secretion of epithelial uterine steroids can be altered by endocrine disruptors acting on embryos even before its implantation, resulting in aberrant methylation in this latter [103]. The alteration of fetal methylation patterns, following the changes in preimplantation intrauterine environment, affects nonimprinted and imprinted genes too, as detected by Wu et al. [118]. The second way used by EDCs to reach the embryo is crossing placenta [119]. This possibility was reported after transplacental exposure to endocrine disruptors like 17α-ethinyl estradiol, bisphenol A, and genistein during the gestation days 11–20 in rat, which was associated with changes in several genes' expression.

3.3 Effect of EDC Epigenetics Modification on Gene Expression in Human Reproduction

EDCs could regulate gene expression in many different ways [120, 121], inducing alterations in DNA methylation patterns [122]. In fact, DNA methylation in key genes that occurs after EDC exposure can be followed by transcriptional changes, leading to cellular abnormalities that may cause functional perturbation of tissues or organs [104]. Barrett et al. first reported an association between EDC exposure and cell transformation [123], laying the bases to speculate, by applying the current knowledge of epigenetic mechanisms, that such transformations could be the result of an epigenetic process. This was also supported by the evidence that individuals exposed to a secondary environmental exposure presented an increased susceptibility in terms of aberrant DNA methylation, changes in the transcription of key genes, and the consequent tumorigenic processes [104]. The research group of Li et al. [124], after the neonatal administration of DES, observed abnormalities in the demethylation of the lactoferrin promoter. In addition, the administration to newborn mice of some phytoestrogens, such as coumestrol and equol, showed an increased methylation that implies the silencing of the proto-oncogene H-Ras [125]. Lastly, Day et al. [126] individuated alteration in methylation patterns in 8-week-old mice, caused by the consumption of genistein. The association between EDCs and methylation status of genes has been recently confirmed by new scientific findings, for example, the discovery that 2,3,7,8-tetrachlorodibenzodioxin (TCDD), DES, or polychlorinated biphenyl-153 (PCB153) influences DNMT activity in early embryos [127].

EDCs' epigenetic effect may also be due to histone modification. With regard to EDCs on histone acetylation, Hong et al. [128] revealed that the chemicals genistein and equol produce this kind of epigenetic modification through the stimulation of the histone acetyltransferase activity, mediated by either estrogen receptors alpha and beta (ERα and ERβ). Again, Singleton et al. [129] showed that treating breast cancer cells overexpressing ERα with bisphenol A (BPA) or estradiol leads to differential expression of a set of histone-related genes. Moreover, BPA upregulated histone H2B and downregulated histone H1, while it had no effect on histone deacetylase, showing a completely opposite effect compared with estradiol [129]. Interestingly, from an epigenetic perspective, these histones have implications for chromatin condensation, which means gene silencing. Other findings about genetic expression regulation mediated by histone modifications demonstrate that gene silencing was associated with histone H3 trimethylation at lysine9 (H3K9me3) and with histone H3 acetylation at lysine 4 and di- or tri-lysine methylation (H3K4me2/3), and these were very common modifications related to changes in gene expression [130]. The gene expression alteration observed was due to the effect of these histone modifications, as well as others, on transcription regulation. However, no known histone code was related to the regulation processes mediated by hormones, and neither these modifications have been directly associated with EDCs.

In conclusion, changes in gene expressions due to epigenetics and without any modification in nucleotide sequence could be determined by both DNA methylation and chromatin state modifications. The epigenetic and epigenome regulation has been studied to identify the genes involved in the endocrine reproductive signaling and their relationship with emerging toxicants in the environment.

The theory that environmental factors can influence physiological phenotype and in particular the reproductive system was first derived from the observations of a wildlife biologist in the field [76], for example, the observation of reproductive dysfunction in many species (fish, birds, reptiles, mammals, etc.) living in areas contaminated with environmental toxicants [63, 131–133]. Then, chemical contaminant levels were increasingly being detected in humans in hormonally active tissues and in breast milk [57], which became a hot topic of discussion in the 1990s, highlighting the impact of environmental chemicals on human reproduction. After the discovery of the huge variety of EDCs, a growing body of literature confirmed the link between the increasing contamination in the environment and the parallel increasing incidence of breast cancer [134], decreasing sperm counts and increasing incidence of testicular cancer [135], which together had adverse effects including birth defects, reproductive failures, and sexual abnormalities. As the genetic background in human populations was essentially static, while disease disorders and infertility were dramatically increasing [136], it has been understood that environmental exposures must act primarily through epigenetic mechanisms to promote reproductive diseases [137, 138].

In fact, EDC exposure has the main effect of causing negative epigenetic changes, like alterations in DNA methylation or histone modification patterns, both inducing changes in normal gene expression that is associated with a wide range of diseases, including various reproductive disorders [139]. The reason for this type of EDCs' molecular mechanism of action on human health is that the majority of endocrine disruptors are not actually able to alter DNA sequence, but their action appears to be related to alterations in the epigenome, where they can affect normal reproductive physiological development and functions by acting as weak estrogenic, antiestrogenic, or antiandrogenic compounds. Females exposed to an excess of androgens early in gestation exhibit increased susceptibility to diseases such as polycystic ovaries in adult life [140], while, in adult male, perinatal or pubertal exposure to compounds such as estradiol and BPA alters the prostate epigenome [141].

Based on the findings supporting the close relationship between EDCs, epigenetic changes, and reproductive system, it is clear that the most vulnerable period for EDC exposure is embryogenesis, due to the high level of cell division characterized by specific epigenetic marks and critical modifications [142] that consequently can be transmitted over consecutive mitotic divisions and affect more cells than those occurring in adults during postnatal development. In addition, the placenta and its functions can be altered or influenced by the environment as well, which may result in pregnancy problems such as early pregnancy loss, preterm birth, intrauterine growth restriction (IUGR), congenital syndromes, and preeclampsia, which have all been linked again to epigenetic alterations [143].

Among the different epigenetic modifications induced by EDCs, DNA methylation is the most frequent and consequently the most studied one, due to its heritable nature, stability, and ease of measurement. Nevertheless, there are not yet many publications examining EDC effects on chromatin state, more precisely on chromatin condensation. However, DNA methylation has been studied extensively in particular in relation to reproductive biology, because the main methylation reprogramming occurs in germ cells formation (primordial germ cells) [48] and during the early stages of embryo development after fertilization [49]. Thus, the alteration of the methylation process has also been related to various disorders, such as those linked to imprinting [144], as mentioned above referring to Angelman syndrome and PWS. In these terms, a special attention has been paid, for example, to the chemicals vinclozolin and methoxychlor; the two pesticides that act as antiandrogenic endocrine and as estrogenic endocrine disruptors, respectively [145], reported to exert a specific effect on DNA methylation correlated to an aberrant phenotype of the reproductive tract.

3.4 Effect of EDC-Induced DNA Methylation on Male and Female Reproductive Tract

The endocrine-disrupting effects of many EDCs can be interpreted as interference with the normal regulation of reproductive processes by steroid hormones. Several evidences indicate that xenobiotics such as EDCs can bind to androgen and estrogen receptors on target tissues, to androgen-binding protein and to sex hormone–binding globulin [146]. Although environmental chemicals have a weak hormonal activity, their ability to interact with more than one steroid-sensitive pathway provides a mechanism by which their nature can be augmented. A given toxicant may be present in low concentration in the environment, and therefore, it can be harmless. However, we are not exposed to one toxicant at a time, but rather to all of the xenobiotics present in the environment. Therefore, numerous potential agonists/antagonists working together through several steroid-dependent signaling pathways could prove to be dangerous to human reproductive health.

During embryogenesis, the genital tract in males and females is first set up, but it is fully differentiated only after puberty, when sex hormone levels rise [147]. Normal physiology can be altered by EDCs exactly during the initiation of the functional activation of male/female reproductive system. Indeed, the most risky periods for xenobiotic exposure are represented by embryonic, neonatal, and pubertal periods, when the reproductive systems undergo to a finely tuned modulation by steroid hormones [76]. Physiological effects due to EDC exposure have been reported to occur in germ line in both males and females during the critical stages of development such as sex determination.

For example, embryo exposure to methoxychlor (MTX) or vinclozolin (Table 3.3) during sex determination period affects embryonic testis cellular composition and germ cell number and survival [109, 148]. In fact, the transient exposure to these EDCs can induce reprogram or imprint changes that show an effect in the adult reproductive physiology. The proof of concept that the effect observed on reproductive system was due to MTX or vinclozolin exposure came from the evidence that exposition of pregnant rats to both EDCs during the critical period for gonadal sex differentiation and testis morphogenesis (days 8–14 of pregnancy) produced transgenerational defects in spermatogenic capacity, which are transmitted through four generations (F1 to F4) [109]. This event was found to be due to an epigenetic mechanism involving altered DNA methylation that led to a permanent reprogramming of the male germ line. The causal effect of EDC involvement in reproductive tract morphogenesis derived from several findings demonstrates that a transient embryonic in utero exposure to an endocrine disruptor influences the embryonic testis transcriptome by epigenetic effects like DNA methylation. These epigenetic alterations resulted in abnormal testis development and in an increased adult spermatogenic cell apoptosis with decreased sperm concentration [109]. In addition to this alteration in the male reproductive tract, vinclozolin exposure has also been reported to induce transgenerational phenotypes in these animals, including adult onset diseases like male infertility [109, 110], increased frequencies of tumors, prostate disease, kidney diseases, and immune abnormalities [151]. Moreover, vinclozolin also induced changes in behavior and learning capacity [152–156], including transgenerational changes in mate preference [153] and anxiety behavior [156]. Transgenerational effects on tissue transcriptomes have also been observed. For example, in the embryonic testis transcriptome, a subset of genes presented a significantly altered expression in males from the F1 through the F3 generation, after vinclozolin exposure [165]. This transgenerational modified phenotype appears to be due to epigenetic changes, particularly due to alterations in DNA methylation of the male germ line [109, 166, 167]. After these first observations on MTX and vinclozolin, other agents that may promote transgenerational phenotypes associated with reproductive tract alterations have been identified.

In male testes, the expression of several genes is regulated via epigenetic modifications, underlining once again the direct influence of epigenetics on the process of spermatogenesis and how epigenetic aberrations (epimutations) can cause male infertility. Genes like MTHFR, PAX8, NTF3, SFN, HRAS, JHM2DA, IGF2, H19, RASGRF1, GTL2, PLAG1, D1RAS3, MEST, KCNQ1, LIT1, and

Table 3.3 Some endocrine-disrupting chemicals and their effects of induced DNA methylation on male/female reproductive systems and germ line differentiation

Endocrine-disrupting chemicals	Class	Effect
Methoxychlor (MTX)	Pesticide, estrogenic	Transgenerational defects in spermatogenic capacity, increase in adult sperm cell apoptosis and decrease in concentration, abnormal testis development [109, 148], and aberrant folliculogenesis [149]
Vinclozolin (VCZ)	Fungicide, antiandrogenic	Transgenerational defects in spermatogenic capacity, increase in adult sperm cell apoptosis and decrease in concentration, abnormal testis development [109, 150], male infertility [109, 110], predisposition to tumors, prostate disease, kidney diseases and immune abnormalities [151], changes in behavior and learning capacity [152–156], transgenerational changes in mate preference [153] and anxiety behavior [156], and aberrant development of PGCs [150]
Bisphenol A (BPA)	Plastic, estrogenic	Female fertility problems, polycystic ovary syndrome and endometriosis, decrease in antral follicle counts, and decrease in oocytes number [157–159]
Genistein	Plant hormones, estrogenic/antiestrogenic	Inhibition of oocytes maturation [108]
2,3,7,8-Tetrachlorodibenzodioxin (TCDD)	Dioxin, estrogenic	Aberrant folliculogenesis [149]
Phthalates	Plastic, antiandrogenic	Aberrant folliculogenesis [149]
Diethylstilbestrol (DES)	Drug, estrogenic	Predisposition to uterus epithelial tumors and to vaginal and cervical cancer, reproductive tract abnormalities [104], and susceptibility to tumor in testis and reproductive tract tissues [160]
Decitabine (5-aza-CdR)	Drug	Altered sperm morphology, decrease in sperm motility and capacity, and decrease in embryo survival [161, 162]
Chlorambucil	Drug	Transgenerational increase in deletions and other mutations in germ cells [163, 164]
Melphalan	Drug	Transgenerational increase in deletions and other mutations in germ cells [163, 164]

SNRPN can be often hypermethylated by environmental toxins/drugs and lead to poor semen parameters and male infertility [36]. For example, the anticancer agent decitabine (5-aza-20-deoxycytidine or 5-aza-CdR) (Table 3.3) is able to reduce global DNA methylation [161, 162], causing altered sperm morphology, decreased sperm motility, decreased fertilization capacity, and decreased embryo survival, similar to the effect showed by the EDCs previously mentioned, methoxychlor and vinclozolin. Among all the endocrine toxicants that can induce aberration in the reproductive tract, BPA has been reported to affect both male [168] and female [169] reproductive tracts.

Concerning EDC effect in female reproductive tract, beside BPA, other EDCs, such as genistein (Table 3.3), have shown to have an inhibitory effect on maturation of mammalian oocytes [108]. This evidence is crucial, since the main biological adverse effects of EDCs with regard to the development of female reproductive system are attributed to folliculogenesis [149]. The primordial follicles evolve to primary, preantral, and antral follicles. In particular, it has been reported that toxicity caused by EDCs to the antral follicles can lead to infertility. EDCs, such as BPA, MTX, TCDD, and phthalates (Table 3.3), can interfere with the development of the aforementioned types of follicles. For example, it was found that 3-month-old mice exposed in utero to 250 µg/kg BPA presented an increased percentage of ovarian tissue occupied by antral follicles. BPA has also been associated with female fertility problems, polycystic ovary syndrome, and endometriosis, whereas in women undergoing fertility treatments BPA levels have been associated with decreased antral follicle counts and a reduction in the number of oocytes [157–159].

Other findings reported the effects of female reproductive tract of perinatal exposure to diethylstilbestrol (DES) (Table 3.3). DES exhibits an estrogen agonist with an effect on the development of reproductive organs [37], supporting the epigenetic effect of DES exposure on the methylation pattern promoters controlling several estrogen-responsive genes associated with the development of reproductive tract. Perinatal DES exposure early in life has been found to increase predisposition to uterus epithelial tumors in adulthood, to several reproductive tract abnormalities, and to vaginal and cervical cancer risk in women [104]. Newbold et al. [160] administrated DES to pregnant rats during early postimplantation development and neonatal period, observing in males a grater susceptibility for tumor in rete testis and reproductive tract tissues in F1 and F2, due to epigenetic alterations like DNA methylation transmitted through the germ line.

Taken together, these evidences represent a clear example of how estrogenic xenobiotic exposure during a critical period of development can modify DNA sequence methylation status and consequently change the transcription of key genes involved in organ development, possibly increasing cancer risk later in life.

3.5   Effects of EDC-Induced DNA Methylation During Development and Germ Line Differentiation

In mammals, germ cell differentiation is initiated in the primordial germ cells (PGCs) during fetal development. PGCs are the embryonic precursors of the germ cell lineage (gametes, i.e., sperms and eggs), and their specification consists of global epigenetic reprograming, characterized by epigenetic phenomena such as the erasure of DNA methylation and histone modifications [170]. After the onset of gonadal sex determination, the PGC genome initiates the remethylation process of

DNA accompanied by remodeling of histone modifications in a sex-specific manner [170, 171]. Genetic and epigenetic changes during reprogramming of embryonic germ cell precursors make the prenatal period a sensitive window for potential adverse effects caused by environmental factors like EDCs [150]. In addition, the lack of any metabolic or excretion mechanism in the fetus highlights how much harmful chemicals can be if exposure occurs in this specific period of development.

However, the observation that PGCs and the developing germ line undergo major epigenetic programming, which can be transgenerationally altered by endocrine disruptors, was identified at first in 2015 [64].

During mammalian development, the primordial germ cells migrate down the genital ridge toward the newly formed gonad, prior to sex determination [172–174]. The germ cells develop into a male or female germ cell lineage at the initial stages of gonadal sex determination: The female germ line forms from oogenesis during follicle development that generate oocytes, and the male germ line, in turn, develops from spermatogonial stem cells and undergoes spermatogenesis, which originates spermatozoa in the testis. The critical period for epigenetic regulation of the germ line takes place during the phase of primordial germ cell migration and gonadal sex determination. Permanent alteration in the epigenetic programming of the germ line appears to be the mechanism involved in the transgenerational altered phenotype [109, 166, 167, 175].

Heritable damage can also occur in the zygote at the beginning of the embryonic development and can be transmitted to the next generation through modification occurred during germ line development [176]. Moreover, such heritable damage can be induced while germ line is developing. For example, chlorambucil and melphalan (Table 3.3) are able to induce a high frequency of heritable deletions and other mutations in mouse germ cells [163, 164], thereby producing a transgenerational mutation. Nevertheless, although some endogenous and exogenous agents are frequently associated with DNA mutations and transgenerational transmission, chemically induced epigenetic modifications of DNA may have the same net effect on the phenotype of newly altered cells and on their progeny [177]. Regarding this, Holliday [178] reported that teratogens could target mechanisms that control patterns of DNA methylation on genome of developing embryos, modifying methylation patterns that will be present on somatic cells, leading to a developmental alteration and subsequently to changes in germ line cells.

Modifications transmitted through germ line cells that occurred during the process of differentiation have been studied by Anway et al. [109], as already mentioned earlier. Later, the authors also detected 25 different genes that had altered methylation patterns in the F1 born to mothers subjected to the vinclozolin administration (Table 3.3) [109]. Therefore, the exposure of a gestating mother to EDC during critical periods of sex differentiation and testis morphogenesis triggers to decreased spermatogenic capacity and sperm viability that was transgenerationally transmitted in the male. This alteration appears to be associated with altered DNA methylation of the germ line [179].

Another evidence of ECDs' implication on germ line methylation alteration came from Brieno-Enriquez et al. [150]. In this study, gestating female mice were exposed to vinclozolin (Table 3.3) with the aim to produce epigenetic transgenerational inheritance of testicular cell apoptosis and abnormalities. Then, the observations were extended to PGCs, allowing the identification of alterations in epigenetic programming and gene expression that were critical for PGC development (such as those in Blimp-1) that promoted epigenetic PGC noncoding RNA programming. In fact, Blimp-1 pathway plays a critical role in determining DNA methylation reprogramming and gene expression alterations that occur during normal development of PGCs and the subsequent germ line, providing a major resource for epigenetic alterations during the development of the human germ line epigenome [180]. The importance of epigenomic control in PGCs was confirmed by the identification of specific DNA methylation sites that escaped DNA methylation erasure in PGCs specification, termed "escapees," supporting a role for altered germ line DNA methylation in epigenetic transgenerational inheritance [180].

3.6 Final Considerations

Epigenetics, first described by Waddington [1], involves all those molecular mechanisms that are able to modulate genome expression without modifying DNA nucleotide sequences. Among the main epigenetic processes, we can mention DNA methylation [10, 11], the action of small noncoding RNA like miRNA and siRNA [13–16], and a large variety of histone modifications [17]. In fact, in eukaryotic organism genome is compacted by basic proteins named histones that allow the organization of DNA into chromatin [17], whose conformation influences gene expression and can be modulated by these epigenetic mechanisms, which have the ability to induce chromatin condensation and/or chromatin relaxation to respectively silence and/or activate gene expression (Table 3.1). Because of their effect on gene expression related to the chromatin conformation (condensed or uncondensed), epigenetic modifications take part in many human biological processes, mostly involving reproductive system, and in some genetic diseases [45]. Their action appears to be crucial during embryo development, from gametogenesis to organogenesis, controlling gene expression to guarantee the correct development of tissues and organs [41, 44]. During oogenesis [27], epigenetics allows the establishment of the physiological epigenome and the gaining of meiotic and developmental potential of the oocyte via chromatin modifications. Differential methylation patterns are established during gametogenesis [31–33], and the processes of DNA methylation at specific loci during oogenesis have been investigated. During spermatogenesis, the expression of many genes in the testes is controlled in the same way by epigenetic mechanisms, like DNA methylation, histone modifications, and chromatin remodeling. Testicular DNA has a unique pattern of methylation established before meiosis [35], that is, much more hypomethylated than somatic tissues. Therefore, it is clear that methylation pattern and the consequent methylation phenomena occurring during both male and female germ line development are crucial [34].

Epigenetics is responsible not only for proper regulation of gametogenesis (oogenesis and spermatogenesis), but also for cellular specialization during embryo development [41]. In fact, cells of the early embryo must be specialized from totipotency to a specific cell population, depending on lineage-specific pattern of gene expression based on modifications of chromatin structure and function. During the entire embryo development, there are two major rounds of epigenetic reprogramming [48, 49]: a global DNA methylation loss in cells of the early embryo [49], followed instead by a process of methylation during implantation [50]. These events of reprogramming can be influenced by external stresses and represent a period of increased sensitivity of the embryo toward potential environmental toxicants, because this latter can transmit the alteration resulting from the stresses through the germ line, with the consequence of abnormal developmental outcomes in adulthood [54, 55].

Nowadays, it is well known that environmental factors have a significant impact on biology [57] since they have the capability to modulate biological systems and to influence physiology, even promoting disease states. Among all the different toxicants, there is a group of growing interest which is known as endocrine-disrupting chemicals (EDCs) that act on human epigenome without modifying DNA sequence. As the name itself suggests (coined in 1991 by Colborn T. [63]), endocrine disruptors are a set of endocrine-active substances that can interfere with human endocrine system mimicking endogenous hormones and other signaling molecules and affect the reproductive system [57]. EDCs can be classified according to their endocrine effect as shown in Fig. 3.1, and they can be found in a huge variety of sources. Therefore, EDCs may influence human health in two principal ways: acting directly on endocrine system as hormonal agonists/antagonists binding with hormonal receptors and inducing epigenetic changes in the genome. In fact, exposure to toxicants during critical periods of the ontogeny of the organism exposed [103, 104] can lead to epigenetic modification even transmissible through offspring by the germ line (transgenerational epigenetic disorders) [114].

After the first observations of how much environmental factors were linked to reproductive dysfunction in many species in wildlife [63, 76, 131–133], during the past decades, it has been reported that endocrine chemicals can seriously affect human reproduction (Table 3.2) [95–98], with a negative influence on fertility [91–94]. In human reproduction, EDCs may change gene expression through DNA methylation and histone modifications of key genes, both acting through the alteration of chromatin state, and it is associated with a wide range of diseases like various reproductive disorders [139].

However, between all the epigenetic mechanisms induced by EDCs, DNA methylation is the most frequent and consequently the most studied one, because of its heritable nature and stability. Normal physiology and subsequent phenotype can be altered by EDCs, and the most sensitive periods for exposure are represented by embryonic, neonatal, and pubertal periods, when there is a finely tuned modulation of hormones [76]. For example, the exposition of pregnant rats to chemicals, such as methoxychlor or vinclozolin (Table 3.3), during the critical period for gonadal sex differentiation and testis morphogenesis, produced transgenerational defects in spermatogenic capacity [109] due to the alteration of DNA methylation in male germ line. Therefore, embryonic testis transcriptome is influenced by DNA methylation, resulting in abnormal reproductive tract morphogenesis and in a general loss of gametes [109]. After these first observations, other agents that may promote transgenerational phenotypes associated with reproductive tract alterations have been identified, like bisphenol A (Table 3.3), which has been reported to affect both male [168] and female [169] reproductive tracts. The main biological adverse effects of EDCs concerning female reproductive system are attributed to folliculogenesis [149]. Besides BPA; MXC; 2,3,7,8-tetrachlorodibenzodioxin; and phthalates (Table 3.3), the chemical genistein, for example, has shown to have an inhibitory effect on maturation of mammalian oocytes [108], [157–159]. Diethylstilbestrol (Table 3.3) exposure early in life is instead associated with several reproductive tract abnormalities and increased vaginal and cervical cancer risk in women [104], once again due to the alteration of methylation pattern controlling several estrogen-responsive genes.

In humans, germ cell differentiation is initiated in PGCs that are the embryonic precursors of the germ cell lineage. Their specification is characterized by a global epigenetic reprogramming [170], involving erasure of DNA methylation and histone modifications in a sex-specific manner [170, 171], which makes prenatal period a particular sensitive window for the harmful effects of EDCs [150]. Alteration in the epigenetic programming of the germ line appears to be the mechanism involved in the transgenerational altered phenotype [109, 166, 167, 175]. Holliday [178] reported the association between teratogens and modification of DNA methylation pattern in particular genomic regions of developing embryos, leading to a developmental alteration and thus to changes in germ line cells. Anway et al. [109] showed that the exposure of a gestating mother rat to vinclozolin or methoxychlor (Table 3.3) during the critical periods was related to transgenerational defects in the spermatogenic capacity as the result of altered DNA methylation of the germ line [179]. Again, the exposure of gestating female mice to vinclozolin (Table 3.3) promotes epigenetic transgenerational inheritance of abnormalities in male reproductive tract [150], confirming that PGC alterations in epigenetic programming and gene expression were critical for their development [180].

Therefore, we can assume that epigenetics could represent an innovative frontier of scientific investigation to identify the molecular basis of alterations of normal epigenome during the sensitive periods of embryo and germinal line development associated with diseases in adulthood. The increase of pollutants in the environment, including EDCs, with their direct effect on human endocrine and reproductive systems, as well as their epigenetic mechanisms of action that induce the abovementioned aberrant phenotypes and diseases, represents a growing health concern that underlines the need of decreasing the release of contaminants in the environment and searching for new therapies acting through epigenetic mechanisms too.

References

1. Speybroeck V. From epigenesis to epigenetics: the case of CH Waddington. Ann N Y Acad Sci. 2002;981(1):61–81.
2. Neilson JR, Sharp PA. Small RNA regulators of gene expression. Cell. 2008;134(6):899–902.
3. Portnoy V, et al. Small RNA and transcriptional upregulation. Wiley Interdiscip Rev. 2011;2(5):748–60.
4. Grunstein M. Histone acetylation in chromatin structure and transcription. Nature. 1997;389(6649):349–52.
5. Bannister AJ, Schneider R, Kouzarides T. Histone methylation: dynamic or static? Cell. 2002;109(7):801–6.
6. Rossetto D, Avvakumov N, Cote J. Histone phosphorylation: a chromatin modification involved in diverse nuclear events. Epigenetics. 2012;7(10):1098–108.
7. Shiio Y, Eisenman RN. Histone sumoylation is associated with transcriptional repression. Proc Natl Acad Sci U S A. 2003;100(23):13225–30.
8. Jason LJ, et al. Histone ubiquitination: a tagging tail unfolds? BioEssays. 2002;24(2):166–74.
9. Martinez-Zamudio R, Ha HC. Histone ADP-ribosylation facilitates gene transcription by directly remodeling nucleosomes. Mol Cell Biol. 2012;32(13):2490–502.
10. Baylin SB. DNA methylation and gene silencing in cancer. Nat Clin Pract Oncol. 2005;2 Suppl 1(1):S4–11.
11. Singal R, Ginder GD. DNA methylation. Blood. 1999;93(12):4059–70.
12. Holmes R, Soloway PD. Regulation of imprinted DNA methylation. Cytogenet Genome Res. 2006;113(1–4):122–9.
13. Bartel DP. MicroRNAs: genomics, biogenesis, mechanism, and function. Cell. 2004;116(2):281–97.
14. Kim VN. Small RNAs just got bigger: Piwi-interacting RNAs (piRNAs) in mammalian testes. Genes Dev. 2006;20(15):1993–7.
15. Esquela-Kerscher A, Slack FJ. Oncomirs—microRNAs with a role in cancer. Nat Rev Cancer. 2006;6(4):259–69.
16. Wall NR, Shi Y. Small RNA: can RNA interference be exploited for therapy? Lancet. 2003;362(9393):1401–3.
17. Margueron R, Trojer P, Reinberg D. The key to development: interpreting the histone code? Curr Opin Genet Dev. 2005;15(2):163–76.
18. Craig JMJB. Heterochromatin—many flavours, common themes. BioEssays. 2005;27(1):17–28.
19. Jenuwein T, Allis CD. Translating the histone code. Science. 2001;293(5532):1074–80.
20. Beiter T, et al. Antisense transcription: a critical look in both directions. Cell Mol Life Sci. 2009;66(1):94–112.
21. Bernstein BE, Meissner A, Lander ES. The mammalian epigenome. Cell. 2007;128(4):669–81.
22. Gibney ER, Nolan CM. Epigenetics and gene expression. Heredity (Edinb). 2010;105(1):4–13.
23. Hartley PD, Madhani HD. Mechanisms that specify promoter nucleosome location and identity. Cell. 2009;137(3):445–58.
24. Jia D, et al. Structure of Dnmt3a bound to Dnmt3L suggests a model for de novo DNA methylation. Nature. 2007;449(7159):248–51.
25. Klose RJ, Bird AP. Genomic DNA methylation: the mark and its mediators. Trends Biochem Sci. 2006;31(2):89–97.
26. Bromfield J, Messamore W, Albertini DF. Epigenetic regulation during mammalian oogenesis. Reprod Fertil Dev. 2008;20(1):74–80.
27. De La Fuente R, Baumann C, Viveiros MM. Epigenetic modifications during mammalian oogenesis: emerging roles of chromatin structure during oocyte growth and meiotic maturation. In: Epigenetics in human reproduction and development. World Scientific; 2017. p. 35–58.

28. De La Fuente R, Baumann C, Viveiros MM. Chromatin structure and ATRX function in mouse oocytes. In: Mouse development. Cham: Springer; 2012. p. 45–68.
29. Hackett JA, Surani MA. DNA methylation dynamics during the mammalian life cycle. Philos Trans R Soc Lond Ser B Biol Sci. 2013;368(1609):20110328.
30. Weaver JR, Bartolomei MS. Chromatin regulators of genomic imprinting. Biochim Biophys Acta. 2014;1839(3):169–77.
31. Lucifero D, et al. Gene-specific timing and epigenetic memory in oocyte imprinting. Hum Mol Genet. 2004;13(8):839–49.
32. Obata Y, Kono T. Maternal primary imprinting is established at a specific time for each gene throughout oocyte growth. J Biol Chem. 2002;277(7):5285–9.
33. Smallwood SA, et al. Dynamic CpG Island methylation landscape in oocytes and preimplantation embryos. Nat Genet. 2011;43(8):811–4.
34. Marchal R, et al. DNA methylation in mouse gametogenesis. Cytogenet Genome Res. 2004;105(2–4):316–24.
35. Oakes CC, et al. Developmental acquisition of genome-wide DNA methylation occurs prior to meiosis in male germ cells. Dev Biol. 2007;307(2):368–79.
36. Rajender S, Avery K, Agarwal A. Epigenetics, spermatogenesis and male infertility. Mutat Res. 2011;727(3):62–71.
37. Huang Y, et al. Differential methylation of TSP50 and mTSP50 genes in different types of human tissues and mouse spermatic cells. Biochem Biophys Res Commun. 2008;374(4):658–61.
38. Govin J, et al. Pericentric heterochromatin reprogramming by new histone variants during mouse spermiogenesis. J Cell Biol. 2007;176(3):283–94.
39. Yan W, et al. HILS1 is a spermatid-specific linker histone H1-like protein implicated in chromatin remodeling during mammalian spermiogenesis. Proc Natl Acad Sci U S A. 2003;100(18):10546–51.
40. Iguchi N, et al. Isolation and characterization of a novel cDNA encoding a DNA-binding protein (Hils1) specifically expressed in testicular haploid germ cells. Int J Androl. 2003;26(6):354–65.
41. O'Neill C. The epigenetics of embryo development. Anim Front. 2015;5(1):42–9.
42. Athanasiadou R, et al. Targeting of de novo DNA methylation throughout the Oct-4 gene regulatory region in differentiating embryonic stem cells. PLoS One. 2010;5(4):e9937.
43. Wallace JA, Orr-Weaver TLJC. Replication of heterochromatin: insights into mechanisms of epigenetic inheritance. Chromosoma. 2005;114(6):389–402.
44. Yi H, et al. Gene expression atlas for human embryogenesis. FASEB J. 2010;24(9):3341–50.
45. Adams J. Imprinting and genetic disease: Angelman, Prader-Willi and Beckwith-Weidemann syndromes. Nature Education. 2008;1(1):129.
46. Magenis RE, et al. Is Angelman syndrome an alternate result of del(15)(q11q13)? Am J Med Genet. 1987;28(4):829–38.
47. Kim Y, Wang SE, Jiang YH. Epigenetic therapy of Prader-Willi syndrome. Transl Res. 2019;208:105–18.
48. Kobayashi H, et al. High-resolution DNA methylome analysis of primordial germ cells identifies gender-specific reprogramming in mice. Genome Res. 2013;23(4):616–27.
49. Smith ZD, et al. A unique regulatory phase of DNA methylation in the early mammalian embryo. Nature. 2012;484(7394):339–44.
50. Wang L, et al. Programming and inheritance of parental DNA methylomes in mammals. Cell. 2014;157(4):979–91.
51. Iqbal K, et al. Reprogramming of the paternal genome upon fertilization involves genome-wide oxidation of 5-methylcytosine. Proc Natl Acad Sci U S A. 2011;108(9):3642–7.
52. Salvaing J, et al. 5-Methylcytosine and 5-hydroxymethylcytosine spatiotemporal profiles in the mouse zygote. PLoS One. 2012;7(5):e38156.
53. Bachman M, et al. 5-Hydroxymethylcytosine is a predominantly stable DNA modification. Nat Chem. 2014;6(12):1049–55.

54. Markey CM, et al. In utero exposure to bisphenol a alters the development and tissue organization of the mouse mammary gland. Biol Reprod. 2001;65(4):1215–23.
55. Rogers MB, Glozak MA, Heller LC. Induction of altered gene expression in early embryos. Mutat Res. 1997;396(1–2):79–95.
56. Guerrero-Bosagna C, Valladares L. Endocrine Disruptors, Epigenetically Induced Changes, and Transgenerational Transmission of Characters and Epigenetic States. In: Endocrine-Disrupting Chemicals. New Jersey: Humana Press; 2007. p. 175–89.
57. Jacobs MN, et al. Marked for life: epigenetic effects of endocrine disrupting chemicals. Annu Rev Environ Resour. 2017;42(1):105–60.
58. Jablonka E, et al. The genome in context: biologists and philosophers on epigenetics. BioEssays. 2002;24(4):392–4.
59. Surani MAJN. Reprogramming of genome function through epigenetic inheritance. Nature. 2001;414(6859):122–8.
60. Heindel JJ. The developmental basis of disease: update on environmental exposures and animal models. Basic Clin Pharmacol Toxicol. 2019;125(Suppl 3):5–13.
61. Papalou O, et al. Endocrine disrupting chemicals: an occult mediator of metabolic disease. Front Endocrinol (Lausanne). 2019;10:112.
62. In SJ, et al. Benzophenone-1 and nonylphenol stimulated MCF-7 breast cancer growth by regulating cell cycle and metastasis-related genes via an estrogen receptor alpha-dependent pathway. J Toxicol Environ Health A. 2015;78(8):492–505.
63. Colborn T, Clement C. Chemically-induced alterations in sexual and functional development: the wildlife/human connection. Princeton: Princeton Scientific Pub. Co.; 1992.
64. Blumberg B, Evans RM. Orphan nuclear receptors–new ligands and new possibilities. Genes Dev. 1998;12(20):3149–55.
65. Robinson-Rechavi M, et al. How many nuclear hormone receptors are there in the human genome? Trends Genet. 2001;17(10):554–6.
66. Lamba J, Lamba V, Schuetz E. Genetic variants of PXR (NR1I2) and CAR (NR1I3) and their implications in drug metabolism and pharmacogenetics. Curr Drug Metab. 2005;6(4):369–83.
67. Hong H, Yang L, Stallcup MR. Hormone-independent transcriptional activation and coactivator binding by novel orphan nuclear receptor ERR3. J Biol Chem. 1999;274(32):22618–26.
68. Giguére V. To ERR in the estrogen pathway. Trend Endocrinol Metabol. 2002;13(5):220–5.
69. Kraus RJ, et al. Estrogen-related receptor alpha 1 actively antagonizes estrogen receptor-regulated transcription in MCF-7 mammary cells. J Biol Chem. 2002;277(27):24826–34.
70. Greschik H, et al. Structural and functional evidence for ligand-independent transcriptional activation by the estrogen-related receptor 3. Mol Cell. 2002;9(2):303–13.
71. Horard B, Vanacker JM. Estrogen receptor-related receptors: orphan receptors desperately seeking a ligand. J Mol Endocrinol. 2003;31(3):349–57.
72. Sekine Y, et al. Cross-talk between endocrine-disrupting chemicals and cytokine signaling through estrogen receptors. Biochem Biophys Res Commun. 2004;315(3):692–8.
73. Zhang W, et al. PCB 126 and other dioxin-like PCBs specifically suppress hepatic PEPCK expression via the aryl hydrocarbon receptor. PLoS One. 2012;7(5):e37103.
74. Rickenbacher U, et al. Structurally specific binding of halogenated biphenyls to thyroxine transport protein. J Med Chem. 1986;29(5):641–8.
75. Davis DL, et al. Medical hypothesis: xenoestrogens as preventable causes of breast cancer. Environ Health Perspect. 1993;101(5):372–7.
76. Danzo BJ. The effects of environmental hormones on reproduction. Cell Mol Life Sci. 1998;54(11):1249–64.
77. Cheek AO, McLachlan JA. Environmental hormones and the male reproductive system. J Androl. 1998;19(1):5–10.
78. Council NR. Hormonally active agents in the environment. Washington DC: National Academies Press; 2000.
79. Cheek AO, et al. Environmental signaling: a biological context for endocrine disruption. Environ Health Perspect. 1998;106(suppl 1):5–10.

80. Clode SA. Assessment of in vivo assays for endocrine disruption. Best Pract Res Clin Endocrinol Metab. 2006;20(1):35–43.
81. Daughton CG, Ternes TA. Pharmaceuticals and personal care products in the environment: agents of subtle change? Environ Health Perspect. 1999;107(suppl 6):907–38.
82. Sultan C, et al. Environmental xenoestrogens, antiandrogens and disorders of male sexual differentiation. Mol Cell Endocrinol. 2001;178(1–2):99–105.
83. Sunami O, et al. Evaluation of a 5-day Hershberger assay using young mature male rats: methyltestosterone and p, p'-DDE, but not fenitrothion, exhibited androgenic or antiandrogenic activity in vivo. J Toxicol Sci. 2000;25(5):403–15.
84. Lambright C, et al. Cellular and molecular mechanisms of action of linuron: an antiandrogenic herbicide that produces reproductive malformations in male rats. Toxicol Sci. 2000;56(2):389–99.
85. McIntyre BS, et al. Effects of in utero exposure to linuron on androgen-dependent reproductive development in the male Crl:CD(SD)BR rat. Toxicol Appl Pharmacol. 2000;167(2):87–99.
86. Wynne-Edwards KE. Hormonal changes in mammalian fathers. Horm Behav. 2001;40(2):139–45.
87. Benassayag C, Perrot-Applanat M, Ferre F. Phytoestrogens as modulators of steroid action in target cells. J Chomatogr. 2002;777(1–2):233–48.
88. Kuiper GG, et al. Comparison of the ligand binding specificity and transcript tissue distribution of estrogen receptors alpha and beta. Endocrinology. 1997;138(3):863–70.
89. Barkhem T, et al. Differential response of estrogen receptor alpha and estrogen receptor beta to partial estrogen agonists/antagonists. Mol Pharmacol. 1998;54(1):105–12.
90. Kuiper GG, et al. Interaction of estrogenic chemicals and phytoestrogens with estrogen receptor beta. Endocrinology. 1998;139(10):4252–63.
91. Menezo Y, Dale B, Elder K. The negative impact of the environment on methylation/ epigenetic marking in gametes and embryos: a plea for action to protect the fertility of future generations. Mol Reprod Dev. 2019;86(10):1273–82.
92. Perheentupa A. Male infertility and environmental factors. Global Reproductive Health. 2019;4(2):e28.
93. Olea N, Fernandez MF. Chemicals in the environment and human male fertility. Occup Environ Med. 2007;64(7):430–1.
94. Foster WG. Environmental toxicants and human fertility. Minerva Ginecol. 2003;55(5):451–7.
95. Safe S. Endocrine disruptors and falling sperm counts: lessons learned or not! Asian J Androl. 2013;15(2):191–4.
96. Slutsky M, Levin JL, Levy BS. Azoospermia and oligospermia among a large cohort of DBCP applicators in 12 countries. Int J Occup Environ Health. 1999;5(2):116–22.
97. Perry MJ, et al. Environmental pyrethroid and organophosphorus insecticide exposures and sperm concentration. Reprod Toxicol. 2007;23(1):113–8.
98. Jouannet P, et al. Semen quality and male reproductive health: the controversy about human sperm concentration decline. APMIS. 2001;109(S103):S48–61.
99. Munoz-de-Toro M, et al. Perinatal exposure to bisphenol-A alters peripubertal mammary gland development in mice. Endocrinology. 2005;146(9):4138–47.
100. Fowler WC Jr, Edelman DA. In utero exposure to DES. Evaluation and followup of 199 women. Obstet Gynecol. 1978;51(4):459–63.
101. Herbst AL, Ulfelder H, Poskanzer DC. Adenocarcinoma of the vagina. Association of maternal stilbestrol therapy with tumor appearance in young women. N Engl J Med. 1971;284(15):878–81.
102. Pandiyan N. Medical drugs impairing fertility. In: Reproductive health and the environment. Springer; 2007. p. 187–205.
103. Guerrero-Bosagna C, Sabat P, Valladares L. Environmental signaling and evolutionary change: can exposure of pregnant mammals to environmental estrogens lead to epigenetically induced evolutionary changes in embryos. Evol Dev. 2005;7(4):341–50.

104. Li S, et al. Environmental exposure, DNA methylation, and gene regulation: lessons from diethylstilbesterol-induced cancers. Ann N Y Acad Sci. 2003;983(1):161–9.
105. Sekaran S, Jagadeesan A. In utero exposure to phthalate downregulates critical genes in Leydig cells of F1 male progeny. J Cell Biochem. 2015;116(7):1466–77.
106. Chao HH, et al. Bisphenol a exposure modifies methylation of imprinted genes in mouse oocytes via the estrogen receptor signaling pathway. Histochem Cell Biol. 2012;137(2):249–59.
107. Zhang XF, et al. Diethylhexyl phthalate exposure impairs follicular development and affects oocyte maturation in the mouse. Environ Mol Mutagen. 2013;54(5):354–61.
108. Jung T, et al. Effects of the protein phosphorylation inhibitor genistein on maturation of pig oocytes in vitro. J Reprod Fertil. 1993;98(2):529–35.
109. Anway MD, et al. Epigenetic transgenerational actions of endocrine disruptors and male fertility. Science. 2005;308(5727):1466–9.
110. Anway MD, et al. Transgenerational effect of the endocrine disruptor vinclozolin on male spermatogenesis. J Androl. 2006;27(6):868–79.
111. Hong J, et al. Exposure of preimplantation embryos to low-dose bisphenol a impairs testes development and suppresses histone acetylation of StAR promoter to reduce production of testosterone in mice. Mol Cell Endocrinol. 2016;427:101–11.
112. Choi JS, et al. miRNA regulation of cytotoxic effects in mouse Sertoli cells exposed to nonylphenol. Reprod Biol Endocrinol. 2011;9(1):126.
113. Lu H, et al. miRNA-200c mediates mono-butyl phthalate-disrupted steroidogenesis by targeting vimentin in Leydig tumor cells and murine adrenocortical tumor cells. Toxicol Lett. 2016;241:95–102.
114. Skinner MK, Manikkam M, Guerrero-Bosagna C. Epigenetic transgenerational actions of endocrine disruptors. Reprod Toxicol. 2011;31(3):337–43.
115. Reik W, Dean W, Walter J. Epigenetic reprogramming in mammalian development. Science. 2001;293(5532):1089–93.
116. Del-Mazo J, et al. Endocrine disruptors, gene deregulation and male germ cell tumors. Int J Dev Biol. 2013;57(2–4):225–39.
117. McEvoy TG, et al. Feed and forage toxicants affecting embryo survival and fetal development. Theriogenology. 2001;55(1):113–29.
118. Wu Q, et al. Exposure of mouse preimplantation embryos to 2,3,7,8-tetrachlorodibenzo-p-dioxin (TCDD) alters the methylation status of imprinted genes H19 and Igf2. Biol Reprod. 2004;70(6):1790–7.
119. Naciff JM, Daston GP. Toxicogenomic approach to endocrine disrupters: identification of a transcript profile characteristic of chemicals with estrogenic activity. Toxicol Pathol. 2004;32(2_suppl):59–70.
120. Lonard DM, Smith CL. Molecular perspectives on selective estrogen receptor modulators (SERMs): progress in understanding their tissue-specific agonist and antagonist actions. Steroids. 2002;67(1):15–24.
121. Nilsson S, et al. Mechanisms of estrogen action. Physiol Rev. 2001;81(4):1535–65.
122. Wachsman JT. DNA methylation and the association between genetic and epigenetic changes: relation to carcinogenesis. Mutat Res–Fundam Mol Mech Mutagen. 1997;375(1):1–8.
123. Barrett JC, Wong A, McLachlan JA. Diethylstilbestrol induces neoplastic transformation without measurable gene mutation at two loci. Science. 1981;212(4501):1402–4.
124. Li S, et al. Developmental exposure to diethylstilbestrol elicits demethylation of estrogen-responsive lactoferrin gene in mouse uterus. Cancer Res. 1997;57(19):4356–9.
125. Lyn-Cook BD, et al. Methylation profile and amplification of proto-oncogenes in rat pancreas induced with phytoestrogens. Proc Soc Exp Biol Med. 1995;208(1):116–9.
126. Day JK, et al. Genistein alters methylation patterns in mice. J Nutr. 2002;132(8 Suppl):2419S–23S.
127. Wu Q, Zhou ZJ, Ohsako S. Effect of environmental contaminants on DNA methyltransferase activity of mouse preimplantation embryos. Wei Sheng Yan Jiu. 2006;35(1):30–2.

128. Hong T, et al. Isoflavones stimulate estrogen receptor-mediated core histone acetylation. Biochem Biophys Res Commun. 2004;317(1):259–64.
129. Singleton DW, et al. Gene expression profiling reveals novel regulation by bisphenol-A in estrogen receptor-alpha-positive human cells. Environ Res. 2006;100(1):86–92.
130. Hiragami-Hamada K, et al. The molecular basis for stability of heterochromatin-mediated silencing in mammals. Epigenetics Chromatin. 2009;2(1):14.
131. Colborn T, vom Saal FS, Soto AM. Developmental effects of endocrine-disrupting chemicals in wildlife and humans. Environ Health Perspect. 1993;101(5):378–84.
132. Toppari J, et al. Male reproductive health and environmental xenoestrogens. Environ Health Perspect. 1996;104(suppl 4):741–803.
133. Giesy JP, et al. Contaminants of fishes from Great Lakes-influenced sections and above dams of three Michigan rivers: III. Implications for health of bald eagles. Arch Environ Contam Toxicol. 1995;29(3):309–21.
134. Jenkins S, et al. Endocrine-active chemicals in mammary cancer causation and prevention. J Steroid Biochem Mol Biol. 2012;129(3–5):191–200.
135. Giwercman A, et al. Evidence for increasing incidence of abnormalities of the human testis: a review. Environ Health Perspect. 1993;101(suppl 2):65–71.
136. Vos T, et al. Global, regional, and national incidence, prevalence, and years lived with disability for 301 acute and chronic diseases and injuries in 188 countries, 1990–2013: a systematic analysis for the global burden of disease study 2013. Lancet. 2015;386(9995):743–800.
137. Skinner MK. Environment, epigenetics and reproduction. Mol Cell Endocrinol. 2014;398(1–2):1–3.
138. Crews D, et al. Nature, nurture and epigenetics. Mol Cell Endocrinol. 2014;398(1–2):42–52.
139. Cortessis VK, et al. Environmental epigenetics: prospects for studying epigenetic mediation of exposure-response relationships. Hum Genet. 2012;131(10):1565–89.
140. Abbott DH, et al. Androgen excess fetal programming of female reproduction: a developmental aetiology for polycystic ovary syndrome? Hum Reprod Update. 2005;11(4):357–74.
141. Prins GS, et al. Perinatal exposure to oestradiol and bisphenol a alters the prostate epigenome and increases susceptibility to carcinogenesis. Basic Clin Pharmacol Toxicol. 2008;102(2):134–8.
142. Dolinoy DC, et al. Maternal genistein alters coat color and protects Avy mouse offspring from obesity by modifying the fetal epigenome. Environ Health Perspect. 2006;114(4):567–72.
143. Robins JC, et al. Endocrine disruptors, environmental oxygen, epigenetics and pregnancy. Front Biosci (Elite Ed). 2011;3:690–700.
144. Sandhu KS. Systems properties of proteins encoded by imprinted genes. Epigenetics. 2010;5(7):627–36.
145. Kelce WR, et al. Environmental hormone disruptors: evidence that vinclozolin developmental toxicity is mediated by antiandrogenic metabolites. Toxicol Appl Pharmacol. 1994;126(2):276–85.
146. Fisher JS. Are all EDC effects mediated via steroid hormone receptors? Toxicology. 2004;205(1–2):33–41.
147. Malasanos TH. Sexual development of the fetus and pubertal child. Clin Obstet Gynecol. 1997;40(1):153–67.
148. Cupp AS, et al. Effect of transient embryonic in vivo exposure to the endocrine disruptor methoxychlor on embryonic and postnatal testis development. J Androl. 2003;24(5):736–45.
149. Sifakis S, et al. Human exposure to endocrine disrupting chemicals: effects on the male and female reproductive systems. Environ Toxicol Pharmacol. 2017;51:56–70.
150. Brieno-Enriquez MA, et al. Exposure to endocrine disruptor induces transgenerational epigenetic deregulation of microRNAs in primordial germ cells. PLoS One. 2015;10(4):e0124296.
151. Anway MD, Leathers C, Skinner MK. Endocrine disruptor vinclozolin induced epigenetic transgenerational adult-onset disease. Endocrinology. 2006;147(12):5515–23.

152. Andre SM, Markowski VP. Learning deficits expressed as delayed extinction of a conditioned running response following perinatal exposure to vinclozolin. Neurotoxicol Teratol. 2006;28(4):482–8.
153. Crews D, et al. Transgenerational epigenetic imprints on mate preference. Proc Natl Acad Sci U S A. 2007;104(14):5942–6.
154. Ottinger MA, et al. Neuroendocrine and behavioral effects of embryonic exposure to endocrine disrupting chemicals in birds. Brain Res Rev. 2008;57(2):376–85.
155. Ottinger MA, et al. Consequences of endocrine disrupting chemicals on reproductive endocrine function in birds: establishing reliable end points of exposure. Domest Anim Endocrinol. 2005;29(2):411–9.
156. Skinner MK, et al. Transgenerational epigenetic programming of the brain transcriptome and anxiety behavior. PLoS One. 2008;3(11):e3745.
157. Kandaraki E, et al. Endocrine disruptors and polycystic ovary syndrome (PCOS): elevated serum levels of bisphenol A in women with PCOS. J Clin Endocrinol Metab. 2011;96(3):E480–4.
158. Caserta D, et al. Bisphenol A and the female reproductive tract: an overview of recent laboratory evidence and epidemiological studies. Reprod Biol Endocrinol. 2014;12(1):37.
159. Souter I, et al. The association of bisphenol-A urinary concentrations with antral follicle counts and other measures of ovarian reserve in women undergoing infertility treatments. Reprod Toxicol. 2013;42:224–31.
160. Newbold RR, et al. Proliferative lesions and reproductive tract tumors in male descendants of mice exposed developmentally to diethylstilbestrol. Carcinogenesis. 2000;21(7):1355–63.
161. Ning Y, et al. 5-Aza-2′-deoxycytidine inhibited PDGF-induced rat airway smooth muscle cell phenotypic switching. Arch Toxicol. 2013;87(5):871–81.
162. Stenzig J, et al. DNA methylation in an engineered heart tissue model of cardiac hypertrophy: common signatures and effects of DNA methylation inhibitors. Basic Res Cardiol. 2016;111(1):9.
163. Russell LB, et al. Chlorambucil effectively induces deletion mutations in mouse germ cells. Proc Natl Acad Sci U S A. 1989;86(10):3704–8.
164. Russell LB, Hunsicker PR, Shelby MD. Melphalan, a second chemical for which specific-locus mutation induction in the mouse is maximum in early spermatids. Mutat Res Lett. 1992;282(3):151–8.
165. Anway MD, Rekow SS, Skinner MK. Transgenerational epigenetic programming of the embryonic testis transcriptome. Genomics. 2008;91(1):30–40.
166. Jirtle RL, Skinner MK. Environmental epigenomics and disease susceptibility. Nat Rev Genet. 2007;8(4):253–62.
167. Guerrero-Bosagna C, et al. Epigenetic transgenerational actions of vinclozolin on promoter regions of the sperm epigenome. PLoS One. 2010;5(9):e13100.
168. Salian S, Doshi T, Vanage G. Impairment in protein expression profile of testicular steroid receptor coregulators in male rat offspring perinatally exposed to bisphenol A. Life Sci. 2009;85(1–2):11–8.
169. Markey CM, et al. Mammalian development in a changing environment: exposure to endocrine disruptors reveals the developmental plasticity of steroid-hormone target organs. Evol Dev. 2003;5(1):67–75.
170. Guibert S, Forne T, Weber M. Global profiling of DNA methylation erasure in mouse primordial germ cells. Genome Res. 2012;22(4):633–41.
171. Seki Y, et al. Cellular dynamics associated with the genome-wide epigenetic reprogramming in migrating primordial germ cells in mice. Development. 2007;134(14):2627–38.
172. Allegrucci C, et al. Epigenetics and the germline. Reproduction. 2005;129(2):137–49.
173. Durcova-Hills G, et al. Influence of sex chromosome constitution on the genomic imprinting of germ cells. Proc Natl Acad Sci U S A. 2006;103(30):11184–8.
174. Trasler JM. Origin and roles of genomic methylation patterns in male germ cells. In: Seminars in cell & developmental biology, vol. 9. Elsevier; 1998. p. 467.

175. Skinner MK, Manikkam M, Guerrero-Bosagna C. Epigenetic transgenerational actions of environmental factors in disease etiology. Trends Endocrinol Metab. 2010;21(4):214–22.
176. Lewis SE. Life cycle of the mammalian germ cell: implication for spontaneous mutation frequencies. Teratology. 1999;59(4):205–9.
177. MacPhee DG. Epigenetics and epimutagens: some new perspectives on cancer, germ line effects and endocrine disrupters. Mutat Res–Fundam Mol Mech Mutagen. 1998;400(1–2): 369–79.
178. Holliday R. The possibility of epigenetic transmission of defects induced by teratogens. Mutat Res. 1998;422(2):203–5.
179. Skinner MK, Anway MD. Seminiferous cord formation and germ-cell programming: epigenetic transgenerational actions of endocrine disruptors. Ann N Y Acad Sci. 2005;1061:18–32.
180. Tang WW, et al. A unique gene regulatory network resets the human germline epigenome for development. Cell. 2015;161(6):1453–67.

Chapter 4
Introduction to Environmental Pollutants and Human Reproduction

Roberto Marci, Giovanni Buzzaccarini, Jean Marie Wenger, and Amerigo Vitagliano

4.1 Introduction

4.1.1 Recent Trends in Human Reproduction

Human reproductive health (defined as a state of physical, emotional, mental, and social well-being in relation to sexuality) is a matter of great concern in the new millennium. Over the past half century, there has been a growing trend toward delayed motherhood in the developed countries. In 2019, 29.4 years was the mean age of women at birth of the first child in European Union. The lowest mean age at birth of the first child was observed in Bulgaria (26.3 years), while the highest values were found in Italy (31.3 years) [1].

Postponement of parenthood has decreased the total fertility rates in almost all the European countries, with a mean fertility rate per woman of 1.53 in 2019 (ranging from 1.86 live births per woman in France to less than 1.3 in Italy, Spain, and Malta). Overall, the number of live births in Europe has decreased by 50% approximately, in the last 50 years. In parallel to the decreased global fertility rates, the chance of a couple remaining involuntarily childless has dramatically increased. A systematic analysis of 277 health surveys conducted by the World Health Organization (WHO) found that 25% of couples were affected by infertility in 2012 [2]. Infertility is defined as "the failure to achieve a pregnancy after 12 months or more of regular unprotected sexual intercourse." The rise in infertility has been accompanied by the rapid diffusion of assisted reproductive technologies (ARTs) worldwide. Over 9 million in vitro fertilization (IVF) children have been born, and

R. Marci (✉) · J. M. Wenger
Department of Translational Medicine, University of Ferrara, Ferrara, Italy
e-mail: mrcrrt@unife.it

G. Buzzaccarini · A. Vitagliano
Department of Women's and Children's Health, University of Padova, Padova, Italy

© The Author(s) 2023
R. Marci (ed.), *Environment Impact on Reproductive Health*,
https://doi.org/10.1007/978-3-031-36494-5_4

over 2.5 million cycles are performed annually, resulting in over 500,000 deliveries every year. Although ARTs have an increasing contribution to the overall birth rate, they can only partially compensate for the drop in fertility rates in the developed countries [2].

The risk of infertility increases with advancing age. The relationship between age and fertility is particularly significant for the female gender, in which a decline in fertility occurs early. By the age of 30, female fertility starts to decline. The decline becomes more rapid once women reach their mid-30s. By the age of 45, the majority of women are infertile. The key reasons for age-related infertility include reduced number and competence of oocytes due to aging insults, resulting in a higher risk of embryo aneuploidies. With respect to males, a significant reduction of fertility is described around the age of 40–45 years, mainly due to worsening of sperm number and motility. Therefore, in general, increasing parental age reduces the overall chances of pregnancy and increases the risk of spontaneous miscarriage [3].

In addition to parental age, there are many causes of infertility including female factors (e.g., ovulation disorders, tubal disease, endometriosis, uterine abnormalities, and reduced ovarian reserve) and male problems (e.g., varicocele, obstructive disorders, and testicle insufficiency), while 20–30% of cases are idiopathic [4]. With respect to age-independent infertility factors, an alarming phenomenon was recently described in males, also defined as "the male infertility crisis." Such a crisis refers to a steady annual decline of 1.4% in sperm counts from 1970 (with an overall decline of 52.4% in the last 40 years) to date. The causes of this phenomenon are partly unexplained. However, it is a common belief among scientists that some environmental factors may have played a role in determining a general worsening of human reproductive health.

Environmental pollutants are chemical, biological, and physical substances introduced in the environment as a result of human activities. The short- and long-term effects of these substances on human reproduction are a present matter of concern, especially in the developed countries. Several mechanisms may be involved in reproductive damage caused by environmental pollutants including hormone-mediated effects, oxidative stress, and direct genetic damage. In this chapter, we introduce the most relevant aspects inherent to the relationship between environment pollutants and human reproductive health.

4.1.2   Environmental Pollutants

Environmental pollution is the contamination of the environment (air, water, and land) with substances originated from man's activities such as urbanization, industrialization, mining, and exploration. It represents the world's greatest problem faced by humans and a major cause of human morbidity and mortality. According to the recent analyses, environmental pollution may account for 9 million deaths each year worldwide.

Over the last decades, numerous environmental pollutants were established as potential risk factors for various acute and chronic diseases in humans. Although the pathogenic effects of pollutants are often evaluated individually, their action is simultaneous and cumulative if one considers the countless possible sources of exposure [5].

The detrimental effects of pollutants on human fertility are supposed to vary based on the age of the exposed subject. Theoretically, the earlier the exposure (i.e., from intrauterine life to adolescence), the greater the resulting reproductive damage. However, the adults are not exempt from sequelae.

In addition, as an individual's intensity of exposure increases, so the severity of the damage will increase.

The mechanisms of interaction between pollutants and reproductive health can be academically divided into three classes:

1. Endocrine-disrupting chemicals (EDCs): The EDCs are "exogenous chemicals, or mixture of chemicals, that interfere with any aspect of hormone action." This action can be distinguished in transitory or permanent depending on the time of exposure. This mechanism is typical of polycyclic aromatic hydrocarbons (PAHs) and heavy metals (Cu, Pb, Zn, etc.) contained in particulate matter (PM), especially from diesel exhaust. EDCs can influence ovarian reserve by acting mainly on the aryl hydrocarbon receptor (AhR) or estrogen receptors (ERs). After binding the exogenous ligand, AhR translocates toward the nucleus; it associates with a nuclear receptor and is able to bind to DNA sequences and modulate gene transcription. AhR induces Bax synthesis, namely a proapoptotic factor contributing to follicular atresia [6]. In addition, diesel exhaust particles contain substances with estrogenic, antiestrogenic, and antiandrogenic activities that can affect gonadal steroidogenesis and gametogenesis. ERs play a crucial role during the early phase of folliculogenesis in humans, as they are increasingly expressed from the primordial stage onward [7]. Moreover, they are consistently expressed by oocytes in human fetal ovaries whatever the follicular stage [8]. In different mammals, estrogens can interfere with primordial follicle formation: in a positive manner in primates and bovines and in a negative manner in mice [8].
2. Induction of oxidative stress: Reactive oxygen species (ROSs) are normally balanced in the organism. However, in case of augmented ROS production or reduced ROS metabolism, oxidative stress occurs. In this situation, the ovarian function can be negatively influenced, because ROS may lead to antral follicle apoptosis [9]. In this respect, solid data found that oxidative stress markers are increased in patients with primary ovarian insufficiency (POI) syndrome [10]. This mechanism is demonstrated for nitrogen oxide (NO_2), ozone (O_3), or PM (through the heavy metals and PAHs they contain) pollutant.
3. Modifications of DNA: Molecules can create alterations to the DNA chain through the formation of DNA adducts. This type of interaction can lead to modifications in gene expression. Moreover, the exposure to pollutants could cause epigenetic modifications on the three-dimensional DNA structure, due to

alterations in DNA methylation. If these modifications affect the germ line in a nonmodifiable way, the mutation will be transmitted to the offspring [11].

4.1.2.1 Air Pollution

Air pollution has been considered for decades as a cause of concern for human fertility. Particulate matter (PM) and ground-level ozone (O_3) are Europe's most troubling pollutants, followed by benzo(a)pyrene (BaP). The main sources of these pollutants are transport and energy chains followed by the industries. The correlation between PM levels in the atmosphere and infertility has been largely investigated. It was found that every increase of 10 µg/m³ in PM2.5 concentration was associated with a 22% fecundability decrease (95% confidence interval [CI] = 6–35%) [12]. A further confirmation comes from an observational study with over 36,000 nurses, which showed a direct association between infertility and the proximity of residence to a main road. Hazard ratio (HR) for infertility when living close to major roads compared with farther was 1.11 (CI = 1.02–1.20). The authors therefore concluded that air pollution has a potentially harmful effect on fertility. In addition, the HR for every 10 µg/m³ increase in cumulative PM2.5–10 among women with primary infertility was 1.10 (CI = 0.96–1.27), and similarly, it was 1.10 (CI: 0.94–1.28) for those with secondary infertility [13].

4.1.2.2 Polycyclic Aromatic Hydrocarbons

Polycyclic aromatic hydrocarbons (PAHs) are a class of common environmental pollutants found in water, air, soil, and plants. They can be released by the natural sources; however, the vast majority derives from vehicular emissions, coal-burning plants, and the production and use of petroleum-derived substances. Exposure to PAHs has been associated with the onset of cancer and other diseases, including infertility. Their deleterious action on female reproductive system comes from their ability to interact with the pituitary–ovarian axis, causing alterations on the ovarian physiology and function [14].

4.1.2.3 Pesticides

Pesticides have wide applications in agriculture, especially with growing mass production for commercial export. At toxic doses, they cause oxidative stress due to a direct damage to antioxidant defense system. These substances have been associated with a variety of reproductive issues in males (e.g., germ cell apoptosis, hypotestosteronemia, and asthenozoospermia) and females (e.g., oligoanovulation, impaired folliculogenesis, follicular atresia, implantation defects, and endometriosis) and with obstetrical complications including spontaneous abortions and fetal malformations. For all these reasons, which will be extensively debated, pesticides directly affect fertility and reproductive physiology of the organism [15].

4.1.2.4 Parabens

Parabens are chemical substances used as preservatives in foods, cosmetics, and pharmaceutical products. These substances are endocrine disruptors that mainly act by mimicking the sex hormones, resulting in reproductive imbalance in both males and females [16]. In males, parabens were shown to affect total sperm count, semen motility, and morphology. In females, toxic exposure to parabens (as assessed through an increase in urinary propylparaben) was associated with short menstrual cycle length, low antral follicle count, and high cycle Day 3 follicle-stimulating hormone (FSH) level. Low antral follicle count and high FSH levels are markers of diminished ovarian reserve and reduced success rates of fertility treatments [17]. Other studies showed a variety of hormonal abnormalities after parabens exposure. In particular, increased levels of butylparaben were associated with lower levels of endogenous estradiol levels, altered thyroid hormone levels, and shifts in estradiol/progesterone ratios in women [18].

4.1.2.5 Perfluorooctanesulfonate (PFOS)/Perfluorooctanoate (PFOA)

Perfluorinated chemicals are substances widely used in everyday items such as food packaging, pesticides, clothing, upholstery, carpets, and personal care products. They have been extensively studied, and recent findings led to the acquaintance that they could be associated with infertility in women and men.

It was found that women who had higher levels of perfluorooctanoate (PFOA) and perfluorooctanesulfonate (PFOS) in their blood took longer to achieve a pregnancy than women with lower levels. The researchers divided the women's levels of PFOS/PFOA into four quartiles and found that, compared with women with the lowest levels of exposure, the likelihood of infertility increased by 70–134% for women in the higher three quartiles of PFOS exposure and by 60–154% for women in the higher three quartiles of PFOA exposure [19]. In men, the recent studies have suggested that PFOA/PFOS exposure can lead to hypotestosteronemia and reduction in semen quality.

The PFOS/PFOA role in male and female infertility is now under the spotlight, deserving an appropriate discussion in a later section.

Considering the wide distribution of different pollutant molecules in the environment and the difficulty in approaching their debate, we consider eligible an academic and a systematic approach. For this reason, here we present two main examples of environmental pollutants and how they could affect human reproductive health.

4.1.3   A Striking Example: The Role of Bisphenol

Bisphenol-A, 2,2-bis(4-hydroxyphenyl) propane (BPA), is one of the most investigated bisphenols. It is largely found in polycarbonate resin mainly used for plastic bags, bottles, baby battles and packaging, coated tins, particularly food and drink cans, and microwave ovenware. More than 90% of the overall exposure to BPA is diet being it a constituent of food containers and packaging, because it can leach into food products, especially after heating. The exposure to BPA occurring through the dermal absorption by the handling of thermal paper or by the application of cosmetics, together with air inhalation and dust and dental material ingestion, represents only the 5% of BPA exposure [20, 21]. In accordance with the Chapel Hill BPA consensus statement, "low BPA doses" have been considered in human epidemiological studies below the reference dose of tolerable daily intake (TDI), corresponding to 0.05 mg/kg (50 µg/kg) body weight/day as established by the United States Environmental Protection Agency (EPA). According to the World Health Organization (WHO) and the Food and Agriculture Organization (FAO) of the United Nations, in Europe, it has been estimated that BPA daily intake is around 0.2 µg/kg bw/day in breast-fed babies and around 11 g/kg bw/day in formula-fed babies for which feeding polycarbonate bottles were used. The estimated daily intake for adults is around 1.5 g/kg bw/day [20, 21].

4.1.3.1   Bisphenol Pathophysiological Impact

BPA acts through a dual mechanism. First, it acts as endocrine-disrupting chemical, thus affecting hormone synthesis, metabolism, and function [22]. In several in vivo and in vitro studies, it has been demonstrated the high affinity of bisphenol-A for estrogen receptors (ERs), having an estrogen-mimicking behavior and consequently stimulating estrogen function [20, 21, 23]. Therefore, BPA has been supposed to be involved in many diseases of female reproductive system [24, 25], due to its property to stimulate ER-dependent gene expression involved in the pathophysiology of female reproductive system [26–29]. Indeed, BPA has a conformational structure that confers the ability to bind both ER alpha (ERα) and ER beta (ERβ), although, according to the in vivo models, the affinity of BPA for ER is 1000-fold to 10,000-fold less than the affinity of 17β-estradiol (E2) [30].

Second, BPA has a slow action on genomic pathways interacting with nuclear ER and regulating several gene expressions; in addition, BPA has a rapid action through nongenomic pathways, activating, for example, the kinase signaling cascades or modulating the calcium flux through the cell's walls [31]. Genomic and nongenomic mechanisms can be triggered by the low- and high-dose exposure of BPA [32].

Taking together these modalities of action, BPA has a different effect if the exposure occurs during prenatal, perinatal, or postnatal period. Deleterious effects are more critical during perinatal exposure, causing dysregulation of hypothalamic–pituitary–ovarian (HPO) axis in babies and adults, with a precocious maturation of the axis through a damage of gonadotropin-releasing hormone (GnRH) pulsatility, gonadotropin signaling, and sex steroid hormone production. Further, a transmission from the pregnant woman to the developing fetus or child through the placenta and breast milk (during gestation and lactation) was also demonstrated, causing BPA-related diseases [33, 34].

4.1.3.2 Bisphenol A and Infertility

Increasing evidence has suggested that BPA might contribute to the pathogenesis of female and male infertility. The hypothetical impact of BPA on natural conception has been investigated in several observational studies [6, 35–37]. The number of subjects with detectable BPA levels (limit of detection [LOD] of assay: 0.5 ng/mL) was higher in infertile than in fertile women [6] and above all in infertile women who live in metropolitan areas [36]. Humans and rodents share the same regulation of reproductive system by the hypothalamic–pituitary–ovarian (HPO) axis. Hypothalamus releases gonadotropin-releasing hormone (GnRH) in rhythmic pulses; the pituitary gland secretes follicle-stimulating hormone (FSH) and luteinizing hormone (LH); the ovary releases sex hormones, including estradiol and progesterone, controlling the function of reproductive system. Most studies were performed on rat models with variation regarding to the exact timing of toxic exposure during their development. The findings of studies conducted on animal models pointed out that the deleterious effect of BPA could vary depending on doses, administration route, window of exposure, and animal models.

Morphological and functional changes in the reproductive system due to BPA can impair female fertility. BPA is able to inhibit androgen function by binding androgen receptors (ARs) [38], resulting in altered ovarian steroidogenesis [39–44] and folliculogenesis [41, 45, 46]. Moreover, BPA can influence endometrial receptivity, resulting in impaired embryo implantation [47–50].

Males are not exempt from reproductive damage due to BPA. In the postpubertal male, BPA is able to interfere with sex hormone synthesis, expression, and function of the respective receptors, resulting in reduced libido and ejaculatory defects. The effects are more detrimental during in utero exposure, as BPA was found to cause a variety of defects including feminization, atrophy of testes and epididymis, increased prostate size, and alteration of adult sperm parameters.

The Reproductive Organ Impairment in Females

The effects of BPA on ovarian, oviduct, and uterus morphology and functions in humans are still unclear, although different authors suggested that BPA exposure can affect in utero morphogenesis and the reproductive function in adults. Most of the experimental studies have been performed on mice; for both humans and rodents, the ovarian development is a dynamic process consisting in the growth of the ovary and establishment of the finite pool of primordial follicles, occurring predominantly during the embryonic period [51]. The functional alterations of ovary, which can be the cause or consequence of alterations of ovarian morphology, include mainly the impairment of folliculogenesis, beyond the impairment of steroidogenesis and sex hormones production.

The BPA action on the ovary and on the sex hormone secretion has been investigated in female pups and adult animals during different phases of the estrus cycle. Indeed, in rat and mouse female offspring perinatally, prenatally, and postnatally exposed to oral, gavage, and subcutaneous administration of low and high BPA doses, increased circulating E2 levels have been recorded [40, 42, 52, 53]. In addition to the regulation of sex hormone synthesis, the exposure to low BPA doses increased messenger and protein expression of FSH receptor (FSHR) too in the ovarian tissue of female adult rats [54].

The Folliculogenesis Impairment

BPA may interfere with multiple molecular processes and pathways involved in folliculogenesis. Indeed, it has been shown that BPA enhances ER messenger expression in ovary and, through the binding to ER, induces epigenetic modifications, in particular DNA hypomethylation, of genes involved in oocyte maturation, with a consequent acceleration of the transformation of the primordial to primary follicles [55]. The administration of low BPA doses in ovaries explanted from mouse female pups inhibited germ cell nest breakdown and enhanced primordial follicle recruitment, by decreasing the expression of Ki-67, Fas, Bac, Bax, and Caspase 3 and 8; increasing the expression of Bcl2; and activating phosphoinositide-3-kinase (PI3K)/Akt pathway. Low BPA doses accelerated follicle development with an increase in antral follicle growth [46], while high BPA doses selectively inhibited antral follicle growth [42, 45, 56]. Low BPA doses' effects on antral follicles were found to be associated with high methylation level of several maternally and paternally imprinted genes [46], whereas high BPA doses' effects on antral follicles were found to be mediated by interference with the expression of genes involved in cell cycle progression (increased expression of cyclin-dependent kinase 4 (CDK4) and cyclin E1 (CCNE1) and decreased expression of cyclin D2 (CCND2), and apoptosis (increased expression of p53, Bcl-2, and Bax) [57].

However, the disruption of folliculogenesis seems to be a reversible process depending on the timing of BPA exposure. Indeed, when BPA exposure occurs in adults, the damage appears to be transient (with a reduction in the number of antral follicles), but a restoration of pre-exposure conditions few weeks later is generally observed. On the contrary, when the exposure to BPA occurs during the postnatal period, a persistent disruption of folliculogenesis is noticed (i.e., a decreased number of primordial follicles and an increased number of atretic follicles persisting in adulthood) [58]. Alarmingly, a recent study on a cohort of women undergoing IVF reported that higher urinary BPA levels were associated with lower antral follicle count, raising concerns for a possible accelerated follicle loss [59].

The Altered Embryo Implantation

Experimental ex vivo studies on uterus confirmed that exposure to BPA affected uterine function, particularly interfering with uterine receptivity and embryo implantation. An ex vivo study, conducted on uterus of female adult mice treated with BPA in the first 3 days of pregnancy, demonstrated that high BPA doses delayed the transfer of embryos to the uterus, damaged blastocyst development before implantation, and inhibited embryo implantation. Regarding the putative mechanism, high BPA dose exposure induces a dose-dependent increase of endothelial nitric oxide synthase (eNOS) protein expression in trophoblast cells, the cells forming the outer layer of a blastocyst, with a consequent induction of excess nitrogen monoxide (NO), which might represent one of the causal factors involved in embryo implantation [47].

Unfavorable embryo implantation was also observed in female adult mice exposed to low BPA dose treatment, in which ex vivo analysis of uterus showed reduced decidual cells surrounding the attached embryo and an increased percentage of intrauterine hemorrhage, due to the shedding and collapse of the endometrium [55].

The impairment of uterine receptivity and the unfavorable embryo implantation can be addressed by the BPA capability to increase uterine luminal area, enhancing the uterine luminal epithelial cell height, and capability to affect E2 and heart and neural crest derivatives expressed 2 (P/HAND2) pathways [56]. Indeed, BPA-exposed uterine epithelial and stromal tissues showed a marked suppression of E2 and P receptor expression and P receptor downstream target gene, HAND2. These factors enhance the activation of fibroblast growth factor and mitogen-activated protein kinase (MAPK) signaling in the epithelium, thus contributing to aberrant proliferation, lack of uterine receptivity, and impaired embryo implantation [47–50].

In conclusion, prenatal, perinatal, and postnatal BPA exposure may influence the following: (1) ovarian development, by reducing the breakdown of germ cell nest, as a consequence of deregulated expression of apoptotic genes; (2) oviduct morphology; (3) folliculogenesis, by downregulating the expression of cell cycle regulatory genes and steroidogenic enzymes, leading to increased follicle apoptosis and premature ovarian insufficiency; and (4) uterine receptivity and embryo implantation.

4.1.3.3 To Strengthen the Concept: The Role of Phthalates

Phthalates are chemical substances that are mainly used as plasticizers in disposable and non-disposable products. Since their widespread use in factories and environment, knowledge on their potential effects on human health is of paramount importance. Notably, different disorders were described in connection to phthalates exposure, including infertility [60].

Chemically, phthalates are esters of 1,2-benzene dicarboxylic acid with a structure that varies according to the number of side chains. Side chains can be composed by dialkyl, alkyl, or aryl groups. Physically, phthalates are colorless, oily, and odorless substances, with a low solubility in water, which is inversely correlated to the chain length. Conversely, phthalic acid derivatives are more soluble in organic solvents [60–62].

Regarding their role in human pathology, phthalates are classified as endocrine disruptors (EDs), namely specific substances hampering the hormonal balance in males and females. Regarding males, phthalates can affect their reproductive function with different mechanisms. First, they can interfere with male reproductive system development [63]. Second, they can induce testicular dysgenesis syndrome (TDS) being responsible for decreased testis weight, spermatogenesis dysfunction, and external genital malformations (shortened anogenital distance, hypospadias, and cryptorchidism) [64]. Third, they can be responsible for male puberty dysfunction [65]. Fourth, phthalates can induce cancer in male reproductive organs [66].

In females, a putative negative effect of phthalates on ovarian function has been hypothesized. Studies on mice showed a considerable reduction of antral follicle development [67] after phthalate exposure, up to follicle exhaustion and premature ovarian failure (POF). Perhaps, the antiestrogenic activity of phthalates may be involved in POF [68]. When the exposure occurs in postnatal age, the natural onset of female puberty can be anticipated or delayed [69, 70]. In addition, exposure to phthalate during pregnancy may somehow result in spontaneous miscarriage and other obstetric complications [71]. Finally, also in women, a role of phthalates in carcinogenesis of the genital tract cannot be excluded [72].

References

1. EUROSTAT. Last accessed 12 Feb 2022. https://ec.europa.eu/eurostat/statistics-explained/index.php?title=Fertility_statistics
2. Mascarenhas MN, Flaxman SR, Boerma T, Vanderpoel S, Stevens GA. National, regional, and global trends in infertility prevalence since 1990: a systematic analysis of 277 health surveys. PLoS Med. 2012;9(12):e1001356. https://doi.org/10.1371/journal.pmed.1001356. Epub 2012 Dec 18. PMID: 23271957; PMCID: PMC3525527.
3. Gnoth C, Godehardt E, Frank-Herrmann P, Friol K, Tigges J, Freundl G. Definition and prevalence of subfertility and infertility. Hum Reprod. 2005 May;20(5):1144–7. https://doi.org/10.1093/humrep/deh870.
4. Bellver J, Donnez J. Introduction: infertility etiology and offspring health. Fertil Steril. 2019;111(6):1033–5. https://doi.org/10.1016/j.fertnstert.2019.04.043. PMID: 31155112
5. Park SK, Tao Y, Meeker JD, Harlow SD, Mukherjee B. Environmental risk score as a new tool to examine multi-pollutants in epidemiologic research: an example from the NHANES study using serum lipid levels. PLoS One. 2014;9(6):e98632. Published 2014 Jun 5. https://doi.org/10.1371/journal.pone.0098632.
6. Caserta D, Bordi G, Ciardo F, Marci R, La Rocca C, Tait S, et al. The influence of endocrine disruptors in a selected population of infertile women. Gynecol Endocrinol. 2013;29(5):444–7.
7. Diamanti-Kandarakis E, Bourguignon JP, Giudice LC, Hauser R, Prins GS, Soto AM, Gore AC. Endocrine-disrupting chemicals: an Endocrine Society scientific statement. Endocr Rev. 2009;30:293–342. https://doi.org/10.1210/er.2009-0002.

8. Fernandez M, Bianchi M, Lux-Lantos V, Libertun C. Neonatal exposure to bisphenol a alters reproductive parameters and gonadotropin releasing hormone signaling in female rats. Environ Health Perspect. 2009;117(5):757–62.
9. Luderer U. Ovarian toxicity from reactive oxygen species. Vitam Horm. 2014;94:99–127. https://doi.org/10.1016/B978-0-12-800095-3.00004-3.
10. Tokmak A, Yildirim G, Sarikaya E, Cinar M, Bogdaycioglu N, Yilmaz FM, et al. Increased oxidative stress markers may be a promising indicator of risk for primary ovarian insufficiency: a cross-sectional case control study. Rev Bras Ginecol Obstet. 2015;37:411–6.
11. Nilsson E, Larsen G, Manikkam M, Guerrero-Bosagna C, Savenkova MI, Skinner MK. Environmentally induced epigenetic transgenerational inheritance of ovarian disease. PLoS One. 2012;7:e36129. https://doi.org/10.1371/journal.pone.0036129.
12. Slama R, Bottagisi S, Solansky I, Lepeule J, Giorgis-Allemand L, Sram R. Short-term impact of atmospheric pollution on fecundability. Epidemiology. 2013:871–879. doi:10.1097/EDE.0b013e3182a702c5.
13. Mahalingaiah S, Hart JE, Laden F, Farland LV, Hewlett MM, Chavarro J, Aschengrau A, Missmer SA. Adult air pollution exposure and risk of infertility in the Nurses' health study II. In: Human Reprod, vol. 31; 2016. p. 638. https://doi.org/10.1093/humrep/dev330.
14. Ramesh A, Archibong AE, Niaz MS. Ovarian susceptibility to benzo[a]pyrene: tissue burden of metabolites and DNA adducts in F-344 rats. J Toxicol Environ Health A. 2010;73(23):1611–25.
15. Kumar BJ, Mittal M, Saraf P, Kumari P. Pesticides induced oxidative stress and female infertility: a review. Toxin Reviews. 2018;39:1–13. https://doi.org/10.1080/15569543.2018.14749266.
16. Jurewicz J, et al. 2017 Human semen quality, sperm DNA damage, and the level of reproductive hormones in relation to urinary concentrations of parabens. J Occup Envrion Med. 2017;59:1034–40.
17. Smith KW, Souter I, Dimitriadis I, Ehrlich S, Williams PL, Calafat AM, Hauser R. Urinary paraben concentrations and ovarian aging among women from a fertility center. Environ Health Perspect. 2013;121(11–12):1299–305.
18. Nishihama Y, Yoshinaga J, Iida A, Konishi S, Imai H, Yoneyama M, Nakajima D, Shiraishi H. Association between paraben exposure and menstrual cycle in female university students in Japan. Reprod Toxicol. 2016;63:107–13.
19. Fei C, McLaughlin JK, Lipworth L, Olsen J. Maternal levels of perfluorinated chemicals and subfecundity. Hum Reprod. 2009;24(5):1200–5. https://doi.org/10.1093/humrep/den490. Epub 2009 Jan 28. PMID: 19176540.
20. Bisphenol A (BPA)-current state of knowledge and future actions by WHO and FAO, in International Food Safety Authorities Network (INFOSAN) Information Note No. 5/2009–Bisphenol A. 2009.
21. Konieczna A, Rutkowska A, Rachon D. Health risk of exposure to bisphenol A (BPA). Rocz Panstw Zakl Hig. 2015;66(1):5–11.
22. Gregoraszczuk EL, Ptak A. Endocrine-disrupting chemicals: some actions of POPs on female reproduction. Int J Endocrinol. 2013;2013:828532.
23. Fenichel P, Chevalier N, Brucker-Davis F. Bisphenol A: an endocrine and metabolic disruptor. Ann Endocrinol (Paris). 2013;74(3):211–20.
24. Sifakis S, Androutsopoulos VP, Tsatsakis AM, Spandidos DA. Human exposure to endocrine disrupting chemicals: effects on the male and female reproductive systems. Environ Toxicol Pharmacol. 2017;51:56–70.
25. Wei M, Chen X, Zhao Y, Cao B, Zhao W. Effects of prenatal environmental exposures on the development of endometriosis in female offspring. Reprod Sci. 2016;23(9):1129–38.
26. Chen X, Wang Y, Xu F, Wei X, Zhang J, Wang C, et al. The rapid effect of Bisphenol-A on long-term potentiation in hippocampus involves estrogen receptors and ERK activation. Neural Plast. 2017;2017:5196958.
27. Richter CA, Birnbaum LS, Farabollini F, Newbold RR, Rubin BS, Talsness CE, et al. In vivo effects of bisphenol A in laboratory rodent studies. Reprod Toxicol. 2007;24(2):199–224.

28. Shi XY, Wang Z, Liu L, Feng LM, Li N, Liu S, et al. Low concentrations of bisphenol A promote human ovarian cancer cell proliferation and glycolysis-based metabolism through the estrogen receptor-alpha pathway. Chemosphere. 2017;185:361–7.
29. Zhang Y, Wei F, Zhang J, Hao L, Jiang J, Dang L, et al. Bisphenol A and estrogen induce proliferation of human thyroid tumor cells via an estrogen-receptor-dependent pathway. Arch Biochem Biophys. 2017;633:29–39.
30. Toxicological and health aspects of bisphenol A, In report of joint FAO/WHO expert meeting. 2010. Ottawa.
31. Shanle EK, Xu W. Endocrine disrupting chemicals targeting estrogen receptor signaling: identification and mechanisms of action. Chem Res Toxicol. 2011;24(1):6–19.
32. Alonso-Magdalena P, Ropero AB, Soriano S, Garcia-Arevalo M, Ripoll C, Fuentes E, et al. Bisphenol-A acts as a potent estrogen via non-classical estrogen triggered pathways. Mol Cell Endocrinol. 2012;355(2):201–7.
33. Cao XL, Zhang J, Goodyer CG, Hayward S, Cooke GM, Curran IH. Bisphenol A in human placental and fetal liver tissues collected from greater Montreal area (Quebec) during 1998–2008. Chemosphere. 2012;89(5):505–11.
34. Lee J, Choi K, Park J, Moon HB, Choi G, Lee JJ, et al. Bisphenol A distribution in serum, urine, placenta, breast milk, and umbilical cord serum in a birth panel of mother-neonate pairs. Sci Total Environ. 2018;626:1494–501.
35. Buck Louis GM, Sundaram R, Sweeney AM, Schisterman EF, Maisog J, Kannan K. Urinary bisphenol A, phthalates, and couple fecundity: the longitudinal investigation of fertility and the environment (LIFE) study. Fertil Steril. 2014;101(5):1359–66.
36. La Rocca C, Tait S, Guerranti C, Busani L, Ciardo F, Bergamasco B, et al. Exposure to endocrine disrupters and nuclear receptor gene expression in infertile and fertile women from different Italian areas. Int J Environ Res Public Health. 2014;11(10):10146–64.
37. Velez MP, Arbuckle TE, Fraser WD. Female exposure to phenols and phthalates and time to pregnancy: the maternal-infant research on environmental chemicals (MIREC) study. Fertil Steril. 2015;103(4):1011–20 e2.
38. Lee HJ, Chattopadhyay S, Gong EY, Ahn RS, Lee K. Antiandrogenic effects of bisphenol a and nonylphenol on the function of androgen receptor. Toxicol Sci. 2003;75(1):40–6.
39. Zhou W, Liu J, Liao L, Han S, Liu J. Effect of bisphenol a on steroid hormone production in rat ovarian theca-interstitial and granulosa cells. Mol Cell Endocrinol. 2008;283(1–2):12–8.
40. Xi W, Lee CKF, Yeung WSB, Giesy JP, Wong MH, Zhang XW, et al. Effect of perinatal and postnatal bisphenol A exposure to the regulatory circuits at the hypothalamus-pituitary-gonadal axis of CD-1 mice. Reprod Toxicol. 2011;31(4):409–17.
41. Peretz J, Gupta RK, Singh J, Hernandez-Ochoa I, Flaws JA. Bisphenol a impairs follicle growth, inhibits steroidogenesis, and downregulates rate-limiting enzymes in the estradiol biosynthesis pathway. Toxicol Sci. 2011;119(1):209–17.
42. Gamez JM, Penalba R, Cardoso N, Bernasconi PS, Carbone S, Ponzo O, et al. Exposure to a low dose of bisphenol A impairs pituitary-ovarian axis in prepubertal rats effects on early folliculogenesis. Environ Toxicol Pharmacol. 2015;39(1):9–15.
43. Mansur A, Adir M, Yerushalmi G, Hourvitz A, Gitman H, Yung Y, et al. Does BPA alter steroid hormone synthesis in human granulosa cells in vitro? Hum Reprod. 2016;31(7):1562–9.
44. Peretz J, Flaws JA. Bisphenol A down-regulates rate-limiting Cyp11a1 to acutely inhibit steroidogenesis in cultured mouse antral follicles. Toxicol Appl Pharmacol. 2013;271(2):249–56.
45. Peretz J, Craig ZR, Flaws JA. Bisphenol A inhibits follicle growth and induces atresia in cultured mouse antral follicles independently of the genomic estrogenic pathway. Biol Reprod. 2012;87(3):63.
46. Trapphoff T, Heiligentag M, El Hajj N, Haaf T, Eichenlaub-Ritter U. Chronic exposure to a low concentration of bisphenol A during follicle culture affects the epigenetic status of germinal vesicles and metaphase II oocytes. Fertil Steril. 2013;100(6):1758–67 e1.

47. Pan X, Wang X, Sun Y, Dou Z, Li Z. Inhibitory effects of preimplantation exposure to bisphenol-a on blastocyst development and implantation. Int J Clin Exp Med. 2015;8(6):8720–9.
48. Xiao S, Diao H, Smith MA, Song X, Ye X. Preimplantation exposure to bisphenol A (BPA) affects embryo transport, preimplantation embryo development, and uterine receptivity in mice. Reprod Toxicol. 2011;32(4):434–41.
49. Berger RG, Foster WG, de Catanzaro D. Bisphenol-A exposure during the period of blastocyst implantation alters uterine morphology and perturbs measures of estrogen and progesterone receptor expression in mice. Reprod Toxicol. 2010;30(3):393–400.
50. Varayoud J, Ramos JG, Bosquiazzo VL, Lower M, Munoz-de-Toro M, Luque EH. Neonatal exposure to bisphenol A alters rat uterine implantation-associated gene expression and reduces the number of implantation sites. Endocrinology. 2011;152(3):1101–11.
51. Grive KJ, Freiman RN. The developmental origins of the mammalian ovarian reserve. Development. 2015;142(15):2554–63.
52. Naule L, Picot M, Martini M, Parmentier C, Hardin-Pouzet H, Keller M, et al. Neuroendocrine and behavioral effects of maternal exposure to oral bisphenol A in female mice. J Endocrinol. 2014;220(3):375–88.
53. Fernandez M, Bourguignon N, Lux-Lantos V, Libertun C. Neonatal exposure to bisphenol A and reproductive and endocrine alterations resembling the polycystic ovarian syndrome in adult rats. Environ Health Perspect. 2010;118(9):1217–22.
54. Zhou J, Qu F, Jin Y, Yang DX. The extracts of Pacific oyster (Crassostrea gigas) alleviate ovarian functional disorders of female rats with exposure to bisphenol a through decreasing FSHR expression in ovarian tissues. Afr J Tradit Complement Altern Med. 2014;11(5):1–7.
55. Li Q, Davila J, Kannan A, Flaws JA, Bagchi MK, Bagchi IC. Chronic exposure to Bisphenol A affects uterine function during early pregnancy in mice. Endocrinology. 2016;157(5):1764–74.
56. Ziv-Gal A, Craig ZR, Wang W, Flaws JA. Bisphenol A inhibits cultured mouse ovarian follicle growth partially via the aryl hydrocarbon receptor signaling pathway. Reprod Toxicol. 2013;42:58–67.
57. Forte M, Mita L, Cobellis L, Merafina V, Specchio R, Rossi S, et al. Triclosan and bisphenol A affect decidualization of human endometrial stromal cells. Mol Cell Endocrinol. 2016;422:74–83.
58. Lopez-Rodriguez D, Franssen D, Sevrin E, Gerard A, Balsat C, Blacher S, et al. Persistent vs transient alteration of folliculogenesis and estrous cycle after neonatal vs adult exposure to bisphenol A. Endocrinology. 2019;160:2558.
59. Souter I, Smith KW, Dimitriadis I, Ehrlich S, Williams PL, Calafat AM, et al. The association of bisphenol-A urinary concentrations with antral follicle counts and other measures of ovarian reserve in women undergoing infertility treatments. Reprod Toxicol. 2013;42:224–31.
60. Benjamin S, Masai E, Kamimura N, Takahashi K, Anderson RC, Faisal PA. Phthalates impact human health: epidemiological evidences and plausible mechanism of action. J Hazard Mater. 2017;340:360–83. https://doi.org/10.1016/j.jhazmat.2017.06.036. Epub 2017 Jun 19. PMID: 28800814
61. Mikula P, Svobodová Z, Smutná M. Phthalates: toxicology and food safety—a review. Czech J Food Sci. 2005;23:217–23. https://doi.org/10.17221/3394-CJFS.
62. Staples CA, Peterson DR, Parkerton TF, Adams WJ. The environmental fate of phthalate esters: a literature review. Chemosphere. 1997;35:667–749. https://doi.org/10.1016/S0045-6535(97)00195-1.
63. Jurewicz J, Radwan M, Sobala W, Ligocka D, Radwan P, Bochenek M. Human urinary phthalate metabolites level and main semen parameters, sperm chromatin structure, sperm aneuploidy and reproductive hormones. Reprod Toxicol. 2013;42:232–41. https://doi.org/10.1016/j.reprotox.2013.10.001.
64. Sharpe RM, Skakkebaek NE. Testicular dysgenesis syndrome: mechanistic insights and potential new downstream effects. Fertil Steril. 2008;89:e33–8. https://doi.org/10.1016/j.fertnstert.2007.12.026.

65. Xie C, Zhao Y, Gao L, Chen J, Cai D, Zhang Y. Elevated phthalates' exposure in children with constitutional delay of growth and puberty. Mol Cell Endocrinol. 2015;407:67–73. https://doi.org/10.1016/j.mce.2015.03.006.
66. Zhu M, Huang C, Ma X, Wu R, Zhu W, Li X. Phthalates promote prostate cancer cell proliferation through activation of ERK5 and p38. Environ Toxicol Pharmacol. 2018;63:29–33. https://doi.org/10.1016/j.etap.2018.08.007.
67. Patiño-García D, Cruz-Fernandes L, Buñay J, Palomino J, Moreno RD. Reproductive alterations in chronically exposed female mice to environmentally relevant doses of a mixture of phthalates and alkylphenols. Endocrinology. 2018;159(2):1050–61. https://doi.org/10.1210/en.2017-00614. PMID: 29300862
68. Gallicchio L, Miller S, Greene T, Zacur H, Flaws J. Premature ovarian failure among hairdressers. Hum Reprod. 2009;24:2636–41. https://doi.org/10.1093/humrep/dep252.
69. Srilanchakon K, Thadsri T, Jantarat C, Thengyai S, Nosoognoen W, Supornsilchai V. Higher phthalate concentrations are associated with precocious puberty in normal weight Thai girls. J Pediatr Endocrinol Metab. 2017;30:1293–8. https://doi.org/10.1515/jpem-2017-0281.
70. Newbold RR. Impact of environmental endocrine disrupting chemicals on the development of obesity. Hormones. 2010;9:206–17. https://doi.org/10.14310/horm.2002.1271.
71. Messerlian C, Wylie BJ, Mínguez-Alarcón L, Williams PL, Ford JB, Souter IC. Urinary concentrations of phthalate metabolites and pregnancy loss among women conceiving with medically assisted reproduction. Epidemiology. 2016;27:879–88. https://doi.org/10.1097/EDE.0000000000000525.
72. Park MA, Hwang KA, Lee HR, Yi BR, Jeung EB, Choi KC. Cell growth of BG-1 ovarian cancer cells is promoted by di-n-butyl phthalate and hexabromocyclododecane via upregulation of the cyclin D and cyclin-dependent kinase-4 genes. Mol Med Rep. 2012 Mar;5(3):761–6. https://doi.org/10.3892/mmr.2011.712.

Chapter 5
Endocrine Disruption in Women: A Cause of PCOS, Early Puberty, or Endometriosis

Jean Marie Wenger and Roberto Marci

5.1 Introduction

A growing number of scientific studies have shown, since the last decade, increasing evidence suggesting that the human health and wildlife could be affected by a wide range of substances broadly disseminated in the environment and also found recurrently in a wide array of everyday products. These products were identified as toxicants with various effects on endocrine processes and functions as neoplasm development, reproductive dysfunctions, and immunological and thyroid disorders [1]. These endocrine-disrupting chemicals (EDCs), which are defined as "an exogenous chemical, or mixture of chemicals, that interferes with any aspect of hormone action" [2], are not rogue pharmaceuticals or rare contaminants.

EDCs enter the human body via food, water, dust by inhalation, and the transdermal route after contact and using cosmetics and creams. Transplacental transfer of these substances to the developing fetus has also been demonstrated [3] and therefore can be found in all body fluids (urine, serum, breast milk, and amniotic fluid). EDCs can accumulate and alter the adipose tissue, pancreas, liver, gastrointestinal tract, muscle, and brain homeostatic and hedonic pathways [4], and the effects of their metabolites can be functional at low doses and can persist for a long time [5].

The US Food and Drug Administration identified more than 1800 chemicals that disrupt at least one of the three endocrine pathways (estrogen, androgen, and thyroid) [6], and 320 of 575 chemicals were screened during the instruction of the European Commission, with either evidence or potential evidence for endocrine

J. M. Wenger (✉)
Department of gynecology, University of Geneva, Geneva, Switzerland
e-mail: jean-marie.wenger@hcuge.ch

R. Marci
Department of Translational Medicine, University of Ferrara, Ferrara, Italy
e-mail: mrcrrt@unife.it

© The Author(s) 2023
R. Marci (ed.), *Environment Impact on Reproductive Health*,
https://doi.org/10.1007/978-3-031-36494-5_5

disruption. Today, medical societies and governmental agencies such as the Endocrine Society [7], the International Federation of Gynecology and Obstetrics [8], the World Health Organization (WHO) and the United Nations Environment Programme (UNEP) [9], and the American Academy of Pediatrics [10] document the rapidly accelerating evidence and implications for human health. EDCs are usually used by the industries, as plastics (bisphenol A [BPA]), plasticizers (phthalates), solvents/lubricants (polybrominated biphenyls [PBBs], polychlorinated biphenyls [PCBs], and dioxins), pesticides (chlorpyrifos, dichlorodiphenyltrichloroethane [DDT], and methoxychlor), fungicides (vinclozolin), and also as flame-retardant additives in manufactured materials and pharmaceutical agents, for example, diethylstilbestrol (DES), a nonsteroidal synthetic estrogen [11]. EDCs may also be made by nature; for example, phytoestrogens, which interfere with endogenous endocrine function, are produced by plants and act primarily through estrogen receptors [12].

One of the most widely discussed and abundant EDCs is bisphenol A (BPA). Bisphenols are found in polycarbonates, epoxy resins, food, cosmetics packaging, and even dental composite materials [13]. BPA is able to interact with estrogen receptors through its phenolic structure: This allows the modification of hormonal homeostasis via a combination of agonist and/or antagonist actions depending on the target tissue. BPA does act not only on estrogens but also on androgen, pregnane X, thyroid, and glucocorticoid receptors [14]. Although in the recent years the use of BPA has been limited, and replaced in some products by its structural analogs such as bisphenol S (BPS), bisphenol F (BPF), and bisphenol AF (BPAF), comparable endocrine-disrupting effects have been observed with these alternative bisphenols, as the metabolism and mechanism of action are similar to BPA [15]. Unfortunately, BPS are not the only EDCs, and our body is subject to a "cocktail effect" as the addition and multiplication of each EDC occurs and can amplify the risks. Many personal care products, foods, and pharmaceuticals contain mixtures of bisphenols, parabens, and other EDCs; esters of p-hydroxybenzoic acid are used as antimicrobial agents and preservatives. In addition to the estrogenic effect, several parabens also possess antiandrogenic activity as they can bind to androgen receptors and thereby inhibit testosterone-induced transcription [16]. Methylparaben (MP) and propylparaben (PP), along with ethylparaben (EP), butylparaben (BP), and benzylparaben (benzylP), are among the most commonly used. In vivo studies indicate that parabens can disrupt reproduction, development, and homeostasis. In humans, they have been detected in serum, urinary cord blood, meconium, milk, amniotic fluid, and placental tissue [17, 18]. Relevant associations of MP and hormones affecting metabolic health and energy were observed, indicating obesogenic potential. Associations of methylparaben and hormones affecting energy balance and metabolic health have been observed, indicating its obesogenic potential [19]. Moreover, their effect seems to be transgenerational, occurring over at least two or three generations. As the window of susceptibility, puberty is considered as the one of the hot spots in the lifetime when EDCs may exert their effects [20], and areas of concern appear to be conditions like polycystic ovaries pathology [21, 22], precocious puberty issues [23], and endometriosis [24–27].

5.2   PCOS

Polycystic ovary syndrome (PCOS) is a complex and heterogeneous endocrine disorder in women of reproductive age [28]. Its prevalence is estimated to be between 5% and 10% and even up to 21%, depending on the diagnostic criteria and the geographic location [29–31]. In 1990, the National Institute of Health proposed the following diagnostic criteria: the presence of clinical and/or biochemical hyperandrogenism and oligomenorrhea/amenorrhea with anovulation [32]. In 2006, the Androgen Excess Society proposed diagnostic criteria: an androgen in excess is a critical element in the development and pathogenesis of PCOS that should be present and accompanied by oligomenorrhea, polycystic ovarian morphology, or both [30]. According to the Rotterdam criteria, the PCOS diagnosis requires meeting two of the three criteria mentioned above [33]. Today, the Rotterdam criteria are used by the medicals professional and researchers [34].

Hyperandrogenism seems to be the key feature of PCOS that contributes to clinical phenotypes and fertility dysregulation [35]. The most common sequelae of hyperandrogenism in the setting of the PCOS phenotype are hirsutism, acne, and alopecia [36]. The hormonal and metabolic alterations may result in reproductive disruption, including menstrual cycle dysfunction, chronic anovulation, and infertility [37], and the majority of women with PCOS have insulin resistance [38, 39], which may lead to the development of obesity [40]. This obesity is characterized by metabolic disturbances similar to metabolic syndrome [41] such atherogenic dyslipidemia and decreased glucose tolerance, which can lead to type 2 diabetes [42], with higher blood pressure values, increased thrombotic activity and several cardiovascular markers [43], and hyperinsulinemia and peripheral insulin resistance, which can occur independently on body weight [44]. Obesity, which is not always found in the ovaries with PCOS, where insulin resistance and compensatory hyperinsulinemia seem to play a vital role in the mechanisms of reproductive disorders by directly affecting the insulin-resistant ovaries with PCOS, has a detrimental impact on the ovulation process [21]. Conceiving difficulties may be due to slightly enlarged ovaries with numerous antral follicles, by two- to three-fold that of normal ovaries causing irregular ovulation and oligomenorrhea/amenorrhea. Other several features of PCOS are an excess of androgen [45], with a correlation with of a two- to three-fold higher anti-Müllerian hormone (AMH) than in ovulatory women with normal ovaries [46].

In addition, a "vicious circle" of hyperandrogenemia is created [47], following an elevated luteinizing hormone (LH) levels that promote androgen production and a reduction in estrogens. The underlying causes of PCOS are unclear, likely both genetic and environmental/nutritional, and the variety of clinical manifestations raises the possibility that multiple etiological factors simultaneously promote the final PCOS phenotype [28]. While geographic location, ethnicity, lifestyle, and environmental factors [48] appear to play a role, the latter along with endocrine-disrupting chemicals (EDCs) in the pathogenetic mechanisms of PCOS has been evoked recently. EDCs are a heterogeneous group of molecules, of natural or synthetic origin, capable of interacting with the endocrine system [28] by affecting hormonal biosynthesis, modifying their genomic and nongenomic effects, modifying the mechanisms of control and regulation and their epigenetic manifestations [18]. EDCs can be found in many everyday products (e.g., plastic bottles, cosmetics, metal cans, flame retardants, detergents, foods, toys, and pesticides) and penetrate in an organism through the ingestion of contaminated food and liquids, the breathing of contaminated air, and transdermal absorption [49]. Although in the recent years the use of BPA has been limited, and replaced with some products by its structural analogs such as bisphenol S (BPS), bisphenol F (BPF), and bisphenol AF (BPAF), comparable endocrine-disrupting effects have been observed with these alternative bisphenols, as the metabolism and mechanism of action are similar to BPA [15, 50]. The serum concentration of BPA is elevated in PCOS and correlates with androgen levels [51, 52]. The data suggest that reproductive function is disturbed directly at the ovary level by affecting ovarian steroid hormone production and the maturation of the follicle or indirectly by interfering with the hypothalamic–pituitary axis [21]. As obesity is associated with PCOS, low-grade inflammation, and increased inflammatory cytokines, several groups have indicated elevated levels of specific cytokines in women with PCOS, pointing out that chronic low-grade inflammation may affect the development of ovarian dysfunction and metabolic derangement [53, 54] The question is if the principal role in low-grade inflammation is due to only obesity or also due to PCOS. It is known from the literature that the interaction between BPA and testosterone is complex. On the one hand, testosterone interacts with BPA metabolism by decreasing uridine diphosphate glucuronosyl transferase activity, which leads to increased levels of BPA. On the other hand, BPA interferes with testosterone metabolism first by the inhibition of testosterone hydroxylases (2- and 6-hydroxylase), which are not that important in the degradation of testosterone as much as oxidoreductases, but still can play a role in its metabolism, and secondary by displacing testosterone on sex hormone–binding globulin (SHBG), which leads to the increase of circulating free androgen concentration [21].These interactions, especially the influence on binding protein, could explain our findings of the correlation between BPA exposure and testosterone only in a healthy control group, unlike in PCOS women, where the testosterone levels are high; thus, a "vicious circle" with BPA is formed.

The higher levels of BPA in PCOS patients were found compared with healthy controls [51, 52] without differences between those with normal-weight and obese ones and higher cytokines levels in obese ones with PCOS [13], which, in complexity, reflect activation and proinflammatory state. Findings in obese women with PCOS (insulin resistance, lousy lipid profile, risk of fatty liver disease, and proinflammatory state) compared with normal-weight PCOS women, which have very similar metabolic profile as healthy control, are confirmation of how obesity could obscure the searching of PCOS etiopathogenesis. The combination of genetic predispositions associated with environmental factors favored PCOS. In this context, being able to interact with the metabolism of testosterone, EDCs constitute one of the causes of PCOS.

We can conclude that these findings confirm that BPA could be one of the essential elements in the PCOS etiopathogenesis [13].

It is important to emphasize that other studies will have to be done because the number of endocrine disruptors continues to increase. For women with PCOS, it is essential that they maintain their body weight within normal range as this may protect them from the metabolic complications associated with this condition.

5.3 Early Puberty and EDCs

Early puberty is defined by the presence of clinical and auxological signs of pubertal development between the age of 8 and 10 years [55], between the age of 7.5 and 8.5 years [56], or between the age of 8 and 9 years [57]. Some authors consider that, when pubertal onset occurs before the age of 8 years, it is considered precocious, and when it occurs after 8 years but before 9 years of age, it is considered early. The mechanism of early pubertal development has not been clarified yet [58]. Puberty begins with the release of the hypothalamic gonadotropin-releasing hormone (GnRH) pulse generator from central nervous system inhibition after a quiescent period during childhood [59]. The age of menarche has definitely decreased from 16 years in the 1800s to 13 years in the 1960s, after which this downward trend seems to have slowed or even stopped [60]. Although genetic factors remain the main determinant of the timing of puberty [61], the trend toward earlier onset of puberty has coincided with improvements in public health and nutrition [62]. At the same time, endocrine-disrupting chemicals (EDCs) have been suggested as affecting the age of pubertal onset, especially in girls. Hence, researchers were led to hypothesize that increasing exposure to EDC had a role in the trend for earlier sexual maturation. Moreover, it was suggested that early puberty manifesting in immigrants from the developing countries was the result of previous exposure to organochlorine pesticides [63]. Constitutional advancement of growth (CAG) is the growth pattern of early growth acceleration, which is present in the majority of girls with idiopathic precocious puberty and in girls with early puberty [64]. While endocrine disruptors are commonly used by the industries, such as plastics (bisphenol A [BPA]), solvents/lubricants (polybrominated biphenyls [PBBs], polychlorinated biphenyls [PCBs]), dioxins, plasticizers (phthalates), pesticides (chlorpyrifos, dichlorodiphenyltrichloroethane [DDT], and methoxychlor), fungicides (vinclozolin), flame-retardant additives in manufactured materials and pharmaceutical agents, for example, diethylstilbestrol (DES), a nonsteroidal synthetic estrogen [11], they can also be natural, for example, phytoestrogens, produced by plants and that act mainly through estrogen receptors [12]. The large quantity of endocrine disruptors and their ability to interact with the endocrine system combined with the tendency toward the early onset of puberty have led many researchers to associate them with precocious puberty, especially since they have estrogenic activity. Several EDCs have been studied, and we will cite the main ones such as phthalates, bisphenol A (BPA), pesticides, flame-retardant chemicals, and PCBs.

5.4   Phthalates

Phthalates are esters of phthalic anhydride, used as liquid plasticizers in plastics, flooring, personal care products, medical devices, and tubing because they increase the flexibility, transparency, durability, and longevity of materials. Their most common use is to soften polyvinyl chloride (PVC). Phthalates can be classified as low- or high-molecular-weight phthalates, and depending on the class and the timing of exposure, different outcomes have been observed.

Their endocrine-disrupting mechanism is not fully clarified, but they act either as estrogen receptor agonists and antagonists or as androgen receptor antagonists and can also disrupt androgen synthesis. Different studies have demonstrated a significant association with premature thelarche and precocious or early puberty [65, 66]. High-molecular-weight phthalate levels several years before puberty are associated with later pubic hair development and younger age of menarche. Low-molecular-weight phthalate levels are related to advanced breast or pubic hair development [67]. In a study of the Danish schoolgirls, high phthalate excretion in urine was associated with delayed pubarche, but not thelarche, which suggests antiandrogenic actions of phthalate [68]. Similar results were obtained in a study of the US girls [69]. In contrast, in another study on the US girls with central precocious puberty (CPP), such an association was not found [70], and furthermore, a recent Korean study showed that phthalate metabolites in girls with central precocious puberty were significantly lower than the prepubertal control girls [71]. Many results are conflicting, and further studies are needed to confirm or refute the effect of phthalate exposure on pubertal timing.

5.5   BPA (Bisphenol A)

BPA is a precursor of plastics, polycarbonates, and epoxy resins coating the inside of beverage, found in plastics (e.g., bottles, Tupperware, food cans, etc.). It is the most commonly found estrogen-like endocrine disruptor that can also act as an antiandrogen in the environment. This chemical is almost ubiquitous, and even if the estrogen receptor agonist activity is weak, its potential should not be underestimated. In some experimental animals, it has been shown that BPA advances puberty [72], but no effect on pubertal timing [73]. Similar to the experimental animals, the results of BPA on human puberty are inconsistent. In a study of the US girls, Wolff et al. reported that BPA had no influence on breast development [67]; however, in studies performed in Turkey and in Thailand, idiopathic central precocious puberty was associated with higher levels of BPA than in control girls [74, 75]. Watkins et al. studied the in utero and peripubertal exposure to phthalates and BPA in relation to sexual maturation and did not find any association between BPA and sexual maturation, although in utero phthalate exposure impacted on earlier timing of sexual maturation [76]. Other studies shown that EDCs are associated with premature thelarche, precocious puberty, and pubertal development [74, 77, 78]. On the other hand, in a recent review, of 19 studies, only seven showed a correlation between BPA and puberty with evidence of the possible disruptive role of BPA in people with central precocious puberty or isolated premature breast development aged from 2 months to 4 years, although the mechanism is not defined. Some studies have also found a close relationship between urinary BPA, body weight, and precocious puberty, which may be explained by the obesogenic effect of BPA itself [79].

5.6 Pesticides

They are classified into various classes, for example, insecticides, herbicides, and fungicides, and can enter the human body through water, air, and food and can pass from mother to fetus via the placenta and to the infant through mother's milk. One of the well-known dichlorodiphenyltrichloroethane (DDT) is an organochlorine, originally developed as an insecticide for use in agriculture. Exposure to DDT is imperceptible, because it is odorless, tasteless, and colorless, and being exposed during fetal life and lactation can affect sexual development. Despite the fact that DDT is still widely used in some low-income countries and has been banned from our markets, it can persist in the environment as a persistent organic pollutant (POP). Dichlorodiphenyldichloroethane (DDE), a metabolite of DDT, has antiandrogenic, antiprogestin, and estrogenic effect and induces aromatase. Vasiliu et al. found an association between the exposure to these chemicals and precocious puberty and earlier age of menarche [80].

A study performed in Denmark, female offspring of mothers exposed to pesticide in a greenhouse showed a decreased age of breast development at 8.9 years, compared with 10.4 years in the unexposed population and 10.0 years in a Danish reference population [81], but the significance of the association disappeared when weight at menarche was controlled for. Pesticide exposure to pesticides has also been suggested in adopted or immigrant girls in Belgium, with central precocious puberty (CPP), following the discovery of higher levels of plasma DDE [63]. Conversely, other studies did not found an association between DDE levels and early puberty [82], but unlike a puberty delay [83].

Flame-retardant chemicals are added to the manufactured materials (plastics, textiles, surface finishes, and coatings) intended to prevent or slow the further development of ignition with their physical and chemical properties. Among them, organohalogen compounds such as polybrominated diphenyl ethers (PBDEs) are lipophilic persistent endocrine disruptors exhibiting estrogenic and androgenic properties. PBDEs might alter pubertal timing, resulting in later menarche in girls [84], but in girls with idiopathic central precocious puberty, particularly those with higher body mass index (BMI) have been found with higher serum concentrations of PBDEs [85]. Thus, the inconsistency of the results of the various studies examining the association of endocrine disruptor chemicals with the onset of puberty [86] makes it imperative that more studies on the subject are performed.

Polychlorinated biphenyl (PCB) is a dioxin-like compound derived from biphenyl, used as a dielectric and coolant fluid in electrical apparatuses. Its mechanism of action is rather similar to that of dioxins, and there is evidence that exposure during the prenatal period leads to early onset of menarche and to delayed pubertal development [58].

The conclusion is that the onset of puberty occurs earlier in girls, and physiological variability and multiple other factors affect the onset of puberty. Exposure to a wide and growing range of known and unknown endocrine disruptors is ubiquitous, and changes in the onset of puberty may be influenced by exposures to endocrine disruptors at critical developmental windows. Endocrine disruptors are hormonally active substances that can act via several mechanisms to disrupt puberty either peripherally on the target organs (adipose tissue or adrenal glands) or centrally via the hypothalamic–pituitary–gonadal (HPG) axis. Nevertheless, the definitive evidence of associations between exposures to endocrine disruptors remains controversial [87, 88]. It seems obvious that some endocrine disruptors modify metabolic parameters: The increase in the latter [10] coincides with the increase in the prevalence of obesity with its risks over the last three decades and suggests that they are one of the major factors of the obesity epidemic [10]. The association between EDC and precocious puberty is subject to a bias that, as we have seen, is constituted by the improvement of health and nutritional conditions and the increase in the prevalence of obesity [89–91], which both can advance the age of puberty. However, current data are insufficient and conflicting to provide sufficient evidence for a causal relationship between exposure to endocrine disruptors and changes in the timing of puberty in humans. Definitive evidence for associations between exposures to endocrine disruptors remains controversial and still insufficient and contradictory to establish sufficient evidence for a causal relationship between exposure to endocrine disruptors and changes in the timing of puberty in humans. Further human epidemiological studies of a prospective and longitudinal nature are needed to determine the combined effect of EDC exposure on puberty and reproduction during critical periods. Furthermore, the underlying mechanisms by which early exposures to endocrine disruptors influence puberty, including epigenetic factors, need to be explored separately.

5.7   Endometriosis

Endometriosis is a common benign condition with potentially significant morbidity such as pelvic pain, dysmenorrhea, dyspareunia, and infertility and is thought to affect 2–50% of women of reproductive age [92, 93]. It is present in 71–87% of women with chronic pelvic pain [94].

The incidence and the prevalence associated with this disease showed an increasing trend in countries with a high sociodemographic index between 1990 and 2017 [92, 93]. Biologically, endometriosis is an estrogen-dependent, inflammatory, potentially chronic gynecological condition characterized by the proliferation of cells resembling functional endometrial tissue and growing outside the uterine cavity [95]. Despite the proposal of many theories, the precise etiology of the disease remains unknown. The oldest and still recognized hypothesis is the theory of retrograde menstruation [96]. Although the attachment of ectopic glands emanating from menstrual debris from reflux remains a plausible mechanistic explanation for the development of endometriosis, it does not explain all the incidences and presentation of the disease. Other theories regarding the development of endometriosis include coelomic metaplasia, activation of remnant stem cells, and inherent epigenetic abnormalities [97–100].

An additional difficulty is associated with the fact that endometriosis may take several different forms (ovarian endometrioma, peritoneal endometriosis, deeply infiltrating endometriosis, and adenomyosis—or endometriosis of the uterine muscle), which not only differ in location but also have different clinical presentations. In some cases, endometriosis remains asymptomatic, and a certain diagnosis can only be established by invasive evaluation (laparoscopy) and histopathological confirmation. Sometimes silent endometriosis is a condition in which the patient does not experience any discomfort resulting from the development of the disease, and symptoms may appear later in life or remain dormant.

Today, it appears that the development of endometriosis is determined by complex interactions between the composite effects of genetic and environmental risk factors. Indeed, families of genes associated with the immune system and inflammatory pathways, cell adhesion, and extracellular matrix remodeling have been described as being differentially expressed when comparing women with and without endometriosis [101, 102]. As a common environmental risk factor, endocrine-disrupting chemicals (EDCs) are ubiquitous in the environment and food chains and can affect the dynamic balance of sex hormones and mediate the innate dysregulation of immune cells, which may therefore play a role important in the pathogenesis of endometriosis [11, 103–106]. Nevertheless, there is a clear lack of well-established and modifiable risk factors for this disease; several existing publications have given conflicting results. There is therefore still no conclusive evidence for these potential risk factors regarding the combinations themselves or their management.

Because of the potential association between exposure to EDCs and the development of endometriosis, many studies have been devoted to this topic. Such studies are difficult to design, as it is difficult to identify both the study group and the control group and to measure the exposure to EDCs and the effects of other factors on the development of this condition.

Of the many EDCs, compounds that are best understood in terms of potential involvement in the pathogenesis of endometriosis are bisphenols [107], dioxin and dioxin-like compounds [25, 104], phthalates [108], and others.

5.8   Bisphenols

Bisphenol A (BPA) was the first to be synthesized, but evidences gathered in 1936 showed a low estrogen effect with affinity for the nuclear estrogen receptor. Its effects depend on dosage, targeted tissue, and tissue development on the site where it acts. The occurrence of estrogenic or antiestrogenic effects depends on the tissue targeted and on their impact on receptors [50]. Global production of BPA has steadily grown in the recent years on account of its multiple applications in the plastic and manufacturing industries, in food packaging, and in toys, causing a constant and permanent poisoning of food, water, and the environment. In 1950, it was found that bisphosphonates could be polymerized, and since then, they have been used to make polycarbonate plastics. These plastics have convenient features such as lightweight, moldability, and impact and heat resistance and are not susceptible to changes over time. About 20% of these plastics are used as a component of epoxy resin, serving as internal coating for plastic containers, bottles, and dental sealants. Therefore, it is a liquid and food contaminant present in abnormal levels in human serum analysis according to the literature. BPA is rapidly metabolized to inactive forms with a mean life cycle of approximately 4–5 h in adults, while in fetuses and children the metabolic rate is relatively low [109]. BPA can easily accumulate in adipose tissue for having lipophilic properties. Measurements of human serum have determined varied and controversial toxicity rates. Currently, the United States Environmental Protection Agency has established a safe level of 50 µg/kg/day, and the European Food Safety Authority has established a tolerable daily intake of less than 4 µg/kg/day. The list of products containing bisphenols available on the market has continued to grow, the most common being bisphenols BPS, BPF, BPB, and BPAF, which nevertheless seem to have the same properties.

Bisphenols are therefore estrogen-mimicking EDCs that are capable of maintaining low levels of progesterone receptors that can lead to disruptions in uterine cyclicity, a potential mechanism for the development of endometriosis [107]. The first, bisphenol A (BPA), previously used in the manufacturing of food cans and dental sealants, is one of the most well-studied and widespread EDCs.

Several previous experimental studies reported that the exposure of prenatal mice to bisphenol A (BPA) can cause endometriosis-like symptoms in offspring [110]. In human, it was abundantly present in sera of women with endometriosis compared with women without disease [111, 112]. A population-based case–control study to determine whether BPA exposure was linked to an increased risk of endometriosis, after measuring total urinary BPA concentrations in 143 cases (women with surgically diagnosed endometriosis) and 287 controls (women without a known endometriosis diagnosis), revealed a statistically significant, positive correlation between urinary BPA concentrations and peritoneal endometriosis, but not ovarian disease [113]. In contrast, in other studies, patients with ovarian endometriomas were found to have significantly higher urinary BPA concentrations than controls [112]. Other studies found no association between urinary [114, 115]. Inconsistencies among human studies likely reflect differences in populations, experimental design variations, and the rigorousness of the control groups [115].

5.9 Dioxins and Dioxin-Like Compounds

Dioxins and dioxin-like compounds are extremely resistant by-products of various industrial processes (e.g., waste incineration and iron/steel industries) or natural, and they represent ubiquitous environmental pollutants, chemically stable and lipophilic [116], and are polycyclic aromatic agents with chloral substituents.

Dioxins and dioxin-like compounds include the following:

(a) Polychlorinated dibenzo-p-dioxins (PCDDs or dioxins): There are 75 PCDDs.
(b) Seven of them are highly toxic polychlorinated dibenzofurans (PCDFs): There are 135 PCDFs. They are not dioxins, but ten of them have dioxin-like properties, the polychlorinated biphenyls (PCBs): There are 209 PCBs, and 12 of them have dioxin-like properties (the so-called coplanar PCBs because of the absence of chlorine substitution in ortho positions that gives the molecule a planar configuration). They have been widely used as dielectric and coolant fluids until they were banned worldwide in the 1980s [104].

PCDDs, PCDFs, and PCBs together form the group of polyhalogenated hydrocarbons and were found, by some authors, to be significantly associated with endometriosis [117, 118].

Dioxin generally enters the environment after accidents like the one in Seveso, Italy, in 1976. Dioxins then get into soil sediments, being carried by weather patterns, and become incorporated into the food chain [119]. They mainly enter the human body through food and, due to their lipophilic nature, accumulate in tissues with high-fat content [116]. Because of this property, it does not surprise to find high levels of dioxin and dioxin-like compounds in older people and reduced levels after delivery or breastfeeding [120]. Ten PCDFs, 12 PCBs (those with dioxin-like properties), and seven PCDDs bind to the aryl hydrocarbon receptor (AhR), an activated ligand transcription factor. AhR could be mostly found in the cytosol (sometimes in the nucleus) and represents the key component of the dioxin pathways [121]. In order to quantify their biological potency, all dioxin-like compounds have received a toxic equivalency factor (TEF) in terms of the most toxic dioxin (2,3,7,8-tetrachlorodibenzo-p-dioxin [TCDD]), which has a TEF of 1. However, the toxicity of a mixture of these compounds is often expressed in pg TEQ (toxic equivalent units)/g lipids, which represents the sum of the product of the concentration of each compound multiplied by its TEF [104]. The concentration is expressed per g lipids because they are mainly stored in adipose tissue [122].

The most toxic dioxin 2,3,7,8-tetrachlorodibenzo-p-dioxin (TCDD), due to its lipophilic nature, has the particularity of being very resistant to degradation and is able to modulate signaling processes mediated by estrogen and progesterone, steroid hormones necessary for the maintenance of normal uterine physiology. Exposure to TCDD has been experimentally linked to the development of reproductive disorders in mammals, most notably in a publication first reported by Rier in 1993, which found a positive correlation between exposure to TCDD and the incidence of endometriosis in a colony of rhesus monkeys [123]. Several studies have since been followed to examine the potential link between exposure to TCDD and the development of endometriosis [117, 124–126].

Concerning PCBs, within the reproductive tract, coplanar PCBs are particularly suited to act in concert with TCDD to disrupt key elements of communication between the immune and endocrine systems ([127, 128], potentially promoting reproductive disorders such as endometriosis. Rier, who had previously linked TCDD and endometriosis [123], subsequently reported a probable coexposure of these animals to significant levels of dioxin-like PCBs following food contaminated with toxic substances [129]. It therefore appears that, even within the framework of a controlled experimental study, it may be difficult to completely exclude additional occult sources of exposure to environmental toxicants via food or water [126, 129].

As with TCDD, although systematic review and meta-analysis results have shown that total PCBs are significantly associated with the risk of endometriosis, epidemiological data remain weak [130], or mixed [131], as for TCDD [126], with a number of studies failing to identify a clear association between TCDD exposure and endometriosis [115], even if certain authors concluded that a bad classification of the disease could have led to underestimating the risk [125].

5.10   Phthalates

Phthalates and their esters consist of a large group of chemical compounds with antiandrogenic and estrogenic activity frequently used in the plastic, coating, cosmetic, and toy industries and medical devices such as syringes and blood bags, and women are generally more at risk than men due to their employment in feminine care products and cosmetics [132]. Phthalates are the by-products of phthalic acid and are used in the plastics industry for their excellent moldability. In the roster of phthalates, three esters are considered endocrine disruptors with estrogenic effects: diethyl-hexyl phthalate (DHEP), benzyl-butyl phthalate (BBP), and dibutyl phthalate (DBP). Phthalates can be found not only in serum and human urine, but also in milk samples. Nevertheless, the mechanisms triggering the development of endometriosis by phthalates remain unclear. Tolerable daily intake ranges between 3 and 30 µg/kg/day [133–135]. In women with advanced endometriosis, significantly higher levels of mono-ethylhexyl phthalate (MEHP) and di-(2-ethylhexyl) phthalate (DEHP) were found in their plasma compared with disease-free women [136, 137]. The results of other studies, the National Health and Nutrition Examination Survey (NHANES), and the Endometriosis, Natural History, Diagnosis, and Outcomes study also revealed a significant association between urinary phthalates and endometriosis [115, 138]. Studies on the association between phthalate exposure and the presence of disease in Taiwanese women revealed a significant increase ($p < 0.05$) in urinary mono-n-butyl phthalate (MBP) and MEHP in patients with endometriosis [139, 140]. Nevertheless, other epidemiological studies failed to validate these findings. Upson [141], in a study including women from the northeast of the United States of America, showed an inverse association between the risk of developing endometriosis and levels of MEHP. These data were confirmed by Itoh [142] in a study of infertile women, although the authors only included 57 cases with endometriosis and 80 controls without endometriosis.

Despite suspicions of causation between phthalates and endometriosis, there are no regulations limiting their use in the United States or Brazil, although the European Community has banned them.

5.11   Medications as Endocrine Disruptors

5.11.1   Diethylstilbestrol

Historically, one of the most well-known pharmaceutical exposures to EDCs was the consequence of the consumption of diethylstilbestrol (DES) by pregnant women, which was originally prescribed with the aim of mitigating the risk of miscarriage, premature delivery, and other pregnancy-related complications [26]. DES is a synthetic, highly potent estrogen that was initially prescribed to women with high-risk pregnancies. Soon after, it was recommended to all pregnant women from the 1940s through the 1970s. In 1971, DES was banned in the United States because, in addition to being completely ineffective in preventing miscarriage, it was shown to increase the risk of serious illness in mothers and their children [143, 144].

Relevant to the current discussion, additional studies revealed an increased incidence of endometriosis in women whose mothers were prescribed DES compared with the daughters of women that were not given DES during pregnancy [145, 146].

5.12   Conclusion

The various studies concerning these three pathologies cited above, which show not only sometimes strong but also weak or contradictory relationships with endocrine disruptors, their involvement in complex metabolic disorders, and the new harmful effects on health of endocrine disruptors frequently used, highlight the full complexity of the problem. Taking this complexity into account in the assessment, management, and attempts to resolve it requires an approach from several points of view: environmental, ethical, scientific, epidemiological, economic, political, strategic, and preventive. Compounds potentially incriminated as endocrine disruptors are ubiquitous, present in our daily life (diet and lifestyle), increasing exponentially, persistent but also sporadic, and capable of producing potentially active metabolites. The scientific challenges are numerous due to the difficulties in dosing the compounds, the confusions, the complex mixtures of exposures and their interrelationships [147], the variability of the distributions of exposure from one study to another that can explain the differences in results, the design of numerous studies, and the imprecision of the exposure assessment methods (dosage, the number of patients, the duration of exposure, statistical bias, and difficulty in assaying the substances in question in the target organs), in particular for the chemicals with short half-life. In addition, biostatistical developments have not yet resulted in an ideal method to manage associated exposures that might exist in the human body [148]. Sometimes the limit values that can be considered toxic are unclear, and the relevance of animal models transferred to humans is questionable. Moreover, with the exception of evidence from accidentally exposed populations, experimental evidence demonstrates that developmental exposure to endocrine disruptors can lead to transgenerational adverse effects with health consequences: Such a concept is difficult to prove in humans because randomized designs of interventions to increase or decrease exposure are generally not applicable due to obvious ethical and logistical considerations.

A recurring theme in the studies reviewed is the appearance on the market of a colossal quantity of new substances, but also of their substitutes, little tested, wrongly assumed to be less toxic [15], and on the contrary revealing new signs of toxicity [26]. What about the recommended doses for BPA by the American Environmental Protection Agency for a safety level of 50 µg/kg/day, while the European Food Safety Authority has established a tolerable daily intake of less than 4 µg/kg/day? or concerning restrictions on phthalates, totally absent in the United States or Brazil, but banned by the European Community [149]? Are there divergences between financial interests and public health?

The otherwise justified terms "possible" or "probable" found in the literature for the risky should not obscure the precautionary principle, in light of reality: It is increasingly clear that endocrine disruptors are involved in diseases that are not transferable. Nevertheless, these synthetic compounds are ignored or at least underestimated as sustainable development goals (SDGs) of 2030, and decreasing exposure to synthetic chemicals with endocrine-disrupting or other harmful properties is not identified as one of the SDGs, although these rightly highlight that air pollution and climate change as global priorities [150] and despite the fact that intervention studies have produced rapid decreases in exposure to organophosphate pesticides, bisphenols, phthalates, parabens, and triclosans [151]. However, the decisions must come not only from the decision-makers, but also from the consumers. Since the majority of exposure to endocrine disruptors occurs through diet, choosing organic foods, lean meats, or a vegetarian lifestyle can help everyone minimize exposure. In addition, reducing the use of canned foods containing a BPA liner, using BPA-/BPS-free products, and avoiding long-term storage or heating of foods in plastic containers will also reduce the accidental exposure to the endocrine disruptors [26].

Therefore, in light of the above, clear-cut strategies and recommendations should be targeted to reduce human exposure to protect future generations from ever-increasing adverse health effects, and regulators should strengthen premarketing toxicological testing [152].

The need for additional further research is evident to further elaborate the effects of endocrine disruptors and other products on human health looking, of course, at causation and actions to reduce exposure to endocrine disruptors, taking into account the evidence and issues involved in decisions [153] and finding alternative manufacturing practices that can be applied to mitigate exposure to endocrine disruptors [24]. The additional costs to society can be weighed against the economic benefits of reduced disease and disability and other societal effects (e.g., ecosystem effects) [24], by always bearing in mind, however, that human health must take precedence over any other interest.

References

1. Carpenter DO. Polychlorinated biphenyls (PCBs): routes of exposure and effects on human health. Rev Environ Health. 2006;21:1–23.
2. Zoeller R, Brown TR, Doan LL, Gore AC, Shakkebaek NE, Soto AM, Woodruff TJ, Vom Saal FS, Endocrine-disrupting. Chemicals and public health protection: a statement of principles from the Endocrine Society. Endocrinology. 2012;153:4097–110.
3. Pivonello C, Muscogiuri G, Nardone A, Garifalos F, Provvisiero DP, Verde N, De Angelis C, Conforti A, Piscopo M, Auriemma RS. Bisphenol A: an emerging threat to female fertility. Reprod Biol Endocrinol. 2020;18:22.
4. Heindel JJ, Blumberg B, Cave M, Machtinger R, Mantovani A, Mendez MA, Nadal A, Palanza P, Panzica G, Sargis R. Metabolism disrupting chemicals and metabolic disorders. Reprod Toxicol. 2017;68:3–33.
5. Wang Y, Zhu Q, Dang X, He Y, Li X, Sun Y. Local effect of bisphenol A on the estradiol synthesis of ovarian granulosa cells from PCOS. Gynecol Endocrinol. 2017;33:21–5.
6. Ding D, Xu L, Fang H, Hong H, Perkins R, Harris S, Bearden ED, Shi L, Tong W. The EDKB: an established knowledge base for endocrine disrupting chemicals. BMC Bioinformatics. 2010;11(Suppl 6):S5.
7. Gore AC, Chappell VA, Fenton SE, et al. EDC-2: the Endocrine Society's second scientific statement on endocrine-disrupting chemicals. Endocr Rev. 2015;36:e1–150.

8. Di Renzo GC, Conry JA, Blake J, et al. International federation of gynecology and obstetrics opinion on reproductive health impacts of exposure to toxic environmental chemicals. Int J Gynaecol Obstet. 2015;131:219–25.
9. WHO. International Programme on chemical safety. Global assessment of state-of-the-science for endocrine disruptors. Geneva: World Health Organization; 2012. https://www.who.int/ipcs/publications/new_issues/endocrine_disruptors/en. Accessed 6 Oct 2014.
10. Trasande L, Shaffer RM, Sathyanarayana S. Food additives and child health. Pediatrics. 2018;142:e20181408.
11. Diamanti-Kandarakis E, Bourguignon JP, Giudice LC, Hauser R, Prins GS, Soto AM, Zoeller RT, Gore AC. Endocrine disrupting chemicals: an Endocrine Society scientific statement. Endocr Rev. 2009;30:293–342.
12. Kuiper GG, Lemmen JG, Carlsson B, Corton JC, Safe SH, Van der Saag PT, Van der Burg B, Gustafsson JA. Interaction of estrogenic chemicals and phytoestrogens with estrogen receptor beta. Endocrinology. 1998;139:4252–63.
13. Simkova M, Vitku J, Kolatorova L, Jana Vrbkova J, Vosatkova M, Vcelak J, Dusskova M. Endocrine disruptors, obesity, and cytokines–how relevant are they to PCOS? Physiol Res. 2020;69(Suppl. 2):S279–93.
14. Žalmanová T, Hošková K, Nevoral J, Adámková K, Kott T, Šulc M, Kotíková Z, Prokešová Š, Jílek F, Králíčková M. Bisphenol S negatively affects the meiotic maturation of pig oocytes. Petr J Sci Rep. 2017;7(1):485.
15. Eladak S, Grisin T, Moison D, Guerquin MJ, N'Tumba-Byn T, Pozzi-Gaudin S, Benachi A, Livera G, Rouiller-Fabre V, Habert RA. New chapter in the bisphenol A story: bisphenol S and bisphenol F are not safe alternatives to this compound. Fertil Steril. 2015;103(1):11–21.
16. Błędzka D, Gromadzińska J, Wąsowicz W. Parabens. From environmental studies to human health. Environ Int. 2014;67:27–42.
17. Azzouz A, Colón LP, Hejji L, Ballesteros E. Determination of alkylphenols, phenylphenols, bisphenol A, parabens, organophosphorus pesticides and triclosan in different ceral based foodstuffs by gas chromatography-mass spectrometry. Anal Bioanal Chem. 2020;412(11):2621–31.
18. Kolatorova L, Duskova M, Vitku J, Starka L. Prenatal exposure to bisphenols and parabens and impacts on human physiology. Physiol Res. 2017;66(Suppl 3):S305–15.
19. Kolatorova L, Vitku J, Hampl R, Adamcova K, Skodova T, Simkova PA, Starka LL, Duskova M. Exposure to bisphenols and parabens during pregnancy and relations to steroid changes. Environ Res. 2018;163:115–22.
20. Casey BJ, Tottennham N, Liston C, Durston S. Imaging the developing brain: what have we learned about cognitive development? Trends Cogn Sci. 2005;9:104–10.
21. Palioura E, Diamanti-Kandarakis E. Polycystic ovary syndrome (PCOS) and endocrine disrupting chemicals (EDCs). Rev Endocr Metab Disord. 2015;16(4):365–71.
22. Kawa IA, Masood A, Fatima Q, Mir SA, Jeelani H, Manzoor S, Rashid F. Endocrine disrupting chemical bisphenol A and its potential effects on female health. Diabetes Metab Syndr. 2021;15(3):803–11.
23. Papadimitriou A, Papadimitriou DT. Endocrine-disrupting chemicals and early puberty in girls. Children. 2021;8:492.
24. KahnL L, Philippat C, Nakayama SF, Slama R, Trasande L. Endocrine-disrupting chemicals: implications for human health. Lancet Diabetes Endocrinol. 2020;8(8):703–18.
25. Polak G, Banaszewska B, Filip M, Radwan M, Wdowiak A. Environmental factors and endometriosis. Int J Environ Res Public Health. 2021;18:11025.
26. Rumph JT, Stephens VR, Archibong AE, Osteen KG, Bruner-Tran KL. Environmental endocrine disruptors and endometriosis. Adv Anat Embryol Cell Biol. 2020;232:57–78.
27. Stephens VR, Jelonia T, Rump JT, Amel S, Bruner-Tran KL. Osteen KG the potential relationship between environmental Endocrine disruptor exposure and the development of endometriosis and Adenomyosis. Front Physiol. 2022;12:1–15.

28. Diamanti-Kandarakis E. Polycystic ovarian syndrome: pathophysiology, molecular aspects and clinical implications. Expert Rev Mol Med. 2008;10:e3.
29. Asuncion M, Calvo RM, San Millan JL, Sancho J, Avila S, Escobar-Morreale HF. A prospective study of the prevalence of the polycystic ovary syndrome in unselected Caucasian women from Spain. J Clin Endocrinol Metab. 2000;85:2434–8.
30. Azziz R, Woods KS, Reyna R, Key TJ, Knochenhauer ES, Yildiz BO. The prevalence and features of the polycystic ovary syndrome in an unselected population. J Clin Endocrinol Metab. 2004;89:2745–9.
31. Franks S. Polycystic ovary syndrome. N Engl J Med. 1995;333:853–61.
32. Zawadzki J, Dunaif A. Current issues in endocrinology and metabolism: polycystic ovary syndrome. Cambridge MA: Blackwell Scientific Publications; 1992.
33. Rotterdam ESHRE/ASRM-Sponsored PCOS Consensus Workshop Group. Revised 2003 consensus on diagnostic criteria and long-term health risks related to polycystic ovary syndrome. Fertil Steril. 2004;81:19–25.
34. Wang R, Mol BWJ. The Rotterdam criteria for polycystic ovary syndrome: evidence-based criteria? Hum Reprod. 2017;32:261–4.
35. Lizneva D, Gavrilova-Jordan L, Walker W, Azziz R. Androgen excess: investigations and management. Best Pract Res Clin Obstet Gynaecol. 2016;37:98.
36. Livadas S, Pappas C, Karachalios A, Marinakis E, Tolia N, Drakou M, et al. Prevalence and impact of hyperandrogenemia in 1,218 women with polycystic ovary syndrome. Endocrine. 2014;47:631–8.
37. Norman RJ, Dewailly D, Legro RS, Hickey TE. Polycystic ovary syndrome. Lancet. 2007;370:685–97.
38. Diamanti-Kandarakis E, Livadas S, Katsikis I, Piperi C, Mantziou A, Papavassiliou AG, et al. Serum concentrations of carboxylated osteocalcin are increased and associated with several components of the polycystic ovarian syndrome. J Bone Miner Metab. 2011;29:201–6.
39. Diamanti-Kandarakis E. Insulin resistance in PCOS. Endocrine. 2006;30:13–7.
40. Rachon D, Teede H. Ovarian function and obesity—interrelationship, impact on women's reproductive lifespan and treatment options. Mol Cell Endocrinol. 2010;316:172–9.
41. Sam S, Dunaif A. Polycystic ovary syndrome: syndrome XX? Trends Endocrinol Metab. 2003;14:365–70.
42. Legro RS, Blanche P, Krauss RM, Lobo RA. Alterations in low-density lipoprotein and high-density lipoprotein subclasses among Hispanic women with polycystic ovary syndrome: influence of insulin and genetic factors. Fertil Steril. 1999;72:990–5.
43. Wild RA, Carmina E, Diamanti-Kandarakis E, Dokras A, Escobar-Morreale HF, Futterweit W. Assessment of cardiovascular risk and prevention of cardiovascular disease in women with the polycystic ovary syndrome: a consensus statement by the androgen excess and polycystic ovary syndrome (AE-PCOS) society. J Clin Endocrinol Metab. 2010;95:2038–49.
44. Dunaif A. Insulin resistance and the polycystic ovary syndrome: mechanism and implications for pathogenesis. Endocr Rev. 1997;18:774–800.
45. Cadagan D, Khan R, Amer S. Thecal cell sensitivity to luteinizing hormone and insulin in polycystic ovarian syndrome. Reprod Biol. 2016;16(1):53–60.
46. Laven JS, Mulders AG, Visser JA, Themmen AP, De Jong FH, Fauser BC. Anti-Müllerian hormone serum concentrations in normoovulatory and anovulatory women of reproductive age. J Clin Endocrinol Metab. 2004;89(1):318–23.
47. Burt Solorzano CM, Beller JP, Abshire MY, Collins JS, McCartney CR, Marshall JC. Neuroendocrine dysfunction in polycystic ovary syndrome. Steroids. 2012;77(4):332–7.
48. Wijeyaratne CN, Seneviratne Rde A, Dahanayake S, Kumarapeli V, Palipane E, Kuruppu N, Yapa C, Seneviratne Rde A, Balen AH. Phenotype and metabolic profile of South Asian women with polycystic ovary syndrome (PCOS): results of a large database from a specialist Endocrine Clinic. Hum Reprod. 2011;26(1):202–13.
49. Darbre PD. Overview of air pollution and endocrine disorders. Int J Gen Med. 2018;11:191–207.

50. Rochester JR, Bolden AL. Bisphenol S and F: a systematic review and comparison of the hormonal activity of Bisphenol a substitutes. Environ Health Perspect. 2015;123:643–50.
51. Kandaraki E, Chatzigeorgiou A, Livadas S, Palioura E, Economou F, Koutsilieris M, Palimeri S, Panidis D, Diamanti-Kandarakis E. Endocrine disruptors and polycystic ovary syndrome (PCOS): elevated serum levels of bisphenol A in women with PCOS. J Clin Endocrinol Metab. 2011;96(3):E480–4.
52. Takeuchi T, Tsutsumi O, Ikezuki Y, Takai Y, Taketani Y. Positive relationship between androgen and the endocrine disruptor, bisphenol A, in normal women and women with ovarian dysfunction. Endocr J. 2004;51(2):165–9.
53. Amato G, Conte M, Mazziotti G, Lalli E, Vitolo G, Tucker AT, Bellastella A, Carella C, Izzo A. Serum and follicular fluid cytokines in polycystic ovary syndrome during stimulated cycles. Obstet Gynecol. 2003;101(6):1177–82.
54. Ebejer K, Calleja-Agius J. The role of cytokines in polycystic ovarian syndrome. Gynecol Endocrinol. 2013;29(6):536–40.
55. Mul D, De Muinck K-SSMPF, Oostdijk W, Drop SLS. Auxological and biochemical evaluation of pubertal suppression with the GnRH agonist leuprolide acetate in early and precocious puberty. Horm Res. 1999;51:270–6.
56. Cassio A, Cacciari E, Balsamo A, Bal M, Tassinari D. Randomised trial of LHRH analogue treatment on final height in girls with onset of puberty aged 7.5–8.5 years. Arch Dis Child. 1999;81:329–32.
57. Lebrethon MC, Bourguignon JP. Management of central isosexual precocity: diagnosis, treatment, outcome. Curr Opin Pediatr. 2000;12:394–9.
58. Den Hond E, Roels HA, Hoppenbrouwers K, Nawrot T, Thijs L, Vandermeulen C, Winneke G, Vanderschueren D, Staessen JA. Sexual maturation in relation to polychlorinated aromatic hydrocarbons: Sharpe and Skakkebaek's hypothesis revisited. Environ Health Perspect. 2002;110:771–6.
59. Abreu AP, Kaiser UB. Pubertal development and regulation. Lancet Diabetes Endocrinol. 2016;4:254–64.
60. Parent AS, Franssen D, Fudvoye J, Gérard A, Bourguignon JP. Developmental variations in environmental influences including endocrine disruptors on pubertal timing and neuroendocrine control: revision of human observations and mechanistic insight from rodents. Front Neuroendocrinol. 2015;38:12–36.
61. Zhu J, Kusa TO, Kusa TO, Chan YM. Genetics of pubertal timing. Curr Opin Pediatr. 2018;30:532–40.
62. Cheng G, Buyken AE, Shi L, Karaolis-Danckert N, Kroke A, Wudy SA, et al. Beyond overweight: nutrition as an important lifestyle factor influencing timing of puberty. Nutr Rev. 2012;70:133–52.
63. Krstevska-Konstantinova M, Charlier C, Crae M, Du Caju M, Heinrichs C, de Beaufort C, Plomteux G, Bourguignon JP. Sexual precocity after immigration from developing countries to Belgium: evidence of previous exposure to organochlorine pesticides. Hum Reprod. 2001;16:1020–6.
64. Theodoropoulou S, Papadopoulou A, Karapanou O, Priftis K, Papaevangelou V, Papadimitriou A. Study of XbaI and PvuII polymorphisms of estrogen receptor alpha (ERα) gene in girls with precocious/early puberty. Endocrine. 2021;73(2):455–62.
65. Golestanzadeh M, Riahi R, Kelishadi R. Association of phthalate exposure with precocious and delayed pubertal timing in girls and boys: a systematic review and meta-analysis. Environ Sci Process Impacts. 2020;22:873–94.
66. Hashemipour M, Kelishadi R, Amin MM, Ebrahim K. Is there any association between phthalate exposure and precocious puberty in girls? Environ Sci Pollut Res Int. 2018;25:13589–96.
67. Wolff MS, Pajak A, Pinney SM, Windham GC, Galvez M, Rybak M, Silva MJ, Ye X, Calafat AM, Kushi LH, et al. Associations of urinary phthalate and phenol biomarkers with menarche in a multiethnic cohort of young girls. Reprod Toxicol. 2017;67:56–64.

68. Frederiksen H, Sørensen K, Mouritsen A, Aksglaede L, Hagen CP, Petersen JH, Skakkebaek NE, Andersson AM, Juul A. High urinary phthalate concentration associated with delayed pubarche in girls. Int J Androl. 2012;35:216–26.
69. Wolff MS, Teitelbaum SL, McGovern K, Windham GC, Pinney SM, Galvez M, Calafat AM, Kushi LH, Biro FM. Phthalate exposure and pubertal development in a longitudinal study of US girls. Hum Reprod. 2014;29:1558–66.
70. Lomenick JP, Calafat AM, Melguizo Castro MS, Mier R, Stenger P, Foster MB, Wintergerst KA. Phthalate exposure and precocious puberty in females. J Pediatr. 2010;156:221–5.
71. Jung MK, Choi HS, Suh J, Kwon A, Chae HW, Lee WJ, Yoo EG, Kim HS. The analysis of endocrine disruptors in patients with central precocious puberty. BMC Pediatr. 2019;19:323.
72. Howdeshell KL, Hotchkiss AK, Thayer KA, Vandenbergh JG, Vom Saal FS. Exposure to bisphenol A advances puberty. Nature. 1999;401:763–4.
73. Rya BC, Hotchkiss AK, Crofton KM, Le G Jr. In utero and lactational exposure to bisphenol A, in contrast to ethinyl estradiol, does not alter sexually dimorphic behavior, puberty, fertility, and anatomy of female LE rats. Toxicol Sci Off J Soc Toxicol. 2010;114:133–48.
74. Durmaz E, Asci A, Erkekoglu P, Balcı A, Bircan I, Koçer-Gumusel B. Urinary bisphenol A levels in Turkish girls with premature thelarche. Hum Exp Toxicol. 2018;37:1007–16.
75. Supornsilchai V, Jantarat C, Nosoognoen W, Pornkunwilai S, Wacharasindhu S, Soder O. Increased levels of bisphenol A (BPA) in Thai girls with precocious puberty. J Pediatric Endocrinol Metab JPEM. 2016;29:1233–9.
76. Watkins DJ, Téllez-Rojo MM, Ferguson KK, Lee JM, Solano-Gonzalez M, Blank-Goldenberg C, Peterson KE, Meeker JD. In utero and peripubertal exposure to phthalates and BPA in relation to female sexual maturation. Environ Res. 2014;134:233–41.
77. Chen Y, Wang Y, Ding G, Tian Y, Zhou Z, Wang X, Shen L, Huang H. Association between bisphenol A exposure and idiopathic central precocious puberty (ICPP) among school-aged girls in Shanghai. China Environ Int. 2018;115:410–6.
78. Watkins DJ, Sánchez BN, Téllez-Rojo MM, Lee JM, Mercado-García A, Blank-Goldenberg C, Peterson KE, Meeker JD. Phthalate and bisphenol A exposure during in utero windows of susceptibility in relation to reproductive hormones and pubertal development in girls. Environ Res. 2017;159:143–51.
79. Leonardi A, Cofini M, Rigante DD, Lucchetti L, Cipolla C, Penta L. Esposito S the effect of bisphenol A on puberty: a critical review of the medical literature. Int J Environ Res Public Health. 2017;14:1044.
80. Vasiliu O, Muttineni J, Karmaus W. In utero exposure to organochlorines and age at menarche. Hum Reprod. 2004;19:1506–12.
81. Wohlfahrt-Veje C, Andersen HR, Schmidt IM, Aksglaede L, Sørensen K, Juul A, Jensen TK, Grandjean P, Skakkebaek NE, Main KM. Early breast development in girls after prenatal exposure to non-persistent pesticides. Int J Androl. 2012;35:273–82.
82. Denham M, Schell LM, Deane G, Gallo MV, Ravenscroft J, De Caprio AP. Relationship of lead, mercury, mirex, dichlorodiphenyldichloroethylene, hexachlorobenzene, and polychlorinated biphenyls to timing of menarche among Akwesasne Mohawk girls. Pediatrics. 2005;115:e127–34.
83. Windham GC, Pinney SM, Voss RW, Sjödin A, Biro FM, Greenspan LC, Stewart S, Hiatt RA, Kushi LH. Brominated flame retardants and other persistent organohalogenated compounds in relation to timing of puberty in a longitudinal study of girls. Environ Health Perspect. 2015;123:1046–52.
84. Harley KG, Rauch SA, Chevrier J, Kogut K, Parra KL, Trujillo C, Lustig RH, Greenspan LC, Sjödin A, Bradman A, et al. Association of prenatal and childhood PBDE exposure with timing of puberty in boys and girls. Environ Int. 2017;100:132–8.
85. Tassinari R, Mancini FR, Mantovani A, Busani L, Maranghi F. Pilot study on the dietary habits and lifestyles of girls with idiopathic precocious puberty from the city of Rome: potential impact of exposure to flame retardant polybrominated diphenyl ethers. J Pediatr Endocrinol Metab. 2015;28:1369–72.

86. Alyssa Huang A, Thomas Reinehr T, Christian L, Roth CL. Connections between obesity and puberty: invited by manuel tena-sempere, cordoba. Curr Opin Endocr Metab Res. 2020;14:160–8.

87. Lucaccioni L, Trevisani V, Marrozzini L, Bertoncelli N, Predieri B, Lugli L, Berardi A, Iughetti L. Endocrine-disrupting chemicals and their effects during female puberty: a review of current evidence. Int J Mol Sci. 2020;21:2078.

88. Papadimitriou A, Papadimitriou DT. Endocrine-disrupting chemicals and early puberty in girls. Children (Basel). 2021;8(6):492.

89. Elobeid MA, Allison DB. Putative environmental-endocrine disruptors and obesity: a review. Curr Opin Endocrinol Diabetes Obes. 2008;15:403–8.

90. Heindel JJ, Newbold R, Schug TT. Endocrine disruptors and obesity. Nat Rev Endocrinol. 2015;11:653–61.

91. Reinehr T, Roth CL. Is there a causal relationship between obesity and puberty? Lancet Child Adolesc Health. 2019;3:44–54.

92. Kuznetsov L, Dworzynski K, Davies M, Overton C. Diagnosis and management of endometriosis: summary of NICE guidance. BMJ. 2017;358:j3935.

93. Moradi Y, Shams-Beyranvand M, Khateri S, Ghrahjeh S, Tehrani S, Varse F, Najmi Z. A systematic review on the prevalence of endometriosis in women. J Med Res. 2021;154(3):446–54.

94. Carpinello OJ, Sundheimer LW, Alford CE, Taylor RN, DeCherney AH. Endometriosis. In: Endotext. South Dartmouth (MA): MDText.com, Inc; 2017. 2000.

95. Koninckx PR, Ussia A, Adamyan L, Wattiez A, Donnez J. Deep endometriosis: definition, diagnosis, and treatment. Fertil Steril. 2012;98(3):564–71.

96. Sampson JA. Metastatic or embolic endometriosis, due to the menstrual dissemination of endometrial tissue into the venous circulation. Am J Pathol. 1927;3:93–11043.

97. Bulun SE, Yilmaz BD, Sison C, Miyazaki K, Bernardi L, Liu S, et al. Endometriosis. Endocr Rev. 2019;40:1048–79.

98. Baranova H, Canis M, Ivaschenko T, Albuisson E, Bothorishvilli R, Baranov V, et al. Possible involvement of arylamine N-acetyltransferase 2, glutathione S-transferases M1 and T1 genes in the development of endometriosis. Mol Hum Reprod. 1999;5:636–41.

99. Sourial S, Tempest N, Hapangama DK. Theories on the pathogenesis of endometriosis. Int J Reprod Med. 2014;2014:1.

100. Figueira PGM, Abrao MS, Krikun G, Taylor HS. Stem cells in endometrium and their role in the pathogenesis of endometriosis. Ann N Y Acad Sci. 2011;1221:10.

101. Eyster KM, Klinkova O, Kennedy V, Hansen KA. Whole genome deoxyribonucleic acid microarray analysis of gene expression in ectopic versus eutopic endometrium. Fertil Steril. 2007;88:1505–33.

102. Wren JD, Wu Y, Guo SW. A system-wide analysis of differentially expressed genes in ectopic and eutopic endometrium. Hum Reprod. 2007;22:2093–102.

103. Caserta D, Maranghi L, Mantovani A, Marci R, Maranghi F, Moscarini M. Impact of endocrine disruptor chemicals in gynaecology. Hum Reprod Update. 2008;14:59–72.

104. Soave I, Caserta D, Wenger JM, Dessole S, Perino A, Marci R. Environment and endometriosis: a toxic relationship. Eur Rev Med Pharmacol Sci. 2015;19:1964–72.

105. Street ME, Angelini S, Bernasconi S, Burgio E, Cassio A, Catellani C, et al. Current knowledge on endocrine disrupting chemicals (EDCs) from animal biology to humans, from pregnancy to adulthood: highlights from a National Italian Meeting. Int J Mol Sci. 2018;19:1647.

106. Sutton P, Woodruff TJ, Perron J, Stotland N, Conry JA, Miller MD, et al. Toxic environmental chemicals: the role of reproductive health professionals in preventing harmful exposures. Am J Obstet Gynecol. 2012;207:164–73.

107. Aldad TS, Rahmani N, Leranth C, Taylor HS. Bisphenol-A exposure alters endometrial progesterone receptor expression in the nonhuman primate. Fertil Steril. 2011;96:175–9.

108. Reddy BS, Rozati R, Reddy BV, Raman NV. Association of phthalate esters with endometriosis in Indian women. BJOG. 2006;113:515–20.

109. Sartain CV, Hunt PA. An old culprit but a new story: bisphenol A and "NextGen" bisphenols. Fertil Steril. 2016;106:820–6.
110. Signorile PG, Spugnini EP, Citro G, Viceconte R, Vincenzi B, Baldi F, et al. Endocrine disruptors in utero cause ovarian damages linked to endometriosis. Front Biosci. 2012;4:1724–30.
111. Brotons JA, Olea-Serrano MF, Villalobos M, Pedraza V, Olea N. Xenoestrogens released from lacquer coatings in food cans. Environ Health Perspect. 1995;103:608–12.
112. Rashidi BH, Amalou M, Lak TB, Ghazizadeh M, Eslami B. A case-control study of bisphenol A and endometrioma among subgroup of Iranian women. J Res Med Sci. 2017;22:7.
113. Upson K, Sathyanarayana S, De Roos AJ, Koch HM, Scholes D, Holt VL. A population-based case-control study of urinary bisphenol A concentrations and risk of endometriosis. Hum Reprod. 2014;29:2457–64.
114. Itoh H, Iwasaki M, Hanaoka T, Sasaki H, Tanaka T, Tsugane S. Urinary bisphenol-A concentration in infertile Japanese women and its association with endometriosis: a cross-sectional study. Environ Health Prev Med. 2007;12:258–64.
115. Buck Louis GM, Peterson CM, Chen Z, Croughan M, Sundaram R, Stanford J, et al. Bisphenol A and phthalates and endometriosis: the endometriosis: natural history, diagnosis and outcomes study. Fertil Steril. 2013;100:e1–2.
116. Van Den Berg M, Birnbaum L, Denison M, De Vito M, Farland W, Feeley M, Fiedler H, Hakansoson H, Hanberg A, Haws L, Rose M, Safe S, Schrenk D, Tohyama C, Tritscher A, Tuomisto J, Tysklind M, Walker N, Peterson RE. The 2005 World Health Organization reevaluation of human and mammalian toxic equivalency factors for dioxins and dioxin-like compounds. Toxicol Sci. 2006;93:223–41.
117. Heilier JF, Nackers F, Verougstaete V, Tonglet R, LiSon D, Donnez J. Increased dioxin-like compounds in the serum of women with peritoneal endometriosis and deep endometriotic (adenomyotic) nodules. Fertil Steril. 2005;84:305–12.
118. Porpora MG, Ingelido AM, Di Domenico A, Ferro A, Crobu M, Pallante D, Cardelli CEV, De Felipe E. Increased levels of polychlorobiphenyls in Italian women with endometriosis. Chemosphere. 2006;63:1361–7.
119. Kanematsu M, Shimuzu Y, Sato K, Kim S, Suzuki T, Park B, Hattori K, Nakamura M, Yabishita H, Yokota K. Distribution of dioxins in surface soils and river-mouth sediments and their relevance to watershed properties. Water Sci Technol. 2006;53:11–21.
120. Uemura H, Arisava K, Hikoshi M, Satoh H, Sumiyoshi Y, Morinaga K, Kodama K, Suzuki T, Nagai M, Suzuki T. PCDDs/PCDFs and dioxin-like PCBs: recent body burden levels and their determinants among general inhabitants in Japan. Chemosphere. 2008;73:30–7.
121. Mimura J, Fujii-Kuriyama Y. Functional role of AhR in the expression of toxic effects by TCDD. Biochim Biophys Acta. 2003;1619:263–8.
122. Carver LA, Bradfield CA. Ligand-dependent interaction of the aryl hydrocarbon receptor with a novel immunophilin homolog in vivo. J Biol Chem. 1997;272:11452–6.
123. Rier SE, Martin DC, Bowman RE, Dmowski WP, Becker JL. Endometriosis in rhesus monkeys (Macaca mulatta) following chronic exposure to 2,3,7,8-tetrachlorodibenzo-p-dioxin. Fundam Appl Toxicol. 1993;21:433–41.
124. Bois FY, Eskenazi B. Possible risk of endometriosis for Seveso, Italy, residents: an assessment of exposure to dioxin. Environ Health Perspect. 1994;102:476–7.
125. Eskenazi B, Mocarelli P, Warner M, Samuels S, Vercellini P, Olive D, et al. Serum dioxin concentrations and endometriosis: a cohort study in Seveso, Italy. Environ Health Perspect. 2002;110:629–34.
126. Matta K, Lefebvre T, Vigneau E, Cariou V, Marchand P, Guitton Y. Associations between persistent organic pollutants and endometriosis: a multiblock approach integrating metabolic and cytokine profiling. Environ Int. 2021;158:106926.
127. Aoki Y. Polychlorinated biphenyls, polychlorinated dibenzo-p-dioxins, and polychlorinated dibenzofurans as endocrine disrupters: what we have learned from Yusho disease. Environ Res. 2001;86(1):2–11.

128. Bruner-Tran KL, Osteen KG. Dioxin-like PCBs and Endometriosis. Syst Biol Reprod Med. 2010;56(2):132–46.
129. Rier SE, Turner WE, Martin DC, Morris R, Lucier GW, Clark GC. Serum levels of TCDD and dioxin-like chemicals in Rhesus monkeys chronically exposed to dioxin: correlation of increased serum PCB levels with endometriosis. Toxicol Sci. 2001;59(1):147–59.
130. Zhang Y, Zheng X, Wang P, Zhang Q, Zhang Z. Occurrence and risks of PCDD/Fs and PCBs in three raptors from North China. Ecotoxicol Environ Saf. 2021;223:112541.
131. Cano-Sancho G, Ploteau P, Matta K, Adoamnei E, Buck Louis G, Mendiola J, Darai E, Squifflet J, Le Bizec B, Antignac JP. Human epidemiological evidence about the associations between exposure to organochlorine chemicals and endometriosis: systematic review and meta-analysis. Environ Int. 2019;123:209–23.
132. Duty SM, Ackerman RM, Calafat AM, Hauser R. Personal care product use predicts urinary concentrations of some phthalate monoesters. Environ Health Perspect. 2005;113(11):1530–5.
133. Hannon PR, Flaws JA. The effects of phthalates on the ovary. Front Endocrinol (Lausanne). 2015;6:8.
134. Fromme H, Gruber L, Seckin E, Raab U, Zimmermann S, Kiranoglu M, Schlummer M, Schwegler U, Smolic S, Völkel W, HBMnet. Phthalates and their metabolites in breast Milk results from the Bavarian monitoring of breast milk (BAMBI). Environ Int. 2011;37:715–22.
135. Hines EP, Calafat AM, Silva MJ, Mendola P, Fenton SE. Concentrations of phthalate metabolites in milk, urine, saliva, and serum of lactating North Carolina women. Environ Health Perspect. 2009;117:86–92.
136. Reddy BS, Rozati R, Reddy S, Kodampur S, Reddy P, Reddy R. High plasma concentrations of polychlorinated biphenyls and phthalate esters in women with endometriosis: a prospective case control study. Fertil Steril. 2006;85:775–9.
137. Hartmann G. Are your personal care products disrupting hormonal balance? https://georgia-hartmann.com/are-your-personal-care-products-disrupting-hormonal-balance. 2020.
138. Weuve J, Hauser R, Calafat AM, Missmer SA, Wise LA. Association of exposure to phthalates with endometriosis and uterine leiomyomata: findings from NHANES, 1999-2004. Environ Health Perspect. 2010;118:825–32.
139. Huang PC, Tsai EM, Li WF, Liao PC, Chung MC, Wang YH, et al. Association between phthalate exposure and glutathione S-transferase M1 polymorphism in adenomyosis, leiomyoma and endometriosis. Hum Reprod. 2010;25:986–94.
140. Huang PC, Li WF, Liao PC, Sun CW, Tsai EM, Wang SL. Risk for estrogen-dependent diseases in relation to phthalate exposure and polymorphisms of CYP17A1 and estrogen receptor genes. Environ Sci Pollut Res Int. 2014;21:13964–73.
141. Upson K, Sathyanarayana S, De Roos AJ, Thompson ML, Scholes D, Dills R, Holt VL. Phthalates and risk of endometriosis. Environ Res. 2013;126:91–7.
142. Itoh H, Iwasaki M, Hanaoka T, Sasaki H, Tanaka T, Tsugane S. Urinary phthalates monoesters and endometriosis in infertile Japanese women. Sci Total Environ. 2009;408:37–42.
143. Reed CE, Fenton SE. Exposure to diethylstilbestrol during sensitive life stages: a legacy of heritable health effects. Birth Defects Res C Embryo Today. 2013;99(2):134–46.
144. Harris RM, Waring RH. Diethylstilboestrol—a long term legacy. Maturitas. 2012;72(2):108–12.
145. Benagiano G, Brosens I. In utero exposure and endometriosis. J Matern Fetal Neonatal Med. 2014;27(3):303–8.
146. Upson K, Sathyanarayana S, Scholes D, Holt VL. Early-life factors and endometriosis risk. Fertil Steril. 2015;104(4):964–71.
147. Pollack AZ, Krall J, Kannan K, Buck Louis GM. Adipose to serum ratio and mixtures of persistent organic pollutants in relation to endometriosis: Findings from the ENDO Study. Environ Res. 2021;195:110732.
148. Bellavia A, James-Todd T, Williams PL. Approaches for incorporating environmental mixtures as mediators in mediation analysis. Environ Int. 2019;123:368–74.

149. Piazza MJ, Urbanetz AA. Environmental toxins and the impact of other endocrine disrupting chemicals in women's reproductive health. JBRA Assist Reprod. 2019;23(2):154–64.
150. Figueres C, Landrigan PJ, Fuller R. Tackling air pollution, climate change, and NCDs: time to pull together. Lancet. 2018;392:1502–3.
151. Rudel RA, Gray JM, Engel CL, et al. Food packaging and bisphenol A and bis(2-ethylhexyl) phthalate exposure: findings from a dietary intervention. Environ Health Perspect. 2011;119:914–20.
152. Kassotis CD, Vandenberg LN, Demeneix BA, et al. Endocrine-disrupting chemicals: economic, regulatory, and policy implications. Lancet Diabetes Endocrinol. 2020;8:719–30.
153. Hill AB. The environment and disease: association or causation? Proc R Soc Med. 1965;58:295–300.

Chapter 6
Endocrine Disruptors and Cancers in Women

Lea Scharschmidt, Florence Scheffler, Albane Vandecandelaere, Dorian Bosquet, Elodie Lefranc, Jean Bouquet De La Jolinière, Moncef Benkhalifa, Anis Feki, and Rosalie Cabry-Goubet

6.1 Introduction

Breast cancer is one of the most common cancers among women in industrialized countries and is one of the deadliest. In 2020, 2.26 million new cases of breast cancer have been reported worldwide and 685,000 deaths [1]. In France, there has been a steady increase of +1.1% each year between 1990 and 2018 in the number of new cases diagnosed. However, in contrast to the number of new cases per year, there has been a progressive decrease of −1.6% per year between 2010 and 2018 [2] in the number of deaths.

The epidemiology of cancers in women has continuously changed over the last 30 years, including a progressive increase in the frequency of advanced breast cancer in young women [3].

This change in the profile of patients with breast cancer has motivated the exploration of numerous avenues to identify new and previously unknown environmental risk factors.

Numerous studies have examined the potential carcinogenic effect of endocrine disruptors: molecules present in food, water, ambient air, industrial products, cosmetics, and many everyday objects.

This carcinogenic effect in humans has been proven for certain organs such as the prostate, liver, and blood [4, 5].

L. Scharschmidt · F. Scheffler · A. Vandecandelaere · D. Bosquet · E. Lefranc · M. Benkhalifa
R. Cabry-Goubet (✉)
Medicine and Biology of Reproduction, CECOS Picardie, Amiens University Hospital, Amiens, France
e-mail: cabry.rosalie@chu-amiens.fr

J. Bouquet De La Jolinière · A. Feki
Gynaecology and obstetrics, Endocrinology and Medicine of Reproduction, HFR Fribourg, Fribourg, Switzerland

Diethylstilbestrol is one of the endocrine disruptors whose carcinogenic effect in women has been widely demonstrated for several years, notably for breast cancer in mothers who used it and vaginal cancer and cervical cancer in daughters exposed in utero [6]. Considering the large number of studies and reviews on the subject, we have not retained this substance in this research.

The main objective of this review is to identify the scientific studies concerning other endocrine disruptors and their potential impacts on the risk of developing breast cancer.

The secondary objective is to study the potential increase in the risk of developing endometrial cancer or ovarian cancer in the presence of these disruptors.

6.2   The Importance of Studying Endocrine Disruptors When Studying the Genesis of Breast Cancer

The human endocrine system is a complex communication system involving many organs producing different hormones.

Hormones are chemical mediators that circulate in the blood to their target organs to exert a specific function. They are secreted by different glands in the human body such as the thyroid, thymus, liver, pancreas, pituitary, hypothalamus, stomach, adrenals, ovaries, kidneys, and testes.

These hormones circulate systemically until they bind to their specific receptors present on the cells of the target tissue. After binding, the hormone–receptor complex is internalized within the cell nucleus and then binds to a specific region of the hormone-dependent gene promoter, triggering gene expression.

From this definition of a hormone comes the definition of an endocrine disruptor (ED). According to the World Health Organization (WHO), endocrine disruptors are "chemical substances of natural or synthetic origin, foreign to the organism and likely to interfere with the functioning of the endocrine system, i.e. the cells and organs involved in the production of hormones and their effect on target cells via receptors" [7].

These endocrine disruptors are thus at risk of inducing harmful effects on the organism or on its descendants.

Indeed, EDs can have an impact on physiological hormonal functioning in different ways. They can bind to hormone receptors naturally present in the target organs (the direct effect of EDs) or interfere with the mechanisms of production or regulation of hormones or their receptors (the indirect effect of EDs).

EDs can act directly via membrane or nuclear hormone receptors, resulting in either an agonist effect by mimicking the effects of hormones or an antagonist effect by blocking these effects.

They can also act indirectly through different mechanisms:
– By degrading the molecular structure of natural hormones (enzymatic interference, via cytochrome P450 in particular)

- By invading hormone receptors, which reduces the number of receptors available to bind natural hormones
- By short-circuiting the transport of natural hormones (interference with the internalization or deoxyribonucleic acid (DNA) binding of the hormone–receptor complex)
- By maintaining high levels of natural hormones (interference with elimination by altering plasma clearance)
- By modifying gene expression, that is, by causing epigenetic modification without changing the nucleotide sequence

The endocrine disruptors currently identified are very numerous, and their list is constantly growing. They can be of natural or synthetic origin (Table 6.1). Synthetic endocrine disruptors are found in products from the pharmaceutical industry (ethinylestradiol and diethylstilbestrol); in products used in everyday life such as food packaging, plastics, and cosmetics (bisphenol A, phthalates, and parabens); in food (dioxins); in products from the construction industry such as paints, carpets, solvents, and flame retardants (organochlorines and polychlorinated biphenyls); in air pollution (polycyclic aromatic hydrocarbons), etc.

These disrupting agents can be absorbed by humans via the respiratory tract, the cutaneous or mucous membrane tract, or the digestive tract. They can be absorbed in low doses in a chronic and repeated manner (accumulation in the body) or in an acute manner, which occurs mainly during workplace accidents, namely in industrial settings.

During chronic and repeated exposure to low doses, integrated endocrine disruptors can be stored in the body in a variable manner depending on its eliminated half-life, ranging from a few days to a few years.

Table 6.1 Examples of families of molecules with endocrine-disrupting effects and their potential sources of diffusion in the environment (INSERM) [8]

Chemical family	Potential sources	Examples
Phthalates	Plastics and cosmetics	Dibutyl phthalate
Alkylphenols	Detergents, plastics, and pesticides	Nonylphenol
Flame retardants	Foams for furniture, carpets, and electronic equipment	Polybrominated diphenyl ethers (PBDE)
Polycyclic aromatic hydrocarbons	Combustion sources: cigarette smoke, diesel engine emissions, and fires	Benzo(a)pyrene
Polychlorinated biphenyls	Electrical transformers	Polychlorinated biphenyls (PCBs), Aroclor
Obsolete pesticides	Residuals from storage and persistent pollution	Dichlorodiphenyltrichloroethane (DDT), lindane, dieldrin, and chlordane
Current pesticides	Agriculture, urban cleaning, and private gardens	Chlorpyrifos, acetochlor, fenbuconazole 56.I list
Phenolic derivatives	Disinfectants, plastics, and cosmetics	Bisphenol A, parabens, and halogenated phenols

There are four key points about the mode of action of these endocrine disruptors on human health:

- The most critical period of exposure seems to be during embryonic life, but effects may not manifest until adulthood; this is the **mechanism of delayed programmed toxicity**.
- The effects are mainly manifested in the next generation and not in the exposed subjects; this is the **transgenerational effect**.
- As the quantity of hormones necessary for the normal functioning of the endocrine system is extremely small, disruption can result from a very small quantity of disruptive substances, this is, a **nonmonotonic dose–response relationship** with a toxicity threshold that is difficult to define.
- There are interactions between different endocrine disruptors acting via various mechanisms (synergistic and antagonistic); therefore, **possible potentiated effects** may be suspected.

The large number of women affected by hormone-sensitive breast cancer has motivated many years of research into its risk factors, genesis, and therapeutic management.

Classically, oncogenesis is divided into three key stages:

- Initiation, which represents a rapid, irreversible, and transmissible lesion of the DNA, induced by a carcinogenic factor (physical, chemical, viral, etc.)
- Promotion, which corresponds to a prolonged, repeated, or continuous exposure to a substance that maintains and stabilizes the initiated lesion (mitogenic stimuli such as cytokines and growth factors). This leads to clonal expansion of pretumor cells.
- Progression, characterized by the acquisition of proliferation capacities.

This oncogenesis develops through different mechanisms within healthy cells such as the release of growth factors, escape from tumor suppressor genes, facilitation of cell movement capacity, induction of neoangiogenesis, and the ability to resist apoptosis mechanisms.

Once modified, tumor cells are dedifferentiated, have highly developed motility capabilities, and respond very sensitively to chemoreceptors. In female breast cancer, these cells are mostly hormone-dependent, with an overexpression of estrogen receptor (ER) and progesterone receptor (PR), resulting in a high sensitivity to the latter.

Several risk factors have been studied and are known to increase its occurrence, such as age, parity, alcohol consumption, body mass index, physical activity, the use of menopausal hormone therapy, personal history of atypical breast hyperplasia, lobular/ductal carcinoma in situ or thoracic irradiation, and the presence of genetic mutations such as breast cancer gene 1 (BRCA1), breast cancer gene 2 (BRCA2), and partner and localizer of BRCA2 (PALB2). By looking at the pathophysiology and oncogenesis of the breast, it is inferred that both endogenous and exogenous estrogens may play a major role in tumor proliferation.

Other factors remain poorly studied in terms of their involvement in the development of breast cancer, and endocrine disruptors are among them.

The search for a cause-and-effect relationship between a woman's exposure to endocrine disruptors during her life and the development of breast cancer seems coherent when we know their potential impacts on the hormonal system.

Indeed, certain molecules of the endocrine disruptor family could be responsible for a promoter effect on mammary hormone receptors and thus favor the clonal expansion of tumor cells previously modified by DNA lesions.

6.3 Endocrine Disruptors and Breast Cancer

6.3.1 Organochlorines (DDT, dichlorodiphenyldichloroethyle ne [DDE], and PCB)

Organochlorines are a large family of molecules with an endocrine-disrupting effect. They are neurotropic toxins that alter the functioning of the sodium channels essential for the transmission of nerve impulses. They are synthetic organic compounds in which one or more hydrogen atoms are substituted by one or more chlorine atoms.

Organochlorines are used as solvents, pesticides, insecticides, fungicides, refrigerants, and intermediary molecules in the chemistry and in the pharmaceutical industry.

These molecules are therefore used in agriculture, where they are administered to animals or to plants as growth regulators. They are also used as defoliants (herbicides), desiccants (water removal), and fruit thinners or used to prevent premature fall of fruit from the trees.

Among these compounds, the most studied are dichlorodiphenyltrichloroethane (DDT), dichlorodiphenyldichloroethylene (DDE), dieldrin, hexachlorobenzene (HCB), polychlorinated biphenyls (PCBs) derived from dioxin, chlordecone, and hexachlorocyclohexane (HCH).

A large number of in vitro and in vivo studies have explored the potential impacts of organochlorines on cancer cell proliferation, some of which are listed below.

One of the hypotheses of action of endocrine disruptors on breast cancer cell proliferation is the interaction of these compounds with membrane or intracellular proteins of breast cells, potentially responsible for tumor proliferation.

6.3.1.1 In Vitro Studies and Organochlorines

The study by Montes-Crajales et al. in 2016 explored this avenue by analyzing in vitro the affinity between certain endocrine-disrupting compounds and widely studied cellular proteins such as estrogen receptors (ESR1), progesterone receptors

(PGR), human epidermal growth factor 2 receptors called HER 2 (ERBB2), BRCA susceptibility proteins type 1 (BRCA1), BRCA susceptibility proteins type 2 (BRCA 2), and sex hormone–binding globulin (SHBG).

This study targeted the affinity between these proteins and different forms of dioxin (belonging to the organochlorine family) and bisphenol A via a high-throughput virtual screening technique, followed by an experimental validation in silico by spectroscopy of the protein/ligand affinity suspected during the screening.

This work highlighted the potential for several endocrine disruptors, including some dioxin and bisphenol A derivatives, to bind to breast cancer–associated proteins, not just hormone receptor proteins [9].

Other in vitro study techniques have been implemented, including the analysis of human breast cell lines grown in culture.

A study by MA Garcia et al. in 2010 evaluated the in vitro effects of different doses of an organochlorine pesticide called hexachlorobenzene (HCB) on human cell cultures MCF-7 and MDA-MB-231.

MCF-7 is one of the estrogen receptor (ER)-positive breast tumor cell lines and is the most widely used line in breast cancer research laboratories. MDA-MB-231 represents an ER-negative tumor cell line.

This study showed an impact of HCB on MCF-7 cell proliferation, but not on the MDA-MB-231 line. It was also shown that exposure to certain doses of HCB induces cytochrome P450 gene expression and stimulates the insulin-like growth factor 1 (IGF-1) signaling pathway, but only on ER-positive cells (MCF-7).

This study raises the potential impact of this pesticide on the proliferation of ER-positive tumor cells [10].

The work of J. Payne et al. in 2001 also investigated the effect of several types of organochlorines on cell lines, such as o,p'-DDT, p,p'-DDE, β-HCH, and p,p'-DDT, which are persistent compounds found in human tissues. The objective of this study is to analyze the impact of these endocrine disruptors on human MCF-7 cell cultures after a standardized exposure to one or a mixture of several of these compounds, thus avoiding any exposure bias found in the general population due to the multiple possible uncontrollable sources of exposure.

The regression analysis showed combined effects even when each component of the mixture was present at or below its individual no-effect concentration. Assessments of the proliferation induced by the individual components of the mixture revealed that the effects of the combination of several components were stronger than the effects of the most potent component of the combination. These combined effects of organochlorines can therefore be described as synergistic on human cells.

In addition, comparisons with the expected effects as predicted by the summation of concentrations and independent action showed a strong agreement between prediction and observation. The effects of organochlorines can therefore be described as additive [11].

The analysis of these various in vitro studies supports the hypothesis that organochlorines are carcinogenic via their endocrine-disrupting effect, but the data on cell cultures is not sufficient to conclude that there is a probable relationship between these compounds and the development of breast cancer in women.

The proof of a possible adverse effect requires human studies, but the ethical issue strictly prohibits any randomized interventional study.

The best way to explore the subject is therefore epidemiological studies with a retrospective analysis of cohorts of women with breast cancer or a prospective analysis of women exposed to endocrine disruptors in their environment.

6.3.1.2   In Vivo Studies and Serum Organochlorine Levels

Two Danish (Hoyer et al.) and Norwegian (Ward et al.) teams explored, respectively, in 1998 (Danish team [12]), in 2000 (Norwegian team [13]), and in 2001 (Danish team [14]) via case–control studies the serum levels of certain organochlorine compounds in women with breast cancer and those in women from a control group.

These three studies showed discordant results regarding the correlation between high serum levels of compounds and the presence of breast cancer, which are presented in Table 6.2.

The first study by Hoyer et al. found a significant increase in the number of women with elevated serum levels of dieldrin in the study group compared with women in the control group, irrespective of the immunohistochemical profile of the breast cancers in the cases. No other significant correlation was found for the other organochlorine compounds [12].

The second study by Ward et al. found no significant increase in serum organochlorine levels in the study group compared with the control patients, despite

Table 6.2 Comparison of three studies

Author (year)	Cases/ controls	Organochlorine	High serum levels significantly linked to breast cancer risk are bold values
Hoyer et al. (1998)	240/477	PCB	Odds ratio (OR) 1.11 (95% confidence interval [CI] 0.70–1.77, $p = 0.77$)
		DDT	OR 0.84 (95% CI 0.49–1.45, $p = 0.65$)
		HCH	OR 1.36 (95% CI 0.79–2.33, $p = 0.24$)
		Dieldrin	**OR 2.05 (95% CI 1.17–3.57, $p = 0.01$)**
Ward et al. (2000)	150/150	DDT	OR 1.1 (95% CI 0.5–2.5, p = not significant (NS))
		Dieldrin	OR 1.0 (95% CI 0.4–2.6, p = NS)
Hoyer et al. (2001)	161/318	HCB	Estrogen Receptor (ER)+: relative risk (RR) 1.2 (95% CI 0.7–2.1, $p > 0.2$) ER−: OR 0.4 (95% CI 0.1–1.4, $p > 0.2$)
		PCB	ER+: RR 1.3 (95% CI 0.8–2.2, $p > 0.2$) ER-: OR 0.8 (95% CI 0.3–2.6, $p > 0.2$)
		DDE	ER+: RR 0.9 (95% CI 0.6–1.5, $p > 0.2$) ER−: OR 0.6 (95% CI 0.2–1.7, $p > 0.2$)
		Dieldrin	ER+: RR 1.4 (95% CI 0.8–2.5, $p > 0.2$) **ER−: OR 7.6 (95% CI 1.3–46.1, $p = 0.01$)**

stratification by age, the presence of ER and PR on immunohistochemical analysis of breast tumors, and time between serum level measurement and the diagnosis of breast cancer [13].

Finally, in their second study in 2001, the team of Hoyer et al. explored the relationship between serum organochlorine levels and the presence of breast cancer, according to the amount of serum levels and according to the RE+ or RE− status of the breast tumors in the cases. No significant association was found except for a higher number of women with ER− breast cancer at high dieldrin exposure [14].

In the 1990s to 2000, several other US studies showed nonsignificant results for the association between organochlorines and breast cancer. However, it appeared that in several of these studies associations tended to be significant in certain subgroups, but the lack of power in each study limited interpretation. This is why the team of Laden et al., in 2001, published a meta-analysis of five of these American studies in order to increase their power [15].

This meta-analysis examined retrospectively collected data from 2042 patients regarding DDE and PCB concentrations in adipose tissue in women with breast cancer compared with controls.

No significant association was observed between elevated adipose levels of DDE (OR 0.99, 95% CI 0.77–1.27, p = NS) or PCB (OR 0.94, 95% CI 0.73–1.21, p = NS) and the presence of breast cancer.

Although some of the individual studies suggested an increased risk of PCB-associated breast cancer in certain stratified subgroups, these findings were reversed in this meta-analysis.

The teams of Millikan et al. (the USA) in 2000 [16] and Charlier et al. (Belgium) in 2004 [17] carried out two case–control studies with large cohorts and found contradictory results.

The Millikan et al. study included 748 cases (292 African American and 456 White women) and 659 controls (270 African American and 389 White women). No statistically significant association was found between elevated serum DDE or PCB levels and the presence of breast cancer in this study, even when subgrouped by ethnicity with adjustment for age, BMI, parity, history of breastfeeding, menopausal status, the use of hormone replacement therapy, and annual income.

The only slight increases in OR found were for elevated serum PCB and DDE levels in African American women in some subgroups (including BMI) and without p-value calculations, which further limits interpretation [16].

In contrast, the study by Charlier et al. retrospectively compared the presence of DDE and HCB in 231 cases and 290 controls. The variables were reported both continuously and in a binary manner (the presence or absence of DDE and HCB) according to a serum level below or above the limit of quantification set by the team.

In this study, there was a significant increase in the number of patients with DDE (OR 2.21, 95% CI 1.41–3.48, p = 0.0006) and HCB (OR 4.99, 95% CI 2.95–8.43, p < 0.0001) in the study group compared with the control group. This difference is also significant when analyzing DDE and HCB as continuous variables.

However, there is a potential selection bias due to the significant decrease in the number of postmenopausal patients in the control group. There is also a significant decrease in the number of patients receiving hormone replacement therapy in the case group, which reinforces the suspected role of the organochlorines studied [17].

Among the most recent studies on DDE and DDT are those conducted by Cohn et al. in 2007, 2015, and 2019. Dr. Cohn looked at the potential impact of organochlorines on the development of breast cancer depending on the age at which the exposure to these compounds began. She assumed that the risks are higher when exposure occurs when the mammary gland is still developing. For this research, she used data from the Child Health and Development Studies (CHDS), among others.

The 2007 nested case–control study measured serum levels of two forms of DDT and DDE in young women, dividing them into four age-groups at the time of peak exposure to these endocrine disruptors in the United States.

Statistical analyses revealed a significant increase in the risk of breast cancer when exposed to p,p'-DDT at any age, especially at an age below 14 years (OR 5.4, 95% CI 1.7–17.1, $p < 0.01$). However, there was no significant difference in each age-group (< 4 years, 4–7 years, 8–13 years) [18].

The 2015 nested case–control study is very interesting because it is one of the first to study the impact of in utero exposure to certain organochlorines, notably DDT.

Indeed, 9300 girls had been prospectively followed since their in utero growth started in the 1960s. Their mothers had blood samples taken for future analysis of serum levels of endocrine disruptors in the perinatal period. Of these, 103 cases who had developed invasive and/or noninvasive breast cancer by the age of 52 were selected, and 354 controls were matched. Perinatal maternal serum levels of DDE and two forms of DDT were measured and classified into four groups according to four quartiles.

The statistical analysis showed a significant increase in the risk of developing cancer in the highest quartile of o,p'-DDT compared with the lowest quartile in girls exposed in utero (OR 3.7, 95% CI 1.5–9, $p = 0.04$), irrespective of the mothers' cancer status [19].

Finally, the 2019 study looked at the risk of breast cancer after DDT exposure according to exposure before or after puberty and according to the onset of cancer before or after menopause [20].

This study found a significant increase in breast cancer risk after menopause for postpubertal exposure and a significant increase before menopause for prepubertal exposure.

Two prospective cohort studies have analyzed the potential link between dioxin exposure and the development of breast cancer. However, these two studies contain cohorts of different sizes, which limits their comparison.

The Italian team of Warner et al. in 2011 analyzed data from a cohort of 833 women and showed an increased risk of breast cancer with serum dioxin levels, but not significantly (hazard ratio [HR] = 1.44, 95% CI 0.89–2.33) [21].

The French team of Danjou et al. in 2015 looked at the potential effect of dioxin through a prospective cohort of 63,830 women and found no significant increase in breast cancer risk associated with dioxin exposure (HR = 1.0, 95% CI 0.96–1.05). It should be noted that the degree of exposure is measured via a food questionnaire [22].

Table 6.3 Comparison of nine studies

Author (year)	Cases/ controls	Compounds	Results (compound and breast cancer)
Romieu et al. (2000)	120/126	DDT and DDE	Significant increase with high levels of DDT and DDE [23]
Pavuk et al. (2003)	24/88	DDT, DDE, PCB, and HCB	Discordant results between subgroups [24]
Ibarluzea et al. (2004)	198/260	DDE, aldrin, and lindane	Significant increase with aldrin and lindane [25]
Gatto et al. (2007)	355/327	DDE and PCB	No significant increase [26]
Itoh et al. (2008)	403/403	DDT, DDE, and PCB	No significant increase [27]
Recio-Vega et al. (2011)	70/70	PCB	Significant increase with some groups of PCBs in postmenopausal women [28]
Cohn et al. (2012)	123/117	PCB	Discordant results according to PCB subgroups [29]
Arrebola et al. (2015)	69/54	DDE, PCB, HCB, HCH, and heptachlor	Significant increase with DDE and HCH; no significant increase with others [30]
Kaur et al. (2019)	42/42	DDE, HCH, heptachlor, aldrin, and dieldrin	Significant increase with DDE, β-HCH, heptachlor, and dieldrin [31]

Finally, there are a few case–control studies with small cohorts giving discordant results for DDE, DDT, or PCB (Table 6.3).

6.3.1.3   In Vivo Studies and Fat Levels of Organochlorines

The analysis of the levels of endocrine disruptors in fat tissue is interesting because they remain stored for a longer period without being eliminated. The measurement of this level therefore makes it possible to study more reliably the cumulative exposure to certain compounds and their potential harmful effects long after the first exposure.

Furthermore, as the breast is predominantly composed of fat, the mammary gland is exposed locally to endocrine disruptors accumulated over the years in the fat cells.

Other studies published between 2000 and 2005 investigated the possible association between organochlorine fat levels and breast cancer through case–control studies in the USA, Canada, and Denmark and also presented conflicting results.

Of these four studies, the study by Stellman et al. (in the USA) in 2000 concluded that there was no significant association between high organochlorine body fat levels and the presence of breast cancer. The same is true for the individual analysis of each of the two organochlorine families studied (DDE and PCBs). The only significant association found concerned a specific subgroup of PCBs called PCB-183 (adjusted OR 2, 95% CI 1.2–3.4, p = not reported (NR)) [32].

In contrast, the Canadian study by Aronson et al. in 2000 showed a significant association between high levels of certain groups of PCBs and the presence of breast cancer.

Indeed, high concentrations of PCB-105 (OR 3.17, 95% CI 1.15–6.68, $p < 0.01$) and PCB-118 (OR 2.31, 95% CI 1.11–3.18, $p < 0.01$) are significantly more frequent in patients in the study group than in the control group. This association was also found in the nonmenopausal subgroup, but not in the menopausal subgroup. The analysis of data for other organochlorines (DDE, DDT, HCB, β-HCH, trans-nonachlor, cis-nonachlor, oxychlordane, and mirex) did not show a significant association [33].

Raashou-Nielsen et al. (Denmark) in 2005 built its study on the same spectrum studied by Aronson et al. and showed very different results.

The team investigated the possible association between given fatty levels of organochlorines and the presence of breast cancer in exclusively postmenopausal women, using a study of 409 cases and 409 controls. The relative risk was calculated according to the quartiles of organochlorine fat levels for HR+ breast cancer, HR− breast cancer, and all breast cancers combined. The results of this study are unexpected, as they show the absence of a significant association between high levels and the presence of breast cancer, but also a decrease in the relative risk of HR− breast cancer for high levels of certain organochlorines, notably DDE, β-HCH, oxychlordane, trans-nonachlor, and HCB. This decrease in RR was also found in the all-cancer group for β-HCH, oxychlordane, and HCB [34].

At the same time, no significant association was found when analyzing the data for the different molecules of the PCB family.

However, these results remain difficult to interpret, given the nonexhaustive analysis of endocrine-disrupting compounds and the absence of data concerning the period of exposure of each patient.

The 2003 study by Muscat et al. (USA) looked at the levels of certain organochlorines found in the fat of surgical specimens after lumpectomy or mastectomy and the potential correlation with disease recurrence. There was an increased relative risk of breast cancer recurrence in patients with high levels of PCBs in fat in general (RR 2.9, 95% CI 1.02–8.2), but none of the results from this study are specified in the article. In addition, some biases were raised by the project team, including the small cohort size and the fact that patients with stage 3 or 4 breast cancer are more likely to have a recurrence, whereas no correlation was found between organochlorine fat levels and disease stage [35].

The recent study by Huang et al. (China) in 2019 also found discordant results, with a nonsignificant decrease in the odds ratio risk of breast cancer in women with average fat levels of certain organochlorines (PCB-28, PCB-52, PCB-101, and DDT by adjusted odds ratio only), but it also showed a significant increase in risk in women with high body fat concentrations of PCB-188, PCB-138, PCB-153, and PCB-180, when calculated with the synthesis of all PCBs, as well as with high body fat levels of DDE. This significant association was found with both the unadjusted and the adjusted ORs.

However, it should be noted that in this study there were significantly fewer post-menopausal women in the control group than in the study group, which represents a potential bias [36].

Finally, we can mention the meta-analysis carried out by Lopez-Cervantes et al. and published in 2001, which brings together 22 studies, some of which are mentioned above, and many others were published before the 2000s and therefore not included in this review. The statistical results of this meta-analysis did not show a significant increase in breast cancer risk associated with serum or adipose levels of DDT or DDE [37].

6.3.2 Bisphenol A and Phthalates

Bisphenol A is a synthetic organic compound found primarily in the manufacture of plastics and resins. It is a molecule that has long been used to create materials used in everyday life, such as cables and adhesives, in the world of childcare and in plastic packaging in contact with foodstuffs. It is also used as a flame retardant or developer in thermal papers.

Bisphenol A is one of the compounds recognized as toxic for reproduction and endocrine disruptors, and its use has been banned in France in the composition of food containers since 2015 [38].

Phthalates are a family of chemical compounds derived from phthalic acid, used in the manufacture of floor coverings, cables, pipes, plastic films, shower curtains, and certain medical devices and childcare equipment. Finally, they are found in certain cosmetics such as nail polish and lacquer for their fixing properties.

Investigations over the last 10 years have led to the classification of certain phthalates as reproductive toxicants and potential endocrine disruptors by the European Chemicals Agency (ECHA) [39].

Unlike organochlorines, there are relatively few studies on the effect of bisphenol A and phthalates on the animal or human population.

The three studies conducted by Vandenberg et al. in 2010 [40], Lozada et al. in 2011 [41], and Acevedo et al. in 2013 [42] have explored the potential mammary carcinogenic risks of bisphenol A in the animal population. All the three studies showed consistent results with a significant increase in breast cancer risk in mice after in utero exposure and during lactation.

A meta-analysis by the team of Liu et al. was published in 2021 and brought together nine case–control studies carried out between 1998 and 2019, allowing the creation of a cohort of 7820 women and the study of certain phthalates and bisphenol A [43].

This meta-analysis found no significant association between urinary or adipose levels of bisphenol A and the development of breast cancer in women of all ages (OR 0.85, 95% CI 0.69–1.05).

Table 6.4 Comparison of four studies

Author (year)	Cases/ controls	Compounds	Results (compound and breast cancer)
Lopez-Carrillo et al. (2010)	233/221	Phthalates	Significant association [45]
Reeves et al. (2019)	2419/838	Phthalates	No significant association [46]
Yang et al. (2009)	70/82	Bisphenol A	Discordant results [47]
Trabert et al. (2015)		Bisphenol A	No significant association with postmenopausal breast cancer [48]

The study also looked at urine and fat levels of phthalates, and the statistical analysis of this cohort did not find a significant increase in breast cancer risk. There was even a slight significant decrease in risk for some phthalate subgroups such as mono-benzyl phthalate (OR 0.73, 95% CI 0.60–0.90, p = NC).

A large study by Ahern et al. published in 2019 prospectively collected data from a cohort of 1,122,042 women. The aim was to compare the risk of breast cancer in 161,737 women exposed to phthalates in pharmaceuticals and 960,305 women considered unexposed [44].

The proportion of women with breast cancer among the unexposed was used as a reference with a hazard ratio of 1.0. After a multivariate analysis, only high exposure to dibutyl phthalate appeared to be associated with a significant increase in breast cancer risk (HR 2.0, 95% CI 1.1–3.6, p = NC).

This study also looked for a possible association between high exposure to phthalates in pharmaceuticals and the presence of hormone-dependent breast cancer, but no significant results were found.

Finally, there are four case–control studies that explored serum or urine levels of endocrine disruptors and breast cancer risk (Table 6.4).

6.4 Endocrine Disruptors and Endometrial/Ovarian Cancer

For endometrial cancer and ovarian cancer, very few studies are available in the scientific literature. One example is the large prospective cohort study led by Donat-Vargas et al. in 2016, which included 36,777 women from 1997 [49]. This surveillance recruited 1593 women with breast cancer, 437 with endometrial cancer, and 195 with ovarian cancer. The proportion of women with cancer was compared between those with high PCB exposure (serum levels in the third tertile) and those with low PCB exposure (serum levels in the first tertile).

This large study showed no significant increase in breast (RR = 0.96, 95% CI 0.75–1.24), endometrial (RR = 1.21, 95% CI 0.73–2.01), or ovarian (RR = 0.90, 95% CI 0.45–1.79) cancer in women with higher serum PCB levels.

6.5 Conclusion

The relationship between endocrine disruptors and cancer in women is not clearly demonstrated and remains potential.

The various studies concerning organochlorines and breast cancer are mainly case–control studies with prospective data collection and retrospective analysis that highlight heterogeneous results that do not allow us to draw conclusions on the role played by these compounds.

Studies on phthalates, bisphenol A, and breast cancer remain few, and their results are also heterogeneous, which does not allow them to be incriminated.

Rare prospective cohort studies increase the power of the results, but the methods for measuring exposure remain disparate and the comparisons of these studies are limited.

Studies concerning endometrial cancer and ovarian cancer are rare, contain very small cohorts [50], or show nonsignificant results.

Finally, many studies have focused on old exposure periods, during the peak of the use of the compounds concerned and before the regulations limiting their use and exposure. It would therefore appear interesting to carry out additional studies with a more recent exposure period and large cohorts and by developing reliable in vivo measurements of endocrine disruptors, representative of the previous exposure.

References

1. WHO incidence cancer du sein [Internet]. 2021. https://www.who.int/fr/news-room/fact-sheets/detail/cancer.
2. Le cancer du sein–Les cancers les plus fréquents [Internet]. 2021. https://www.e-cancer.fr/Professionnels-de-sante/Les-chiffres-du-cancer-en-France/Epidemiologie-des-cancers/Les-cancers-les-plus-frequents/Cancer-du-sein.
3. Johnson RH, Chien FL, Bleyer A. Incidence of breast cancer with distant involvement among women in the United States, 1976 to 2009. JAMA. 2013;309(8):800–5. https://doi.org/10.1001/jama.2013.776.
4. Prins GS. Endocrine disruptors and prostate cancer risk. Endocr Relat Cancer. 2008;15(3):649. https://www.ncbi.nlm.nih.gov/pmc/articles/PMC2822396/
5. Birnbaum LS, Fenton SE. Cancer and developmental exposure to endocrine disruptors. Environ Health Perspect. 2003;111(4):389–94.
6. Glaze GM. Diethylstilbestrol exposure in utero: review of literature. J Am Osteopath Assoc. 1984;83(6):435–8.
7. DGS Anne M. Perturbateurs endocriniens. Ministère des Solidarités et de la Santé. 2021. https://solidarites-sante.gouv.fr/sante-et-environnement/risques-microbiologiques-physiques-et-chimiques/article/perturbateurs-endocriniens.
8. https://www.ipubli.inserm.fr/bitstream/handle/10608/102/Chapitre_57.html.
9. Montes-Grajales D, Bernardes GJL, Olivero-Verbel J. Urban endocrine disruptors targeting Breast Cancer proteins. Chem Res Toxicol. 2016;29(2):150–61.

10. García MA, Peña D, Alvarez L, Cocca C, Pontillo C, Bergoc R, et al. Hexachlorobenzene induces cell proliferation and IGF-I signaling pathway in an estrogen receptor alpha-dependent manner in MCF-7 breast cancer cell line. Toxicol Lett. 2010;192(2):195–205.
11. Payne J, Scholze M, Kortenkamp A. Mixtures of four organochlorines enhance human breast cancer cell proliferation. Environ Health Perspect. 2001;109(4):7.
12. Høyer SH, Zacharewski T. Organochlorine exposure and risk for breast cancer. Prog Clin Biol Res. 1998;396:133–45.
13. Ward EM, Schulte P, Grajewski B, Andersen A, Patterson DG, Turner W, et al. serum organochlorine levels and breast cancer: a nested case-control study of Norwegian women. Cancer Epidemiol Biomark Prev. 2000;12:1357–67.
14. Høyer AP, Jørgensen T, Rank F, Grandjean P. Organochlorine exposures influence on breast cancer risk and survival according to estrogen receptor status: a Danish cohort-nested case-control study. BMC Cancer. 2001;1:8.
15. Laden F, Collman G, Iwamoto K, Alberg AJ, Berkowitz GS, Freuden JL. 1,1-Dichloro-2,2-bis(p-chloro-phenyl)ethylene and poly-chlorinated biphenyls and Breast Cancer: combined analysis of five U.S. Studies J Natl Cancer Inst. 2001;93(10):9.
16. Millikan R, DeVoto E, Duell EJ, Tse C-K, Savitz DA, Beach J, et al. Dichlorodiphenyldichloroethene, polychlorinated biphenyls, and breast cancer among African-American and white women in North Carolina. Cancer Epidemiol Biomark Prev. 2000;9(11):1233–40.
17. Charlier C, Foidart J-M. Environmental dichlorodiphenyltrichlorethane or hexachlorobenzene exposure and breast cancer: is there a risk? Clin Chem Lab Med. 2004;42(2):222–7.
18. Cohn BA, Wolff MS, Cirillo PM, Sholtz RI. DDT and Breast Cancer in young women: new data on the significance of age at exposure. Environ Health Perspect. 2007;115(10):9.
19. Cohn BA, Merrill ML, Krigbaum NY, Yeh G, Park J-S, Zimmermann L, et al. DDT exposure in utero and Breast Cancer. J Clin Endocrinol Metab. 2015;8:2865.
20. Cohn BA, Cirillo PM, Terry MB. DDT and breast cancer: prospective study of induction time and susceptibility windows. J Natl Cancer Inst. 2019;111:8.
21. Marcella W, Paolo M, Steven S, Larry N, Paolo B, Brenda E. Dioxin exposure and cancer risk in the Seveso women's health study. Environ Health Perspect. 2011;119(12):1700–5. https://doi.org/10.1289/ehp.1103720.
22. Danjou AM, Fervers B, Boutron-Ruault M-C, Philip T, Clavel-Chapelon F, Dossus L. Estimated dietary dioxin exposure and breast cancer risk among women from the French E3N prospective cohort. Breast Cancer Res BCR. 2015;17:39. https://www.ncbi.nlm.nih.gov/pmc/articles/PMC4362830/.
23. Romieu I, Hernandez-Avila M, Lazcano-Ponce E, Weber JP, Dewailly E. Breast cancer, lactation history, and serum organochlorines. Am J Epidemiol. 2000;152(4):363–70.
24. Pavuk M, Cerhan JR, Lynch CF, Kocan A, Petrik J, Chovancova J. Case-control study of PCBs, other organochlorines and breast cancer in eastern Slovakia. J Expo Anal Environ Epidemiol. 2003;13(4):267–75.
25. Jm IJ, Fernández MF, Santa-Marina L, Olea-Serrano MF, Rivas AM, Aurrekoetxea JJ, et al. Breast cancer risk and the combined effect of environmental estrogens. Cancer Causes Control. 2004;15(6):591–600.
26. Gatto NM, Longnecker MP, Press MF, Sullivan-Halley J, McKean-Cowdin R, Bernstein L. Serum organochlorines and breast cancer: a case-control study among African-American women. Cancer Causes Control. 2007;18(1):29–39.
27. Itoh H, Iwasaki M, Hanaoka T, Kasuga Y, Yokoyama S, Onuma H, et al. Serum organochlorines and breast cancer risk in Japanese women: a case-control study. Cancer Causes Control. 2009;20(5):567–80.
28. Recio-Vega R, Velazco-Rodriguez V, Ocampo-Gómez G, Hernandez-Gonzalez S, Ruiz-Flores P, Lopez-Marquez F. Serum levels of polychlorinated biphenyls in Mexican women and breast cancer risk. J Appl Toxicol. 2011;31(3):270–8.

29. Cohn BA, Terry MB, Plumb M, Cirillo PM. Exposure to polychlorinated biphenyl (PCB) congeners measured shortly after giving birth and subsequent risk of maternal breast cancer before age 50. Breast Cancer Res treat [Internet]. 2012;136(1):267–275. doi:https://doi.org/10.1007/s10549-012-2257-4.
30. Arrebola JP, Belhassen H, Artacho-Cordón F, Ghali R, Ghorbel H, Boussen H, et al. Risk of female breast cancer and serum concentrations of organochlorine pesticides and polychlorinated biphenyls: a case-control study in Tunisia. Sci Total Environ. 2015;520:106–13.
31. Kaur N, Swain SK, Banerjee BD, Sharma T, Krishnalata T. Organochlorine pesticide exposure as a risk factor for breast cancer in young Indian women. South Asian J Cancer. 2019;8:212.
32. Stellman SD, Djordjevic MV, Britton JA, Muscat JE, Citron ML, Kemeny M, et al. Breast cancer risk in relation to adipose concentrations of organochlorine pesticides and polychlorinated biphenyls in long island, new york. Cancer Epidemiol Biomark Prev. 2000;9(11):1241–9.
33. Aronson KJ, Miller AB, Woolcott CG, Sterns EE, McCready DR, Lickley LA, et al. Breast adipose tissue concentrations of polychlorinated biphenyls and other organochlorines and breast cancer risk. Cancer Epidemiol Biomark Prev. 2000;9(1):55–63.
34. Raaschou-Nielsen O, Pavuk M, LeBlanc A, Dumas P, Weber JP, Olsen A, et al. Adipose organochlorine concentrations and risk of Breast Cancer among postmenopausal Danish women. Cancer Epidemiol Biomark Prev. 2005;14:9.
35. Muscat JE, Britton JA, Djordjevic MV, Citron ML, Kemeny M, Busch-Devereaux E, et al. Adipose concentrations of organochlorine compounds and breast cancer recurrence in long island, New York. Cancer Epidemiol Biomark Prev. 2003;12(12):1474–8.
36. Huang W, He Y, Xiao J, Huang Y, Li A, He M, et al. Risk of breast cancer and adipose tissue concentrations of polychlorinated biphenyls and organochlorine pesticides: a hospital-based case-control study in Chinese women. Environ Sci Pollut Res. 2019;26:32128–36.
37. López-Cervantes M, Torres-Sánchez L, Tobías A, López-Carrillo L. Dichlorodiphenyldichloroethane burden and breast cancer risk: a meta-analysis of the epidemiologic evidence. Environ Health Perspect. 2004;112(2):207–14. https://www.ncbi.nlm.nih.gov/pmc/articles/PMC1241830/.
38. Anses-Agence nationale de sécurité sanitaire de l'alimentation, de l'environnement et du travail. Bisphénol A. 2021. https://www.anses.fr/fr/content/bisph%C3%A9nol
39. All news–ECHA. Cité 24 sept 2021. https://echa.europa.eu/fr/-/endocrine-disrupting-properties-to-be-added-for-four-phthalates-in-the-authorisation-list.
40. Vandenberg LN, Maffini MV, Wadia PR, Sonnenschein C, Rubin S, Soto AM. Exposure to environmentally relevant doses of the xenoestrogen Bisphenol-A alters development of the fetal mouse mammary gland. Endocrinology. 2007;148(1):116–27.
41. Lozada KW, Keri RA. Bisphenol A increases mammary cancer risk in two distinct mouse models of breast cancer. Biol Reprod. 2011;85(3):490–7.
42. Acevedo N, Davis B, Schaeberle CM, Sonnenschein C, Soto AM. Perinatally administered bisphenol a as a potential mammary gland carcinogen in rats. Environ Health Perspect. 2013;121(9):7.
43. Liu G, Cai W, Liu H, Jiang H, Bi Y, Wang H. The association of Bisphenol A and phthalates with risk of breast cancer: a meta-analysis. Int J Environ Res Public Health. 2021;18(5):2375.
44. Ahern TP, Broe A, Lash TL, Cronin-Fenton DP, Ulrichsen SP, Christiansen PM, et al. Phthalate exposure and breast cancer incidence: a Danish Nationwide cohort study. J Clin Oncol. 2019;37(21):1800–9.
45. López-Carrillo L, Hernández-Ramírez RU, Calafat AM, Torres-Sánchez L, Galván-Portillo M, Needham LL, et al. Exposure to phthalates and breast cancer risk in northern Mexico. Environ Health Perspect. 2010;118(4):539–44.
46. Reeves KW, Díaz Santana M, Manson JE, Hankinson SE, Zoeller RT, Bigelow C, et al. Urinary phthalate biomarker concentrations and postmenopausal Breast Cancer risk. J Natl Cancer Inst. 2019;111(10):1059–67.
47. Yang M, Ryu J-H, Jeon R, Kang D, Yoo K-Y. Effects of bisphenol A on breast cancer and its risk factors. Arch Toxicol. 2009;83(3):281–5.

48. Trabert B, Falk RT, Figueroa JD, Graubard BI, Garcia M, Lissowska J, et al. Urinary bisphenol A-glucuronide and postmenopausal breast cancer in Poland. Cancer Causes Control. 2014;25:1587–93.
49. Donat-Vargas C. Dietary exposure to polychlorinated biphenyls and risk of breast, endometrial and ovarian cancer in a prospective cohort. Br J Cancer. 2016;2016:9.
50. Hiroi H, Tsutsumi O, Takeuchi T, Momoeda M, Ikezuki Y, Okamura A, et al. Differences in serum bisphenol a concentrations in premenopausal normal women and women with endometrial hyperplasia. Endocr J. 2004;51(6):595–600.

Chapter 7
Endocrine Disruption in the Male

Andrea Garolla, Andrea Di Nisio, Luca De Toni, Alberto Ferlin, and Carlo Foresta

7.1 Introduction

Reproductive health has emerged as an important healthcare need involving many clinical and public health issues, including sexually transmitted infections (STIs), declining fertility and rising rates of testicular cancer [1–4]. Importantly, it is now recognized that many causes and risk factors for testicular dysfunction and infertility indeed act early during life [5]. Many andrological pathologies that we see in adults actually arose in younger age, due to the strong susceptibility and vulnerability of male gonads to external insults, starting from gestational age and during all growth phases.

Of particular scientific and public interest is the possible contribution of endocrine disruptors to increased incidence of male sexual and reproductive problems, such as infertility, hypogonadism, cryptorchidism, hypospadias and testicular cancer. An endocrine-disrupting chemical (EDC) is defined as "an exogenous chemical, or mixture of chemicals, that interferes with any aspect of hormone action" [6]. Contamination from EDCs is almost inevitable, when such chemicals are used in occupational activities or are widely dispersed across the environment. The daily used products like pesticides, plastic items containing bisphenol A and phthalates, flame retardants, personal care products containing antimicrobials, heavy metals and perfluoroalkyls are regularly being manufactured in industries. These are some of the most potential candidates as testicular disruptors among EDCs. Although the biological effects of many EDCs are well known at the molecular and cellular levels in in vitro studies, their mechanism of action is not readily and easily assessed in vivo, as their effects can appear after prolonged and/

A. Garolla (✉) · A. Di Nisio · L. De Toni · A. Ferlin · C. Foresta
Department of Medicine, Unit of Andrology and Reproductive Medicine, University of Padova, Padova, Italy
e-mail: andrea.garolla@unipd.it; andrea.dinisio@unipd.it; luca.detoni@unipd.it; alberto.ferlin@unipd.it; carlo.foresta@unipd.it

© The Author(s) 2023
R. Marci (ed.), *Environment Impact on Reproductive Health*,
https://doi.org/10.1007/978-3-031-36494-5_7

or continuous exposure to a low dose. Importantly, the effects can be transgenerational and therefore two or three generations are necessary to highlight some modest effects, making epidemiological studies in humans very challenging. Furthermore, these effects are often the result of the simultaneous interaction of several substances (mixture effect) at low doses. On the contrary, in vitro and animal studies often use single compounds at a high dose. Human EDC-related diseases are more likely to be the result of long-term exposure to low concentrations of EDCs mixtures, rather than of acute exposure to single compound at high concentration. Anyway, many EDCs that have been linked to impaired male sexual and reproductive development and function seem to act as antiandrogenic and/or estrogenic compounds, after binding or mimicking the actions of either the androgen receptor (AR) or the oestrogen receptor (ER). EDCs are highly heterogeneous and can be classified according to their origins in: (1) natural and artificial hormones (e.g. fitoestrogens, 3-omegafatty acids, contraceptive pills and thyroid medicines); (2) drugs with hormonal side effects (e.g. naproxen, metoprolol and clofibrate); (3) industrial and household chemicals (e.g. phthalates, alkylphenoletoxilate detergents, plasticizers and solvents) and (4) side products of industrial and household processes (e.g. polycyclic aromatic hydrocarbons, dioxins and pentachlorobenzene). As a consequence, many pathways might be disrupted, depending on the period of life when they act (ranging from impairment of sexual differentiation, organogenesis, spermatogenesis and steroidogenesis) and the cocktail of contaminants involved.

In general, three main phases of a man life are particularly susceptible for subsequent normal testis development and function (Fig. 7.1): the intrauterine

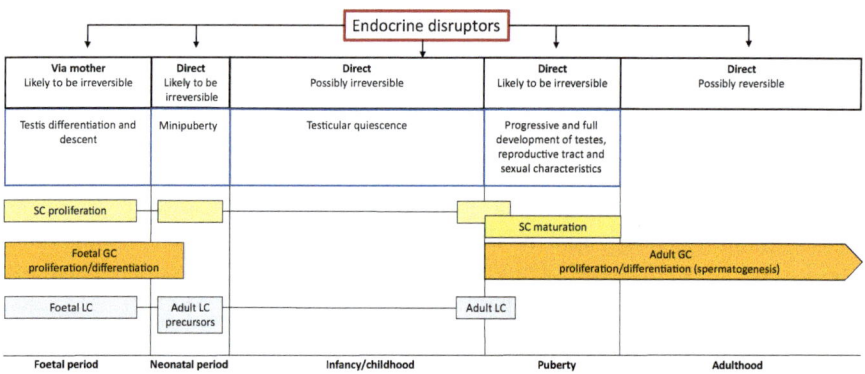

Fig. 7.1 Windows of susceptibility for testicular development and function from the foetal period to adulthood and effects of EDCs. *SC* sertoli cells, *GC* germ cells, *LC* leydig cells. Reproduced with permission from: Ferlin A, Di Nisio A, De Toni L, Foresta C. Impact of Endocrine Disruptors on Male Sexual Development. In: Foresta C, Gianfrilli D (eds), Pediatric and Adolescent Andrology. Trends in Andrology and Sexual Medicine, 2021. Springer, Cham. https://doi.org/10.1007/978-3-030-80015-4_2

phase, the neonatal phase comprising the so called "minipuberty" in the first months of life and puberty. However, even during infancy, when the testes are apparently "sleeping", damaging causes with permanent effects on testicular function can occur. This is, for example, the case of the iatrogenic, devastating effect of chemotherapy in this period of life. Risk factors acting via the mother during pregnancy might compromise definitively testicular function later in life, by disrupting foetal germ cell proliferation and differentiation, Sertoli cell proliferation and establishing of the Leydig cell population (Fig. 7.1). Similarly, risk factors acting directly on minipuberty might compromise germ, Sertoli and Leydig cell differentiation and proliferation. Iatrogenic, environmental and life style risk factors during childhood might interfere above all with the germ cell compartment and those acting during puberty might disrupt Sertoli cell maturation, the establishment of adult Leydig cell population and spermatogenesis (Fig. 7.1) [5, 7]. These fundamental phases of vulnerability are also important when dealing with EDCs, even if the intrauterine, transplacental phase seems to be the most important for future and transgenerational effects. Also adolescence is a vulnerable window for the development and maturation of the genitourinary tract [8]. Risk factors, lifestyles and EDCs effects in adolescence may negatively affect adult health as well as that of future generations, through epigenetics.

A large body of evidence has been published dealing with various molecular and cellular aspects of the action of EDCs and their association with urogenital diseases. However, most studies focused on a single or single class of EDCs. Evidence from epidemiological and clinical studies is less robust for the intrinsic difficulties highlighted above. Indeed, a systematic review and meta-analysis [9] of epidemiological studies reporting association between male reproductive disorders and exposures (documented by biochemical analyses of biospecimens) to chemicals that have been included in the European Commission's list of Category I EDCs showed that there is evidence for a small increased risk following prenatal and postnatal exposure to some persistent environmental chemicals, but the evidence is low, with an overall odds ratio across all exposures and outcomes of 1.11 (95% CI 0.91–1.35). Most studies are focused on bisphenol A and phthalates [10], and more recently on perfluoroalkyl compounds (PFC) (Table 7.1) [11].

Table 7.1 Main endocrine-disrupting chemicals (EDCs) and related mechanisms and effects during the three main phases of male sexual development

EDCs	Pre-natal period Mechanisms	Effects	Neo-natal period/Infancy Mechanisms	Effects	Childhood/Adolescence Mechanisms	Effects
BPA	Foetal Leydig cell dysfunction; germ cell toxicity	Reduced T and/or INSL3; impaired germ cells maturation; impaired HPG maturation	GnRH interference; impaired germ cells maturation	Impaired HPG maturation; congenital malformations (crtyptorchidism, hypospadias)	HPG disruption; reduced steroidogenesis; testicular toxicity on germ cells and Leydig cells	Delayed puberty; reduced spermatogenesis; developmental disorders; delayed sexual maturation; reduced testicular volume
Phthalates	Histological alterations; Leydig cell dysfunction	Reduced T and/or INSL3; impaired germ cells maturation	Impaired germ cells maturation	Congenital malformations (crtyptorchidism, hypospadias)	Reduced steroidogenesis; testicular toxicity	Impaired spermatogenesis
PFAS	Impairment of foetal Leydig cells, germ cells and Sertoli cells	Reduced T and/or INSL3; congenital malformations; impaired germ cells maturation; impaired HPG maturation	GnRH interference; impaired germ cells maturation	Impaired HPG maturation; congenital malformations (crtyptorchidism, hypospadias); reduced AGD	HPG dysruption; reduced steroidogenesis	Delayed puberty; reduced spermatogenesis; developmental disorders; delayed sexual maturation; reduced testicular volume

Abbreviations: *BPA* bisphenol A, *PFAS* perlfuoroalkyl substances, *T* testosterone, *INSL3* insulin-like 3 hormone, *HPG* Hypothalamic–pituitary gland, *GnRH* gonadotropin-releasing hormone, *AGD* anogenital distance

7.2 Mechanism of Action of EDCs on Hypothalamic–Pituitary–Gonadal Axis

As reported above, the interactions involved in gonadal function and hormonal communication are various and complex. Different targets could be impaired by EDCs, and different time points are involved, even at different generations. Various mechanisms can lead to the impairment of endocrine regulation, but mainly involves a reduction in steroid hormones biosynthesis, storage or release and transport, or could involve an antagonistic effect on binding of sex hormones to their receptors, and/or post receptor signal transduction [reviewed in 12].

There is evidence of altered steroid biosynthesis of different hormones from a wide range of toxicological studies on different chemicals [6]. The first reported mechanism of reduced steroidogenesis mainly involves the inhibition of specific enzymatic steps. Another mechanism has been observed by direct inhibition of aromatase activity, therefore blocking the conversion of testosterone to estrogen in the testis. Altogether these chemicals are defined as environmental estrogens and antiandrogens, as they interfere with hormones biosynthesis regulated by gonadal or extragonadal steroids through a series of signals at transcriptional and translational levels [5]. Pituitary hormone synthesis is affected by both estrogen and testosterone, directly or indirectly through changes in the glycosylation of LH and FSH. Therefore, any factor that interferes with the glycosylation has a negative impact on the biological activity of these hormones. As a consequence, any EDCs that mimics or antagonizes the action of these steroid hormones could presumably alter glycosylation.

Another mechanism is the alteration of hormone storage and/or release. For example, after LH stimulation in Leydig cells, the testis synthesizes testosterone. Therefore, any EDC that antagonizes the binding of LH to its receptor on one hand, or that inhibits the activation of the $3',5'$-cyclic AMP (cAMP)-dependent pathway involved in steroidogenesis on the other hand, has the potential to impair testosterone biosynthesis. As for cAMP, second messenger pathways are one of the main routes involved in the release of various hormones. Therefore, any compound that interferes with these processes has the potential to reduce the bioavailability of hormones. For example, disruption of pituitary hormone release has been reported for heavy metals, mainly by interfering with Ca^{2+} flux [12].

Another possible mechanism of hormonal interference on the hypothalamic–pituitary–gonadal (HPG) axis is the perturbation of hormonal transport and clearance. Hormones typically circulate in blood in the free or bound state. Steroids, and androgens in particular, are transported by specific transporters, named steroid hormone-binding globulin (SHBG) or testosterone-estrogen-binding globulin (TEBG). Any modification in the concentration of these carrier proteins in the circulation can lead to an increase or decrease in steroid hormone bioavailability. For example, DDT analogs have been shown to induce an enhancement in the degradation of transport proteins, leading to reduced release of androgen from the testis to the circulation, then limiting its biological systemic activity [10].

Probably the most frequent mechanism of EDCs interference on sex hormones is the altered hormone receptor recognition and/or binding. Since hormones represent a complex system of signal transduction and communication across various body cell types, the correct recognition of hormones with their receptors is fundamental in order to elicit correct responses in target cells. The binding of the physiological ligand to its receptor, which can be either cytoplasmic, nuclear or membrane-bound, is therefore highly specific and represents a crucial step in hormonal signaling. Intracellular receptors, there including steroid receptors, adrenal receptors, thyroid hormones receptors, vitamin D receptor and retinoic acid receptor, normally act by regulating gene transcription upon ligand binding and subsequent nuclear translocation, where they interact with specific DNA target sequences, known as responsive elements that ultimately activate the transcription of target genes. A huge variety of EDCs have been proven to interfere with this process, either by resembling the physiological agonist or acting as an agonist, or even by inhibiting the hormone binding and acting as an antagonist [10–12]. At first, the most studied EDCs were shown to inhibit estrogen receptor activity, such as DDT, some PCBs and BPA. Nonetheless, this interference on steroid hormones is not exclusive of only one compound or towards only one receptor, indeed many EDCs classified as estrogen-like or anti-androgenic compounds have the ability to reduce receptor binding and/or affinity on more than one type of hormone receptor. Classical hormonal receptors are located on and in the cell membrane, upon binding, transduction of a signal across the membrane requires the activation of second messenger signal transduction pathways. Among these, the most frequent involve alterations in G-protein/cAMP-dependent protein kinase A (e.g. after LH stimulation of the Leydig cell), phosphatidylinositol regulation of protein kinase C and inositol triphosphate (e.g. after GnRH stimulation of gonadotrophs; thyrotropin releasing hormone stimulation of thyrotrophs), (c) tyrosine kinase (e.g. after insulin binding to the membrane receptor) and (d) calcium ion flux. Xenobiotics are an example of interference on signaling pathways involving second messengers regulated by peptide hormones. EDCs can also target the cascade of events that follows the hormone binding to its receptor and fundamental to fulfil the physiological response of target cells to hormonal stimulation. Various mechanisms can interfere with activation of steroid hormone receptors. Among these, the most frequent one involves the reduction of receptor sensitivity to its ligand, as observe, for example, after tetrachlorodibenzo-p-dioxin (TCDD) exposure (including the estrogen, progesterone and glucocorticoid receptors). As a result, a wide range of pathways can be altered by EDCs; therefore, any evaluation of their effects on human health should include the possibly largest set of influences on hormonal signaling, receptor function or regulation of feedback [9].

Finally, another possible mechanism relies in the stimulation of oxidative stress, which frequently results in increased apoptosis due to cellular damage as a consequence of oxygen and oxygen-derived formation of free radicals, which is reactive oxygen species (ROS). The generation of ROS has been proven to induce testicular damage after exposure to various chemicals that are associated with hormonal impairment, ultimately leading to infertility. Another target of ROS is the endothelium, where highly reactive radicals can induce cell damage, leading to a reduction of blood flow to the testis with consequent impairment of testicular function. Finally, ROS can also directly damage DNA, by oxidation of DNA bases or by covalent binding that induces strand breaks or cross-linking [12].

7.3   Bisphenol A

Bisphenols, and in particular the phenol compound 2,2 Bis (4-hydroxylphenyl)–propane, universally known as Bisphenol-A (BPA), are widely used as additives for the production of plastic materials, such as polycarbonate, phenol and epoxy resins, polyesters and polyacrylates, as well as antioxidant in foodstuffs and cosmetics [6, 13]. Specifically, nearly 75% of the industrial production of BPA is intended for the manufacture of polycarbonate-based products, which find wide application in food industry such as containers for food and beverages, in plastic dishes, in kitchen utensils, in containers for microwave cooking and, until 2011, in bottles [14]. Of note, BPA is also used in epoxy-resins films used as binary patina: the internal coatings in the cans for canned food [15].

As a result, there is a significant risk of human exposure to BPA through ingestion, skin contact or inhalation [16, 17]. Epidemiological data from the United States have reported detectable levels of BPA in urine samples from more than 90% of general population, resulting in a major problem of exposure to chemical substance [18].

Concerns about BPA issues on the human health date back to 1930s, when severe impact on male sexual development had been suggested. From a mechanistic point of view, the most relevant risks associated with the exposure to BPA are mainly due to its action as an EDC. Available reports in late 1990s first documented a stimulating activity of BPA on ERα [19, 20] confirmed later [21–23]. In addition, unconjugated BPA showed a binding activity to other two receptors: the G protein-coupled oestrogen receptor 30 (GPR30), also known as membrane estrogen receptor alpha (mERα) [24, 25], and the orphan nuclear oestrogen-related receptor gamma (ERR-gamma) [25]. Finally, experimental animal studies demonstrated that BPA binds also to AR, to the peroxisome proliferator-activated receptor gamma (PPAR-gamma) and the thyroid hormone receptor [22].

A wide amount of data from animal studies shows a clear effect of BPA on male reproductive system, even at very low doses. In rodent models, BPA exposure has been associated with reduced sperm count and significant reductions of the absolute weights of the testes and seminal vesicles [26–33]. Furthermore, the exposure to BPA has been associated with the alteration of other non-conventional markers of sperm quality such as the index of DNA fragmentation, suggesting a possible role as mutagen [29, 32, 34–42]. Also, acrosomal integrity, an overall marker of the fertilization potential, was significantly reduced by PBA exposure in murine models [27].

Several studies have been performed to disclose the possible disruption of the hypothalamus–pituitary–testis (HPT) axis associated with BPA exposure in animal models, with the result of a fairly complex picture that invariably leads to the impaired production of testosterone [28, 43], both by direct effects on steroidogenesis of the Leydig cells [39, 44, 45] and indirect effects on HPT. This latter is mediated by indirect suppression of the pituitary LH release through the massive aromatase upregulation in the testes [46]. Importantly, because of its high lipid solubility, BPA undergoes to trans-placental transfer in animal models with a consequent detection in cord blood, an evidence reported also in humans [47–49]. Accordingly, BPA exposure during the prenatal period was associated with the impairment of both foetal development and the endocrine function of the testis, with reduced Leydig cell proliferation and foetal testosterone production [50–52]. Maternal exposure to BPA was associated with reduced sperm count and motility in male offspring and, in turn, with post implantation loss and decreased litter size [53]. Of note, very recent studies disclosed some transgenerational effects associated with BPA exposure [54].

Despite the large availability of data in animal models, fewer studies assessed the possible relationship between BPA exposure and semen quality in humans and a negative association between urinary BPA and sperm concentration [55], motility, morphology and sperm DNA damage [56]. However, two independent studies on male partners from infertile couples attending infertility clinics were not able to retrieve any significant association between BPA urinary concentration and altered semen parameters [57, 58].

Another field of investigation pursued was the possible correlation between exposure to BPA and alteration of the endocrine pattern, but widely varying scenarios can be observed. Lower serum levels of follicle-stimulating hormone (FSH) in exposed workers compared to those non-exposed was found [59], but also a positive and significantly association with serum testosterone levels was observed [60]. Another study found increased serum testosterone, free testosterone, LH and oestradiol in subjects pertaining to higher urinary BPA concentrations quartile, compared with the lowest quartile. Subjects in the highest urinary BPA quartile also showed reduced progressive sperm motility compared with the lowest quartile [61]. On the contrary, urinary BPA concentrations were found positively associated with serum SHBG levels and inversely correlated with free androgen index (FAI) [58].

Finally, few studies aimed to assess the possible impact of BPA exposure on the overall fertility potential in males through the overall evaluation of the relationship between BPA levels and the reproductive outcome in the setting of assisted reproduction facilities. Minimal association between paternal urinary propyl paraben levels and reduced live birth rate in a correlation model corrected by possible confounders has been reported [62]. However, no significant association emerged between paternal urinary BPA and reproductive outcomes after fertility treatments. On the other hand, urinary BPA concentration in either males or females was not associated with increased time to pregnancy [63].

Overall, available data are supportive of detrimental role of BPA on semen parameters, but this is not accompanied by clear data on sex hormones and on fertility outcomes. As suggested by other authors [64], within the limits of the availability of data in humans, a possible reconciling explanation could rely on a greater direct toxicity of BPA on germ line cells, rather than in an albeit important endocrine disruption of the HPT axis.

In conclusion, BPA represents one of the most controversial chemical pollutants, with the typical features of an EDC. Early toxicological evidence on BPA date back to nearly 30 years ago, when major interference with estrogen signaling pathway was claimed. Since that time, a wide range of cell mechanisms of both endocrine and metabolic disruption have been claimed by the use of experimental models. In particular, major impairment of the HPT axis has been recognized as associated with the exposure to BPA during both the foetal and the adult life, resulting in altered testis development, impaired endocrine function and infertility. To this regard, direct disruption of sperm characteristics, such as reduced motility performances and development genetic abnormalities, has been identified. On the other hand, data obtained in humans are actually limited and poorly conclusive to identify a strict causal role of BPA in reduced male fertility potential.

Methodological differences and different study populations are factors that can explain some discrepancies. Moreover, available clinical outcomes, such as semen parameters and time to pregnancy, are likely susceptible of variation related to many different confounding factors. It should be noted that, as for most of chemical pollutants, the identification of a reliable marker of exposure remains a major issue. Specifically, for BPA, urinary concentrations are surely reliable data from an analytical point of view but may not be representative of the real exposure to BPA due to its short half-life. To this regard, Vitku et al. reported that BPA levels in blood plasma were positively correlated with BPA levels in semen, but only seminal BPA was negatively associated with seminal quality [65]. Finally, the cross-sectional design of the available studies surely provides proof of association but limited evidence of causality.

One of the main problems associated with exposure to endocrine disruptors in general, and to BPA in particular, is represented by the potential activity at low concentrations. This represents a critical issue during the development phases, such as embryo/foetal life, newborn or peri-pubertal age, since the effects in these time windows may result irreversible and are generally detected only at adulthood. Accordingly, populations at higher risk include pregnant women, infants and adolescents (Fig. 7.1). On these bases, the current European law restricted the use of BPA in the production of packaging and materials in direct contact with food by limiting migration rate to 0.05 mg/kg of food and prescribing the total absence in products for newborns, from food to food containers and clothes [66]. In addition, based on new toxicological data and methodologies, the European Authorities adjusted the tolerable daily intake from 50 to 4 μg/kg body weight/day with an overall lowering rate of 12 times, highlighting the increasing level of attention for these health concerns.

7.4 Phthalates

Phthalates are employed in virtually all industrial applications and consumer products as additives, used as plasticizers in a broad range of industrial and commercial products [67, 68]. The most commonly used phthalates are di-(2-ethylhexyl) phthalate (DEHP), di-n-butyl phthalate (DBP), diethyl phthalate (DEP) and benzylbutyl phthalate (BzBP). More than 75% of DEHP produced worldwide is used in plastic products. The other phthalates are largely used in personal care products like foams, shampoos, dyes, lubricants and food packaging materials [69]. Since these compounds are not covalently bound polymers, their exposure to heat over time has the potential to favour their migration into food [70]. Indeed, plasticizers such as phthalate esters, because of their anti-androgen and oestrogen-like activity, are indicated as major EDCs. Both in vitro and in vivo toxicology studies have demonstrated their endocrine-disrupting potential in model organisms, with endpoints such as antiandrogen effects, reproductive abnormalities, testicular lesions and reduced sperm production [71]. However, as for other EDCs, dose ranges used for traditional reproductive toxicological studies were much higher than those observed in human epidemiological studies. Therefore, it is not surprising that these studies do not entirely align with the human studies. Nevertheless, in vitro and in vivo toxicology studies with low exposures to phthalates were linked to decreased semen quality and male infertility in animals, as well as to decreased androgen production and steroidogenesis [64, 72–80]. Phthalates have mostly shown the antiandrogen effect on testicular function during steroid formation [81–83]. Furthermore, phthalates as well as their metabolites (e.g. DEHP/MEHP, DBP/MBP) have stimulatory effects at low doses through inducing the production of progesterone, testosterone, steroidogenesis-related proteins and gene expression [64, 74, 75, 77, 78, 80]. The adverse effects of phthalates on sperm quality were confirmed by ex vivo studies, where spermatozoa were exposed to high concentrations of phthalates, showing that sperm motility was affected and that cytotoxicity was caused at long-term exposures (>3 days) to the metabolite DEHP [84]. In parallel, DHEP has been shown to inhibit testosterone production, when cultured in vitro with explants derived from human testes [85].

Epidemiological studies reported an association between phthalates exposure and altered seminal parameters [86]. It is important to note that exposure of infants to phthalates is due to both maternal exposure and breastfeeding. In fact, breast milk levels of the phthalate metabolites are positively associated with maternal diet and water consumption.

Studies in humans corroborated the in vitro findings and suggested that exposure to phthalate metabolites is correlated with lower motility of spermatozoa in men from subfertile couples [87]. The DNA damage induced in spermatozoa, the motility and morphology of the spermatozoa were weakly associated with the exposure to phthalates [88–91], whereas an inverse association between MEHP exposure and testosterone and oestradiol levels was reported [92].

Apart from infertility, data available on the effect of phthalates on male reproductive health are limited [93]. Phthalates are rapidly metabolized and excreted in urine and feces and therefore the assessment of exposure to phthalates in humans relies on the measurement of urinary concentrations of phthalate metabolites. However, little or even no attention is given to the possible accumulation of unmetabolized phthalates in different tissues [94]. This evidence raises some concerns about the appropriateness of parameters employed as index of exposure to contaminants, in particular for those substances like phthalates that, showing specific tissue accumulation, may exert risk associated to long-term exposures [82]. To this regard, quantification of both parent compound and corresponding metabolites in specific body fluids may represent an informative parameter with better correlation with clinical parameters [83].

7.5 Perfluoroalkyl Compounds

Perfluoroalkyl compounds (PFCs) or substances (PFAS) are a class of organic molecules characterized by fluorinated hydrocarbon chains extensively used in industry and consumer products including oil and water repellents, coatings for cookware, carpets and textiles. PFCs possess unique physical chemical properties due to their amphiphilic structures and their strong carbonfluorine bonds. Therefore, long-chain PFCs are non-biodegradable and bioaccumulate in the environment [95, 96]. PFCs have been found in humans and in the global environment and their toxicity, environmental fate and sources of human exposure have been a major subject of research. Currently, 23 PFCs are distinguished, including perfluorooctanoic acid (PFOA) and perfluooctane sulfonate (PFOS), which are the predominant forms in human and environmental samples. Both in vitro and animal studies on PFCs toxicity have shown a detrimental effect of PFOA and PFOS on testicular function, through alteration of steroidogenic machinery and subsequent defect of spermatogenesis [97–101]. Among the endocrine effects of PFOS in particular, it should be emphasized that this compound can affect the HPT axis activity [102, 103]. It is also able to exert its toxicity at testicular level [104], as reported in rats [102, 105] and in testis models [106]. According to a recent study on male rats [107], high doses of PFOS orally administered for 28 days seem to modify the relative gene and protein receptor expressions of several hormones of the reproductive axis (GnRH, LH, FSH and testosterone). Recently, exposure to PFOA was associated to reduction in sperm motility through alteration of sperm membrane fluidity [108].

Various PFCs compounds have been found in human serum [109], seminal fluid [110], breast milk [111] and even umbilical cord [112], suggesting a life-long exposure to PFCs in humans, from foetal stages until the adult life. Indeed, PFCs act as endocrine disruptors on the foetus and newborns, leading to developmental defects [113]. This has led to strict regulation of PFOA and PFOS use in industrial processes, as the compounds were added to the Annex B of the Stockholm

Convention on Persistent Organic Pollutants. In addition to the health concerns related to foetal development, epidemiological studies have focused also on the relationship between PFCs and human fertility. In utero exposure to PFOA was associated later in adult life with lower sperm concentration and total sperm count and with higher levels of LH and FSH [114].

Besides the impact of PFCs on the professionally exposed populations, recent evidence of pollution from chemical industries producing PFCs has emerged also in the general population from at least four different area worldwide: Mid-Ohio valley in the USA, Dordrecht area in Netherlands, Shandong district in China and Veneto region in Italy [115]. Despite strong evidence pointing towards a negative role of PFCs on male reproductive function, to date few evidence is available on the actual effect of these substances on seminal parameters in men, with conflicting results [110, 116, 117]. Two cross-sectional studies reported negative associations of PFOS, or high PFOA and PFOS combined, with the proportion of morphologically normal spermatozoa in adult men [116, 118]. Furthermore, in a study of men attending an in vitro fertilization clinic, Raymer et al. [110] reported that LH and free testosterone significantly and positively correlated with plasma levels of PFOA, although PFOA was not associated with semen quality. Conflicting results are also reported for the association between PFCs and sperm DNA quality, although a significant trend is evident for increased DNA fragmentation in exposed men [117, 119, 120]. In infertile males, PFOS levels were higher than fertile counterparts, together with a higher gene expression of ERα, ERβ and AR [121, 122], suggesting that PFCs activity might be linked also to the genetic expression of sex hormones nuclear receptors. With respect to AR, PFOS and PFOA induce a decrease of the protein expression of this receptor in the hypothalamus and pituitary gland as well as in the testis [123]. These findings clearly suggest an antiandrogenic potential of PFCs. More recently, in a cross-sectional study on 212 exposed males from the Veneto region in Italy, and 171 nonexposed controls, increased levels of PFCs in plasma and seminal fluid positively correlated with circulating testosterone and with a reduction of semen quality, testicular volume, penile length and anogenital distance [124]. Furthermore, the anti-androgenic property of PFOA was related to antagonism on the binding of testosterone to AR [124].

In conclusion, in men, there is little evidence for an association between PFCs exposure and semen quality or levels of reproductive hormones. As is the case for many epidemiological studies, causality cannot be definitively established in these studies, largely because of their cross-sectional design. However, the consistency of findings in preclinical studies strongly suggests a causal relationship for some endpoints.

7.6   Conclusions

EDCs can potentially cause harmful effects to the male reproductive system. In addition to the classical action of EDCs that includes the agonism and/or antagonism with hormone and nuclear receptors, the last decade of scientific research has given significant advances in the field of molecular biology that identified several compounds as endocrine disruptors, by interfering with the cell cycle, the apoptotic machinery and the epigenetic regulation of the target cells [125]. However, action mechanisms should not be generally extrapolated since each chemical has different routes to interfere with endocrine activity. Among the tens of known EDCs, BPA, phthalates and PFCs are particularly intriguing for male sexual and reproductive consequences given the strong experimental evidence of effects on hormone nuclear receptors (AR and/or ER), HPT axis and direct action on spermatogenesis and steroidogenesis [126, 127]. However, epidemiological studies in humans have shown controversial and inconsistent results. This discrepancy can be attributed to several factors that could affect the outcome of the studies, notably to the complexity of the clinical protocols used, the degree of occupational or environmental exposure, the selection of the target group under investigation, the determination of the variables measured and the sample size of the subjects examined. Despite the lack of consistency in the results of the human studies, the overall conclusion points toward a positive association between exposure to EDCs and alteration of the reproductive system.

EDCs and environmental factors can lead to male reproductive alterations at different stages of sexual development and maturation. In addition to the classical mechanism of endocrine disruption by chemicals, there including agonistic and/or antagonistic interference on hormonal and intracellular receptors, the last decades of scientific research have provided new evidence at different experimental levels, from in vitro studies to animal and human studies. Altogether these results have recognized a wide range of chemicals with endocrine-disrupting features that interfere with various biological processes in target cells, such as cell proliferation or apoptosis, and their epigenetic regulation. However, a common mechanism of action cannot be identified, given the very wide range of chemical structures, exposure routes, environmental levels and so on among different chemicals with endocrine activities. Among the tens of known EDCs, BPA, heavy metals, phthalates, organophosphates and PFAS are particularly intriguing for male HPT axis function, given the agreement in experimental studies showing a consistent effect on steroid receptors (AR and/or ER), hormonal metabolism and related enzymes and direct action on steroidogenesis. Although the observed effects may be subtle on an individual level, the biological link between them (i.e. TDS: decreased androgen levels contributing to cryptorchidism, reduced penile length and reduced testicular volume) should raise concerns about the effects of EDCs at population levels in young men. More longitudinal studies performed on a wide number of subjects are clearly needed in order to identify other putative damaging compounds, to clarify new routes of exposure and to replace legacy EDCs with harmless substances. However, epidemiological studies in humans have shown controversial and inconsistent results for different EDCs classes. The lack of consistency across studies and between human and animal studies can be explained by different factors possibly affecting the outcomes of these studies, such as the differences in investigation protocols and study designs, the crude levels of contaminations at different degrees, from occupational to environmental exposure, the different analytical approaches for the quantification of exposure, the selection of different variables of interest as outcomes indicative of endocrine-disrupting features and finally even the wide range of sample sizes of subjects included in the studies. Despite the lack of consistency in the results of the human studies, the overall conclusion points toward a positive association between exposure to EDCs and alteration of the hypothalamus–pituitary–testis axis. Nonetheless, it should be pointed out that humans are not exposed to a single compound during their life, from foetal period to adulthood. Therefore, it is crucial to stress out the importance of toxicological and clinical studies that take into account a cocktail of chemicals rather than the single compound. Although this clearly represent a complicate step, it is fundamental in order to have a more comprehensive view of exposure risk to EDCs in well-define groups, and in particular in those developmental windows more sensitive to hormonal alterations, such as foetus, newborns and adolescents. The collection of new evidence on the cumulative effect of different chemicals with different properties and mechanisms would therefore provide a new avenue for the treatment and prevention of male reproductive alterations.

References

1. Khabbaz RF, Moseley RR, Steiner RJ, Levitt AM, Bell BP. Challenges of infectious diseases in the USA. Lancet. 2014;384:53–63.
2. Slater C, Robinson AJ. Sexual health in adolescents. Clin Dermatol. 2014;32:189–95.
3. Stephen EH, Chandra A, King RB. Supply of and demand for assisted reproductive technologies in the United States: clinic- and population-based data, 1995–2010. Fertil Steril. 2016;105:451–8.
4. Nigam M, Aschebrook-Kilfoy B, Shikanov S, Eggener S. Increasing incidence of testicular cancer in the United States and Europe between 1992 and 2009. World J Urol. 2015;33:623–31.
5. Skakkebaek NE, Rajpert-De Meyts E, Buck Louis GM, Toppari J, Andersson AM, Eisenberg ML, Jensen TK, Jørgensen N, Swan SH, Sapra KJ, Ziebe S, Priskorn L, Juul A. Male reproductive disorders and fertility trends: influences of environment and genetic susceptibility. Physiol Rev. 2016;96:55–97.
6. Gore AC, Chappell VA, Fenton SE, Flaws JA, Nadal A, Prins GS, Toppari J, Zoeller RT. EDC-2: the Endocrine Society's second scientific statement on endocrine-disrupting chemicals. Endocr Rev. 2015;36(6):E1–E150.
7. Sharpe RM. Environmental/lifestyle effects on spermatogenesis. Philos Trans R Soc Lond Ser B Biol Sci. 2010;365:1697–712.
8. Abreu AP, Kaiser UB. Pubertal development and regulation. Lancet Diabetes Endocrinol. 2016;4:254–64.
9. Bonde JP, Flachs EM, Rimborg S, Glazer CH, Giwercman A, Ramlau-Hansen CH, Hougaard KS, Høyer BB, Hærvig KK, Petersen SB, Rylander L, Specht IO, Toft G, Bräuner EV. The epidemiologic evidence linking prenatal and postnatal exposure to endocrine disrupting chemicals with male reproductive disorders: a systematic review and meta-analysis. Hum Reprod Update. 2016;23(1):104–25.
10. Pallotti F, Pelloni M, Gianfrilli D, Lenzi A, Lombardo F, Paoli D. Mechanisms of testicular disruption from exposure to bisphenol A and Phtalates. J Clin Med. 2020;9(2):E471.
11. Di Nisio A, Foresta C. Water and soil pollution as determinant of water and food quality/contamination and its impact on male fertility. Reprod Biol Endocrinol. 2019;17(1):4.
12. Sikka SC, Wang R. Endocrine disruptors and estrogenic effects on male reproductive axis. Asian J Androl. 2008;10(1):134–45. https://doi.org/10.1111/j.1745-7262.2008.00370.x.

13. Murata M, Kang JH. Bisphenol A and all cell signaling pathways. Biotechnol Adv. 2017;36:311–27.
14. Lyons G. Bisphenol A: a known endocrine disruptor. A WWF European toxics Programme report. WWF European Toxics Programme: Godalming, Surrey. Registered Charity No 201707. 2000. http://www.google.it/url?sa=t&rct=j&q=&esrc=s&source=web&cd=1&cad =rja&uact=8&ved=2ahUKEwjQ4uXOnoDoAhWttIsKHeNUAywQFjAAegQIBxAB&url=h ttp%3A%2F%2Fassets.panda.org%2Fdownloads%2Fbisphenol.pdf&usg=AOvVaw1KMvI7 KSfb4MEwV1Ce3KvD.
15. Ehrlich S, Calafat AM, Humblet O, Smith T, Hauser R. Handling of thermal receipts as a source of exposure to bisphenol A. JAMA. 2014;311:859–60.
16. Pivnenko K, Pedersen GA, Eriksson E, Astrup TF. Bisphenol A and its structural analogues in household waste paper. Waste Manag. 2015;44:39–47.
17. Vandenberg LN, Hauser R, Marcus M, Olea N, Welshons WV. Human exposure to bisphenol A (BPA). Reprod Toxicol. 2007;24:139–77.
18. Vandenberg LN, Hunt PA, Myers JP, vom Saal FS. Human exposures to bisphenol a: mismatches between data and assumptions. Rev Environ Health. 2013;28:37–58.
19. Calafat AM, Ye X, Wong LY, Reidy JA, Needham LL. Exposure of the U.S. population to bisphenol A and 4-tertiary-octylphenol: 2003–2004. Environ Health Perspect. 2008;116:39–44.
20. Gould JC, Leonard LS, Maness SC, Wagner BL, Conner K, Zacharewski T, Safe S, McDonnell DP, Gaido KW. Bisphenol A interacts with the estrogen receptor alpha in a distinct manner from estradiol. Mol Cell Endocrinol. 1998;142:203–14.
21. Kuiper GG, Lemmen JG, Carlsson B, Corton JC, Safe SH, van der Saag PT, van der Burg B, Gustafsson JA. Interaction of estrogenic chemicals and phytoestrogens with estrogen receptor beta. Endocrinology. 1998;139:4252–63.
22. Li L, Wang Q, Zhang Y, Niu Y, Yao X, Liu H. The molecular mechanism of bisphenol A (BPA) as an endocrine disruptor by interacting with the nuclear receptors: insights from molecular dynamics (MD) simulations. PLoS One. 2015;10:e0120330.
23. Richter CA, Birnbaum LS, Farabollini F, Newbold RR, Rubin BS, Talsness CE, Vandenbergh JG, Walser-Kuntz DR, vom Saal FS. In vivo effects of bisphenol A in laboratory rodent studies. Reprod Toxicol. 2007;24:199–224.
24. Dong S, Terasaka S, Kiyama R. Bisphenol A induces a rapid activation of Erk1/2 through GPR30 in human breast cancer cells. Environ Pollut. 2011;159:212–8.
25. Wozniak AL, Bulayeva NN, Watson CS. Xenoestrogens at picomolar to nanomolar concentrations trigger membrane estrogen receptor-alpha-mediated Ca^{2+} fluxes and prolactin release in GH3/B6 pituitary tumor cells. Environ Health Perspect. 2005;113:431–9.
26. Okada H, Tokunaga T, Liu X, Takayanagi S, Matsushima A, Shimohigashi Y. Direct evidence revealing structural elements essential for the high binding ability of bisphenol A to human estrogen-related receptor-gamma. Environ Health Perspect. 2008;116:32–8.
27. Al-Hiyasat AS, Darmani H, Elbetieha AM. Effects of bisphenol A onadult male mouse fertility. Eur J Oral Sci. 2002;110:163–7.
28. Wisniewski P, Romano RM, Kizys MML, Oliveira KC, Kasamatsu T, Giannocco G, Chiamolera MI, Dias-da-Silva MR, Romano MA. Adult exposure to bisphenol A (BPA) in Wistar rats reduces sperm quality with disruption of the hypothalamic–pituitary–testicular axis. Toxicology. 2015;329:1–9.
29. Gurmeet K, Rosnah I, Normadiah MK, Das S, Mustafa AM. Detrimental effects of bisphenol A on development and functions of the male reproductive system in experimental rats. EXCLI J. 2014;13:151–60.
30. Dobrzynska MM, Radzikowska J. Genotoxicity and reproductive toxicity of bisphenol A and X-ray/bisphenol A combination in male mice. Drug Chem Toxicol. 2013;36:19–26.
31. Tainaka H, Takahashi H, Umezawa M, Tanaka H, Nishimune Y, Oshio S, Takeda K. Evaluation of the testicular toxicity of prenatal exposure to bisphenol A based on microarray analysis combined with MeSH annotation. J Toxicol Sci. 2012;37:539–48.

32. Tiwari D, Vanage G. Mutagenic effect of bisphenol A on adult rat male germ cells and their fertility. Reprod Toxicol. 2013;40:60–8.
33. Salian S, Doshi T, Vanage G. Neonatal exposure of male rats to bisphenol A impairs fertility and expression of sertoli cell junctional proteins in the testis. Toxicology. 2009;265:56–67.
34. Qiu LL, Wang X, Zhang XH, Zhang Z, Gu J, Liu L, Wang Y, Wang X, Wang SL. Decreased androgen receptor expression may contribute to spermatogenesis failure in rats exposed to low concentration of bisphenol A. Toxicol Lett. 2013;219:116–24.
35. Minamiyama Y, Ichikawa H, Takemura S, Kusunoki H, Naito Y, Yoshikawa T. Generation of reactive oxygen species in sperms of rats as an earlier marker for evaluating the toxicity of endocrine-disrupting chemicals. Free Radic Res. 2010;44:1398–406.
36. Chitra KC, Latchoumycandane C, Mathur PP. Induction of oxidative stress by bisphenol a in the epididymal sperm of rats. Toxicology. 2003;185:119–27.
37. Liu C, Duan W, Li R, Xu S, Zhang L, Chen C, He M, Lu Y, Wu H, Pi H, Luo X, Zhang Y, Zhong M, Yu Z, Zhou Z. Exposure to bisphenol A disrupts meiotic progression during spermatogenesis in adult rats through estrogen-like activity. Cell Death Dis. 2013;4:e676.
38. Rashid H, Ahmad F, Rahman S, Ansari RA, Bhatia K, Kaur M, Islam F, Raisuddin S. Iron deficiency augments bisphenol A-induced oxidative stress in rats. Toxicology. 2009;256:7–12.
39. Wu HJ, Liu C, Duan WX, Xu SC, He MD, Chen CH, Wang Y, Zhou Z, Yu ZP, Zhang L, Chen Y. Melatonin ameliorates bisphenol A-induced DNA damage in the germ cells of adult male rats. Mutat Res. 2013;752:57–67.
40. D'Cruz SC, Jubendradass R, Jayakanthan M, Rani SJ, Mathur PP. Bisphenol A impairs insulin signaling and glucose homeostasis and decreases steroidogenesis in rat testis: an in vivo and in silico study. Food Chem Toxicol. 2012;50:1124–33.
41. Kabuto H, Hasuike S, Minagawa N, Shishibori T. Effects of bisphenol a on the metabolisms of active oxygen species in mouse tissues. Environ Res. 2003;93:31–5.
42. Anjum S, Rahman S, Kaur M, Ahmad F, Rashid H, Ansari RA, Raisuddin S. Melatonin ameliorates bisphenol A-induced biochemical toxicity in testicular mitochondria of mouse. Food Chem Toxicol. 2011;49:2849–54.
43. Fang Y, Zhou Y, Zhong Y, Gao X, Tan T. Effect of vitamin E on reproductive functions and anti-oxidant activity of adolescent male mice exposed to bisphenol A. Wei Sheng Yan Jiu. 2013;42:18–22.
44. El-Beshbishy HA, Aly HA, El-Shafey M. Lipoic acid mitigates bisphenol A-induced testicular mitochondrial toxicity in rats. Toxicol Ind Health. 2013;29:875–87.
45. Lan HC, Wu KY, Lin IW, Yang ZJ, Chang AA, Hu MC. Bisphenol A disrupts steroidogenesis and induces a sex hormone imbalance through c-Jun phosphorylation in Leydig cells. Chemosphere. 2017;185:237–46.
46. Gonçalves GD, Semprebon SC, Biazi BI, Mantovani MS, Fernandes GSA. Bisphenol A reduces testosterone production in TM3 Leydig cells independently of its effects on cell death and mitochondrial membrane potential. Reprod Toxicol. 2018;76:26–34.
47. Xi W, Lee CK, Yeung WS, Giesy JP, Wong MH, Zhang X, Hecker M, Wong CKC. Effect of perinatal and postnatal bisphenol A exposure to the regulatory circuits at the hypothalamus-pituitary-gonadal axis of CD-1 mice. Reprod Toxicol. 2011;31:409–17.
48. Zhang T, Sun H, Kannan K. Blood and urinary bisphenol A concentrations in children, adults, and pregnant women from China: partitioning between blood and urine and maternal and fetal cord blood. Environ Sci Technol. 2013;47:4686–94.
49. Wan Y, Choi K, Kim S, Ji K, Chang H, Wiseman S, Jones PD, Khim JS, Park S, Park J, Lam MHW, Giesy JP. Hydroxylated polybrominated diphenyl ethers and bisphenol A in pregnant women and their matching fetuses: placental transfer and potential risks. Environ Sci Technol. 2010;44:5233–9.
50. Balakrishnan B, Henare K, Thorstensen EB, Ponnampalam AP, Mitchell MD. Transfer of bisphenol A across the human placenta. Am J Obstet Gynecol. 2010;202:393e1.
51. Ben Maamar M, Lesné L, Desdoits-Lethimonier C, Coiffec I, Lassurguère J, Lavoué V, Deceuninck Y, Antignac JP, Le Bizec B, Perdu E, Zalko D, Pineau C, Chevrier C, Dejucq-Rainsford N, Mazaud-Guittot S, Jégou B. An investigation of the endocrine-disruptive effects of bisphenol a in human and rat fetal testes. PLoS One. 2015;10:e0117226.

52. Lv Y, Li L, Fang Y, Chen P, Wu S, Chen X, Ni C, Zhu Q, Huang T, Lian Q, Ge RS. In utero exposure to bisphenol A disrupts fetal testis development in rats. Environ Pollut. 2019;246:217–24.

53. Hong J, Chen F, Wang X, Bai Y, Zhou R, Li Y, Chen L. Exposure of preimplantation embryos to low-dose bisphenol A impairs testesdevelopment and suppresses histone acetylation of StAR promoter toreduce production of testosterone in mice. Mol Cell Endocrinol. 2016;427:101–11.

54. Salian S, Doshi T, Vanage G. Perinatal exposure of rats to bisphenol A affects the fertility of male offspring. Life Sci. 2009;85:742–52.

55. Manikkam M, Tracey R, Guerrero-Bosagna C, Skinner MK. Plastics derived endocrine disruptors (BPA, DEHP and DBP) induce epigenetic transgenerational inheritance of obesity, reproductive disease and sperm epimutations. PLoS One. 2013;8:e55387.

56. Li DK, Zhou Z, Miao M, He Y, Wang J, Ferber J, Herrinton LJ, Gao E, Yuan W. Urine bisphenol-A (BPA) level in relation to semen quality. Fertil Steril. 2011;95:625–30.

57. Meeker JD, Ehrlich S, Toth TL, Wright DL, Calafat AM, Trisini AT, Ye X, Hauser R. Semen quality and sperm DNA damage in relation to urinary bisphenol A among men from an infertility clinic. Reprod Toxicol. 2010;30:532–9.

58. Goldstone AE, Chen Z, Perry MJ, Kannan K, Louis GM. Urinary bisphenol A and semen quality, the LIFE study. Reprod Toxicol. 2015;51:7–13.

59. Mendiola J, Jorgensen N, Andersson AM, Calafat AM, Ye X, Redmon JB, Drobnis EZ, Wang C, Sparks A, Thurston SW, Liu F, Swan SH. Are environmental levels of bisphenol A associated with reproductive function in fertile men? Environ Health Perspect. 2010;118:1286–91.

60. Hanaoka T, Kawamura N, Hara K, Tsugane S. Urinary bisphenol A and plasma hormone concentrations in male workers exposed to bisphenol A diglycidyl ether and mixed organic solvents. Occup Environ Med. 2002;59:625–8.

61. Galloway T, Cipelli R, Guralnik J, Ferrucci L, Bandinelli S, Corsi AM, Money C, McCormak P, Merlzer D. Daily bisphenol A excretion and associations with sex hormone concentrations: results from the InCHIANTI adult population study. Environ Health Perspect. 2010;118:1603–8.

62. Lassen TH, Frederiksen H, Jensen TK, Petersen JH, Joensen UN, Main KM, Skakkebaek NE, Juul A, Jørgensen N, Andersson AM. Urinary bisphenol A levels in young men: association with reproductive hormones and semen quality. Environ Health Perspect. 2014;122:478–84.

63. Dodge LE, Williams PL, Williams MA, Missmer SA, Toth TL, Calafat AM, Hauser R. Paternal urinary concentrations of parabens and other phenols in relation to reproductive outcomes among couples from a fertility clinic. Environ Health Perspect. 2015;123:665–71.

64. Buck Louis GM, Sundaram R, Sweeney AM, Schisterman EF, Maisog J, Kannan K. Urinary bisphenol A, phthalates, and couple fecundity: the longitudinal investigation of fertility and the environment (LIFE) study. Fertil Steril. 2014;101:1359–66.

65. Peretz J, Vrooman L, Ricke WA, Hunt PA, Ehrlich S, Hauser R, Padmanabhan V, Taylor HS, Swan SH, VandeVoort CA, Flaws JA. Bisphenol a and reproductive health: update of experimental and human evidence, 2007–2013. Environ Health Perspect. 2014;122:775–86.

66. Vitku J, Heracek J, Sosvorova L, Hampl R, Chlupacova T, Hill M, Sobotka V, Bicikova M, Starka L. Associations of bisphenol A and polychlorinated biphenyls with spermatogenesis and steroidogenesisin two biological fluids from men attending an infertility clinic. Environ Int. 2016;89–90:66–173.

67. EFSA COMMISSION REGULATION (EU) 2018/213 of 12 February 2018 on the use of bisphenol A in varnishes and coatings intended to come into contact with food and amending regulation (EU) no 10/2011 as regards the use WHO—endocrine disrupting chemicals. 2012.

68. Barr DB, Silva MJ, Kato K, Reidy JA, Malek NA, Hurtz D, et al. Assessing human exposure to phthalates using monoesters and their oxidized metabolites as biomarkers. Environ Health Perspect. 2003;111:1148–51.

69. Guo Y, Wu Q, Kannan K. Phthalate metabolites in urine from China, and implications for human exposures. Environ Int. 2011;37:893–8.

70. Guo Y, Weck J, Sundaram R, Goldstone AE, Louis GB, Kannan K. Urinary concentrations of phthalates in couples planning pregnancy and its association with 8-hydroxy-2'-

deoxyguanosine, a biomarker of oxidative stress: longitudinal investigation of fertility and the environment study. Environ Sci Technol. 2014;48:9804–11.

71. Skinner MK. Endocrine disruptors in 2015: epigenetic transgenerational inheritance. Nat Rev Endocrinol. 2015;12:68–70.

72. Meeker JD, Ferguson KK. Urinary phthalate metabolites are associated with decreased serum testosterone in men, women, and children from NHANES 2011-2012. J Clin Endocrinol Metab. 2014;99:4346–52.

73. Bao A-M, Man X-M, Guo X-J, Dong H-B, Wang F-Q, Sun H, et al. Effects of di-n-butyl phthalate on male rat reproduction following pubertal exposure. Asian J Androl. 2011;13:702–9.

74. Bloom MS, Whitcomb BW, Chen Z, Ye A, Kannan K, Buck Louis GM. Associations between urinary phthalate concentrations and semen quality parameters in a general population. Hum Reprod. 2015;30:2645–57.

75. Fan J, Traore K, Li W, Amri H, Huang H, Wu C, et al. Molecular mechanisms mediating the effect of mono-(2-Ethylhexyl) phthalate on hormone-stimulated steroidogenesis in MA-10 mouse tumor Leydig cells. Endocrinology. 2010;151:3348–62.

76. Gunnarsson D, Leffler P, Ekwurtzel E, Martinsson G, Liu K, Selstam G. Mono-(2-ethylhexyl) phthalate stimulates basal steroidogenesis by a cAMP-independent mechanism in mouse gonadal cells of both sexes. Reproduction. 2008;135:693–703.

77. Han X, Cui Z, Zhou N, Ma M, Li L, Li Y, et al. Urinary phthalate metabolites and male reproductive function parameters in Chongqing general population. China Int J Hyg Environ Health. 2014;217:271–8.

78. Hu Y, Dong C, Chen M, Lu J, Han X, Qiu L, et al. Low-dose monobutyl phthalate stimulates steroidogenesis through steroidogenic acute regulatory protein regulated by SF-1, GATA-4 and C/EBP-beta in mouse Leydig tumor cells. Reprod Biol Endocrinol. 2013;11:72.

79. Li Y, Hu Y, Dong C, Lu H, Zhang C, Hu Q, et al. Vimentin-Mediated Steroidogenesis Induced by Phthalate Esters: Involvement of DNA Demethylation and Nuclear Factor κB. PLoS One. 2016;11:e0146138.

80. Savchuk I, Söder O, Svechnikov K. Mono-2-Ethylhexyl phthalate stimulates androgen production but suppresses mitochondrial function in mouse Leydig cells with different steroidogenic potential. Toxicol Sci. 2015;145:149–56.

81. Chen X, YN L, Zhou QH, Leng L, Chang Y, Tang NJ. Effects of low concentrations of Di-(2-ethylhexyl) and mono-(2-ethylhexyl) phthalate on steroidogenesis pathways and apoptosis in the murine Leydig tumor cell line MLTC-1. Biomed Environ Sci. 2013;26:986–9.

82. Dees JH, Gazouli M, Papadopoulos V. Effect of mono-ethylhexyl phthalate on MA-10 Leydig tumor cells. Reprod Toxicol. 2001;15:171–87.

83. Fiandanese N, Borromeo V, Berrini A, Fischer B, Schaedlich K, Schmidt J-S, et al. Maternal exposure to a mixture of di(2-ethylhexyl) phthalate (DEHP) and polychlorinated biphenyls (PCBs) causes reproductive dysfunction in adult male mouse offspring. Reprod Toxicol. 2016;65:123–32.

84. Wolff MS, Engel SM, Berkowitz GS, Ye X, Silva MJ, Zhu C, et al. Prenatal phenol and phthalate exposures and birth outcomes. Environ Health Perspect. 2008;116:1092–7.

85. Pant N, Pant A, Shukla M, Mathur N, Gupta Y, Saxena D. Environmental and experimental exposure of phthalate esters: the toxicological consequence on human sperm. Hum Exp Toxicol. 2011;30:507–14.

86. Desdoits-Lethimonier C, Albert O, Le Bizec B, Perdu E, Zalko D, Courant F, et al. Human testis steroidogenesis is inhibited by phthalates. Hum Reprod. 2012;27:1451–9.

87. Hauser R, Sokol R. Science linking environmental contaminant exposures with fertility and reproductive health impacts in the adult male. Fertil Steril. 2008;89:e59–65.

88. Duty SM, Silva MJ, Barr DB, Brock JW, Ryan L, Chen Z, et al. Phthalate exposure and human semen parameters. Epidemiology. 2003;14:269–77.

89. Hauser R, Meeker JD, Singh NP, Silva MJ, Ryan L, Duty S, et al. DNA damage in human sperm is related to urinary levels of phthalate monoester and oxidative metabolites. Hum Reprod. 2007;22:688–95.
90. Duty SM, Calafat AM, Silva MJ, Brock JW, Ryan L, Chen Z, et al. The relationship between environmental exposure to phthalates and computer-aided sperm analysis motion parameters. J Androl. 2004;25:293–302.
91. Hauser R, Meeker JD, Duty S, Silva MJ, Calafat AM. Altered semen quality in relation to urinary concentrations of phthalate monoester and oxidative metabolites. Epidemiology. 2006;17:682–91.
92. Liu L, Bao H, Liu F, Zhang J, Shen H. Phthalates exposure of Chinese reproductive age couples and its effect on male semen quality, a primary study. Environ Int. 2012;42:78–83.
93. Meeker JD, Calafat AM, Hauser R. Urinary metabolites of Di(2-ethylhexyl) phthalate are associated with decreased hormone levels in adult men. J Androl. 2009;30(3):287–97.
94. Kay VR, Bloom MS, Foster WG. Reproductive and developmental effects of phthalate diesters in males. Crit Rev Toxicol. 2014;44:467–98.
95. Rusyn I, Peters JM, Cunningham ML. Modes of action and species-specific effects of di-(2-ethylhexyl)phthalate in the liver. Crit Rev Toxicol. 2006;36:459–79.
96. Conder JM, Hoke RA, De Wolf W, Russell MH, Buck RC. Are PFCAs bioaccumulative? A critical review and comparison with regulatory criteria and persistent lipophilic compounds. Environ Sci Technol. 2008;42:995–1003.
97. Steenland K, Zhao L, Winquist A. A cohort incidence study of workers exposed to perfluorooctanoic acid (PFOA). Occup Environ Med. 2015;72:373–80.
98. Biegel LB, Liu RCM, Hurtt ME, Cook JC. Effects of ammonium Perfluorooctanoate on Leydig-cell function: in vitro, in vivo, and ex vivo studies. Toxicol Appl Pharmacol. 1995;134:18–25.
99. Shi Z, Zhang H, Liu Y, Xu M, Dai J. Alterations in gene expression and testosterone synthesis in the testes of male rats exposed to Perfluorododecanoic acid. Toxicol Sci. 2007;98:206–15.
100. Wan HT, Zhao YG, Wong MH, Lee KF, Yeung WSB, Giesy JP, et al. Testicular signaling is the potential target of Perfluorooctanesulfonate-mediated subfertility in male Mice1. Biol Reprod. 2011;84:1016–23.
101. Zhang H, Lu Y, Luo B, Yan S, Guo X, Dai J. Proteomic analysis of mouse testis reveals perfluorooctanoic acid-induced reproductive dysfunction via direct disturbance of testicular steroidogenic machinery. J Proteome Res. 2014;13:3370–85.
102. Kang JS, Choi JS, Park JW. Transcriptional changes in steroidogenesis by perfluoroalkyl acids (PFOA and PFOS) regulate the synthesis of sex hormones in H295R cells. Chemosphere. 2016;155:436–43.
103. López-Doval S, Salgado R, Pereiro N, Moyano R, Lafuente A. Perfluorooctane sulfonate effects on the reproductive axis in adult male rats. Environ Res. 2014;134:158–68.
104. Pereiro N, Moyano R, Blanco A, Lafuente A. Regulation of corticosterone secretion is modified by PFOS exposure at different levels of the hypothalamic–pituitary–adrenal axis in adult male rats. Toxicol Lett. 2014;230:252–62.
105. Qiu L, Zhang X, Zhang X, Zhang Y, Gu J, Chen M, et al. Sertoli cell is a potential target for perfluorooctane sulfonate-induced reproductive dysfunction in male mice. Toxicol Sci. 2013;135:229–40.
106. Jensen AA, Leffers H. Emerging endocrine disrupters: perfluoroalkylated substances. Int J Androl. 2008;31:161–9.
107. Zhang Y, Beesoon S, Zhu L, Martin JW. Biomonitoring of perfluoroalkyl acids in human urine and estimates of biological half-life. Environ Sci Technol. 2013;47:10619–27.
108. López-Doval S, Salgado R, Lafuente A. The expression of several reproductive hormone receptors can be modified by perfluorooctane sulfonate (PFOS) in adult male rats. Chemosphere. 2016;155:488–97.
109. Šabović I, Cosci I, De Toni L, Ferramosca A, Stornaiuolo M, Di Nisio A, Dall'Acqua S, Garolla A, Foresta C. Perfluoro-octanoic acid impairs sperm motility through the alteration of plasma membrane. J Endocrinol Investig. 2020;43(5):641–52.

110. Olsen GW, Lange CC, Ellefson ME, Mair DC, Church TR, Goldberg CL, et al. Temporal trends of Perfluoroalkyl concentrations in American red cross adult blood donors, 2000–2010. Environ Sci Technol. 2012;46:6330–8.
111. Raymer JH, Michael LC, Studabaker WB, Olsen GW, Sloan CS, Wilcosky T, et al. Concentrations of perfluorooctane sulfonate (PFOS) and perfluorooctanoate (PFOA) and their associations with human semen quality measurements. Reprod Toxicol. 2012;33:419–27.
112. Kubwabo C, Kosarac I, Lalonde K. Determination of selected perfluorinated compounds and polyfluoroalkyl phosphate surfactants in human milk. Chemosphere. 2013;91:771–7.
113. Kim S, Choi K, Ji K, Seo J, Kho Y, Park J, et al. Trans-placental transfer of thirteen Perfluorinated compounds and relations with fetal thyroid hormones. Environ Sci Technol. 2011;45:7465–72.
114. Skakkebaek NE, Rajpert-De Meyts E, Main KM. Testicular dysgenesis syndrome: an increasingly common developmental disorder with environmental aspects. Hum Reprod. 2001;16:972–8.
115. Vested A, Ramlau-Hansen CH, Olsen SF, Bonde JP, Kristensen SL, Halldorsson TI, et al. Associations of in utero exposure to perfluorinated alkyl acids with human semen quality and reproductive hormones in adult men. Environ Health Perspect. 2013;121:453–8.
116. Ingelido AM, Abballe A, Gemma S, Dellatte E, Iacovella N, De Angelis G, et al. Biomonitoring of perfluorinated compounds in adults exposed to contaminated drinking water in the Veneto region. Italy Environ Int. 2018;110:149–59.
117. Joensen UN, Bossi R, Leffers H, Jensen AA, Skakkebæk NE, Jørgensen N. Do Perfluoroalkyl compounds impair human semen quality? Environ Health Perspect. 2009;117:923–7.
118. Louis GMB, Chen Z, Schisterman EF, Kim S, Sweeney AM, Sundaram R, et al. Perfluorochemicals and human semen: the LIFE study. Environ Health Perspect. 2015;123(1):57–63.
119. Toft G, Jönsson BAG, Lindh CH, Giwercman A, Spano M, Heederik D, et al. Exposure to perfluorinated compounds and human semen quality in arctic and European populations. Hum Reprod. 2012;27:2532–40.
120. Specht IO, Hougaard KS, Spanò M, Bizzaro D, Manicardi GC, Lindh CH, et al. Sperm DNA integrity in relation to exposure to environmental perfluoroalkyl substances—a study of spouses of pregnant women in three geographical regions. Reprod Toxicol. 2012;33(4):577–83.
121. Governini L, Guerranti C, De Leo V, Boschi L, Luddi A, Gori M, et al. Chromosomal aneuploidies and DNA fragmentation of human spermatozoa from patients exposed to perfluorinated compounds. Andrologia. 2015;47:1012–9.
122. La Rocca C, Alessi E, Bergamasco B, Caserta D, Ciardo F, Fanello E, et al. Exposure and effective dose biomarkers for perfluorooctane sulfonic acid (PFOS) and perfluorooctanoic acid (PFOA) in infertile subjects: preliminary results of the PREVIENI project. Int J Hyg Environ Health. 2012;215:206–11.
123. La Rocca C, Tait S, Guerranti C, Busani L, Ciardo F, Bergamasco B, et al. Exposure to endocrine disruptors and nuclear receptors gene expression in infertile and fertile men from Italian areas with different environmental features. Int J Environ Res Public Health. 2015;12:12426–45.
124. Foresta C, Tescari S, Di Nisio A. Impact of perfluorochemicals on human health and reproduction: a male's perspective. J Endocrinol Investig. 2018;41(6):639–45.
125. Di Nisio A, Sabovic I, Valente U, Tescari S, Rocca MS, Guidolin D, Dall'Acqua S, Acquasaliente L, Pozzi N, Plebani M, Garolla A, Foresta C. Endocrine disruption of androgenic activity by Perfluoroalkyl substances: clinical and experimental evidence. J Clin Endocrinol Metab. 2019;104(4):1259–71.
126. Matsushima A, Kakuta Y, Teramoto T, Koshiba T, Liu X, Okada H, Tokunaga T, Kawabata SI, Kimura M, Shimohigashi Y. Structural evidence for endocrine disruptor bisphenol a binding to human nuclear receptor ERR gamma. J Biochem. 2007;142:517–24.
127. Sifakis S, Androutsopoulos VP, Tsatsakis AM, Spandidos DA. Human exposure to endocrine disrupting chemicals: effects on the male and female reproductive systems. Environ Toxicol Pharmacol. 2017;51:56–70.

Chapter 8
Endocrine-Disrupting Chemicals (EDCs) and Reproductive Outcomes

Arianna D'Angelo (iD) **and Georgina St Pier**

8.1 Introduction

Infertility can be defined as 'a disease characterized by the failure to establish a clinical pregnancy after 12 months of regular and unprotected sexual intercourse'. [1] In the past 50 years, there has been a steady yet undeniable rise of this condition, [2, 3] with one in six couples hoping to conceive being diagnosed as infertile [4]. There are many well-recognised causes of infertility, including advanced age, sexually transmitted infections, endocrine disorders such as polycystic ovarian syndrome and male factor [5–7]. However, in more than 10% of couples, the cause of infertility is unknown, [8] and the prevalence of such unexplained subfertility is increasing [7, 9]. In this condition, women have apparently normal ovulatory cycles, normal hormonal levels and no obvious reproductive disease. Likewise, their male partners have apparently normal semen quality, yet conception does not occur.

One explanation for this increase in unexplained subfertility could be the increasing prevalence of endocrine-disrupting chemicals (EDCs). These are exogenous chemicals that have the ability to disturb the normal endocrine function of humans, [3] thus interfering with normal development, homeostasis and reproduction [10]. EDCs are a diverse group of substances used extensively within the manufacturing, industrial and agricultural industries [11]. The most important EDCs are bisphenol A

A. D'Angelo (✉)
Reproductive Medicine, Wales Fertility Institute, Cardiff, UK

Cardiff University, Cardiff, UK

University Hospital of Wales Heath Park, Cardiff, UK
e-mail: dangeloa@cardiff.ac.uk

G. St. Pier
Cardiff University, Cardiff, UK

Ysbyty Gwynedd, Bangor, UK
e-mail: georgina.stpier2@wales.nhs.uk

© The Author(s) 2023
R. Marci (ed.), *Environment Impact on Reproductive Health*,
https://doi.org/10.1007/978-3-031-36494-5_8

(BPA), phthalates (both used in plastic manufacturing), triclosan (widely used in pharmaceuticals and personal care products) and parabens (used as food preservatives and in cosmetics). Human exposure to EDCs is almost ubiquitous as it takes many forms—mostly via ingestion, but also through inhalation or dermal uptake [12]. Once in the body, most are able to bioaccumulate within adipose tissue; their long half-life results in prolonged exposure to these substances [13]. Furthermore, EDCs are ubiquitous in nature, resulting in a combination of different chemicals accumulating; [14–16] these cumulative effects perhaps worse than those individually.

As the production of synthetic chemicals dramatically rises worldwide, [14, 17, 18] questions have arisen regarding their effects on human health, including reproductive health. It has been widely shown that environmental pollution negatively impacts the fertility of all mammalian species [1, 5]. However, in more recent years, substantial evidence has accumulated showing a negative association specifically between EDC exposure and both male and female fertility and fecundity [11, 15]. Animal studies have demonstrated that in females, ovulation, meiosis, the number of follicles present in the ovaries and embryonic implantation can all be impacted by such chemicals [1]. Whilst comparative effects are also seen in males—effects such as hypospadias and poor semen quality have been noted—the most severe consequences on fertility are seen in females; as the number of oocytes is fixed at birth, there is no possibility for replacement. Whilst the overall pathophysiological mechanisms of EDCs in relation to fertility are not fully understood, [16, 19] studies have shown that they are able to competitively bind with hormone receptors, exerting adverse biological functions. For example, BPA and triclosan have similar structure to 17β-oestradiol [3] and are therefore able to interfere with oestrogen signalling pathways [11, 14].

Perhaps unsurprisingly, as the prevalence of infertility has risen worldwide, so too has demand for services providing assisted reproductive technology (ART), including in vitro fertilisation (IVF) [20–22]. Having established that EDCs negatively impact human fertility, it is prudent to understand their possible impact on fertility treatment. This chapter looks at the effects of specific EDCs on the outcomes of IVF treatment and other forms of ART, as well as their impact on the pregnancy outcomes.

8.2   Impact of EDCs on IVF Outcomes

8.2.1   Bisphenol A

Bisphenol A (BPA) is perhaps the most thoroughly examined EDC. Its extensive presence in everyday products including plastics, medical equipment, the epoxy lining of food and drink containers and dental sealants mean that the general population are widely exposed to its effects [23, 24]. Indeed, one study found up to 95% of patients tested had detectable levels of BPA in urine samples, [11] and it has also been detected in the follicular fluid of most women commencing IVF treatment

[25]. Unlike other EDCs, BPA has a comparably short half-life and is not prone to bioaccumulation, with almost complete urinary excretion within 24 hours, [13, 15] though this does not mitigate its harmful effects; studies show correlations between increased levels of BPA in adult urine samples and increased incidence of obesity, cardiovascular disease and Type II diabetes mellitus [11].

In relation to fertility, it is thought BPA inhibits aromatase activity, thereby inhibiting oestrogen synthesis and disrupting ovarian folliculogenesis and implantation [24]. This therefore directly impacts not only female fertility but also the outcomes of any subsequent fertility treatment. Several studies have shown relationships between increased exposure to BPA in females and poor outcomes of IVF treatment: specifically, serum BPA levels have been negatively associated with oocyte retrieval, oocyte maturation, fertilisation rates and embryo quality [26–28].

One of the seminal studies in this field was the Environmental and Reproductive Health (EARTH) study. Researchers consistently found a correlation between levels of BPA in female urine samples and decreased peak serum oestradiol levels and the number of both total and mature oocytes retrieved [25, 29]. There was also an association between BPA level and the rate of normally fertilised oocytes, with a decrease of 24% and 27%, respectively, for the highest versus lowest quartiles of urinary BPA levels. Additionally, increased urinary BPA concentrations are associated with a reduced successful implantation rate, with these trends showing positive linear dose–response associations [30]. This finding has since been confirmed by multiple other studies [13, 26, 27, 31, 32]. Increased miscarriage rates, post-non successful treatment, have also been reported, [11, 13] alongside increased rates of premature birth [13].

Whilst ethical challenges make it difficult to carry out experimental studies in humans, evidence available from experimental animal studies suggests a similar picture. In female mice, sub-chronic low-dose BPA exposure has been linked to diminished ovarian reserve [33]. Other animal studies have suggested association between BPA and spindle abnormalities, as well as impairment of follicular growth and implantation [11, 32]. Alongside antagonisation of oestrogen pathways, BPA has been shown to alter animal uterine morphology and reduce uterine receptivity preimplantation, [18] as well as impairing steroidogenesis in rat ovarian theca-interstitial and granulosa cells in vitro [34]. The long-term effects of these are unknown, but potentially trans-generational [11].

However, despite the effects of BPA upon significant IVF outcomes including oocyte retrieval rate and implantation rate, the arguably more important outcomes of clinical pregnancy and live birth appear to be unaffected in some studies [15].

8.2.2 Phthalates

Phthalates are a class of synthetic chemicals used extensively as plasticisers and solvents and within food processing and personal care products [11, 24, 35–37]. As a group of substances, there is ever-increasing evidence regarding their effect on human fertility and fecundity, including IVF outcomes. Epidemiological evidence has suggested some women of reproductive age are at particular risk due to daily use of cosmetics and other personal care items [18, 36]. As with BPA, phthalates have a ubiquitous presence, evidenced by the presence of nine out of the 13 phthalate metabolites measured in the urine of more than 99% of people [38]. They are also detectable in amniotic fluid [11] and are able to cross via placental transfer to a developing foetus and via breastmilk to neonates [18]. This may have profound implications for the development of the endocrine system of the affected child, potentially impacting their later reproductive functions.

As with BPA outcomes, in women undergoing fertility treatment, phthalate levels are associated with increased risk of implantation failure [24, 39]. The isotope di-2-exthylhexyl phthalate (DEHP) is the most well-studied phthalate, due to its concerning anti-androgenic properties [24, 35, 37]. Increased urinary DEHP levels have been associated with: decreased antral follicle counts; [11] decreased total and mature oocyte yield during IVF; [11, 15, 23, 24] decreased oocyte quality [24] and reduced probability of clinical pregnancy and live birth following IVF [11, 15, 24]. These studies suggest phthalates have the worrying potential to lower the probability of clinical pregnancy and live birth following IVF treatment—unlike BPA, which only appears to affect earlier ART outcomes. Other phthalate isotypes have also been associated with poor IVF outcomes: monobenzyl phthalate (MBzP) and mono-n-butyl phthalate (MBP) concentrations are associated with decreased blastocyst quality, [37] and urinary monoethyl phthalate (MEP) levels were associated with clinical pregnancy loss and preterm birth in some studies (though other studies found no associations) [23].

Several animal studies provide further evidence of phthalates acting as toxicants to the female reproductive system; [15] they have been shown to disrupt ovarian function and negatively impact folliculogenesis and steroidogenesis [11, 24, 36]. In female rats, DEHP was associated with reduced serum oestradiol and progesterone levels [28, 35]. Likewise, in male rats, exposure to DEHP decreased androgen production and altered sexual differentiation [40]. However, unlike BPA, which has been found in higher concentrations in women requiring infertility treatment compared to those who do not, [26] there have been no associations found between increased urinary concentrations of phthalate metabolites and the need for fertility treatment [35].

A study by Du et al. (2019) found that the follicular fluid concentration of certain phthalate metabolites was associated with altered levels of intrafollicular reproductive hormones—including oestradiol and testosterone—in women undergoing IVF [36]. This is concerning, and the effects of phthalates on theca and granulosa cells within the follicle may explain the adverse outcomes associated with phthalate exposure and IVF, such as decreased rates of implantation, clinical pregnancy and live birth. Interestingly, a separate study by Wu et al. (2017) found that male urinary concentrations of phthalate metabolites—but not female—were associated with reduced blastocyst quality in the IVF setting [37].

In contrast, a study by Alur et al. (2015) found that amongst women with a history of infertility, those who conceived after ART had statistically significantly lower first-trimester urinary concentrations of DEHP metabolites compared to those who conceived without ART [35]. The authors suggest that these slightly surprising results may be due to patients undergoing ART pursuing 'healthier' lifestyle choices and thereby reducing their exposure to phthalates; the majority of DEHP exposure is through via dietary intake of typically unhealthier, processed food high in animal fat, such as beef, pork and cheese [41, 42].

Overall, the potential impact of phthalates to influence both early and late IVF outcomes cannot be ignored. However, all three studies cited above consist of relatively small sample sizes, reducing the robustness of the results. Further research is needed—ideally systematic reviews or meta-analyses—to provide reliable conclusions regarding the unique actions and effect of the individual phthalate subtypes.

8.2.3 Triclosan

Triclosan is a compound with antimicrobial activity, resulting in widespread use within products such as disinfectants, soaps, deodorants and toothpaste, [15, 23, 24, 43, 44] and also within medical devices [44]. Triclosan has been detected in up to in 98% of urine samples from pregnant women [45] and has also been found in the blood, breast milk and adipose tissue of humans; [24] this is unsurprising given its widespread presence. Its bacteriostatic properties mean triclosan has typically been considered safe for human use; however, emerging evidence suggests that it has the potential to detrimentally impact both thyroid and sex hormone homeostasis [11, 24, 46] leading to debate regarding its necessity in household products [44].

There are few epidemiological studies measuring the relationship between triclosan and ART outcomes. The largest and most recent study found that urinary triclosan concentrations were associated with decreased implantation rate. No effect was seen for the other IVF outcomes assessed, namely: metaphase II oocyte count, quality of embryos, rate of fertilisation and rate of clinical pregnancy [46]. Other human studies assessing early IVF outcomes have differing results: Hua et al. (2017) found triclosan affected top quality embryo and implantation rate [47], whereas Lange et al. (2015) found implantation rate to be unaffected, although oocyte yield was seen to decrease with increased triclosan exposure [48]. In addition, the synergistic effects of triclosan and BPA are thought to influence oocyte development and quality, depending on the time of exposure to the mother [18]. Evidence regarding the effects of triclosan on ART outcomes is therefore scarce and inconclusive, with small sample sizes making results hard to compare. Further research is needed amongst larger, diverse populations to provide further evidence regarding the impact of triclosan on reproductive outcomes.

8.2.4   Parabens

Parabens are a class of EDC used primarily as antimicrobial preservatives within pharmaceuticals, personal care products and food [11, 24, 49]. These chemicals exhibit weak oestrogenic properties, [18, 24, 50] and urine sampling has shown women to have a five-fold higher paraben concentration than men [51]—likely due to their extensive presence in cosmetics [52]. However, despite their widespread presence, there are very few studies investigating the association between parabens and fertility, fecundity and fertility treatment.

Existing data has shown that exposure to mixtures of parabens and phenols can alter ovarian hormone levels, [18] likely through their binding to the oestrogen receptors ER-α and ER-β, [53, 54] potentially impacting any IVF outcomes. One study found an association between increasing urinary propyl paraben levels and decreasing antral follicle and oocyte count in women undergoing IVF treatment, suggesting an association between paraben exposure and diminished ovarian reserve [55]. Further studies have shown an association between urinary methyl and propyl paraben levels and poor embryo quality only, [56] with no statistically significant effects seen upon other IVF outcomes, including oocyte retrieval rate, cleavage rate, implantation, [56] or fertilisation and live birth rates [57].

In summary, data surrounding this topic is scarce. There is an as yet unmet need for further human studies examining the potential effects of parabens on fertility and ART, especially given their abundant presence within the general population. In the meantime, patients undergoing IVF may prefer to minimise their exposure to parabens where possible.

8.2.5   Persistent Organic Pollutants

Persistent organic pollutants (POPs) are a large group of EDCs, with various sub-groups including polychlorinated biphenyls (PCBs) and perfluoroalkyl and poly-fluoroalkyl substances (PFAS). As suggested by their name, POPs are incredibly stable molecules that resist degradation and are toxic to both humans and animals. This stability results in their bioaccumulation, usually within the adipose tissue of humans and animals [11, 14]. POPs are known to have oestrogenic and anti-androgenic hormone action, impacting both male and female fertility [11, 24, 58]. Prior to a worldwide ban on PCBs in 1979, [12] they were used in fluids for electrical equipment, lubricants and paints [11]. However, these chemicals have long half-lives of up to 5 years, [59] resulting in persistent environmental exposure to these chemicals, [11, 14] primarily through contaminated food (particularly meat, fish and dairy) but also via inhalation and dermal contact with contaminated soil [24]. In contrast, PFAS are still used today in a wide range of products including textiles, pesticides, personal care products and firefighting foams [11, 18, 58, 60].

A recent review of the evidence by Bjorvang et al. (2020) found conflicting evidence in relation to the effect of POPs on IVF outcomes; [14] some studies found no associations, whilst others found the most lipophilic POPs were associated with poorer oocyte quality [61–63] and reduced fertilisation rate [62–64]. Increasing serum PCB concentration has also been associated with lower rates of high-quality embryos, lower implantation rates and a lower live birth rate in women undergoing IVF treatment [62, 65]. This is in contrast to earlier studies which have suggested chronic exposure to low levels of PCBs does not impact human reproductive [66] or IVF outcomes [67].

The presence of POPs is not restricted to the mother; however, these chemicals are able to cross the placenta, enter foetal circulation and are also present in breast milk [24, 68]. The potential long-term effects of these on the health of the newborn are concerning.

Further experimental and epidemiological studies are clearly needed to clarify the associations seen, which are limited and conflicting. Despite sharing a common chemical structure, each isoform has different biological effects, [69] which limits the extent broad comparisons that can be made across the groups. Efforts should be made to identify exactly which POPs provide the most, if any, risk to human fertility.

8.2.6 Implications for Practice

Studying the impact of EDCs on fertility, fecundity and IVF outcomes is by no means an easy process. Human fertility is dependent on multiple factors, underpinned by the complex underlying mechanisms of processes such as folliculogenesis, spermatogenesis and endometrial receptivity. One of the biggest challenges when examining the effects of EDCs is that these chemicals are vast in number and have varying different biochemical effects—the impact of one cannot be generalised to the impact of them all. Their ubiquitous nature further complicates matters, as these compounds are difficult to avoid and patients will be exposed to many at once. The presence of multiple different chemicals within the human body therefore confounds the effects of any individual chemical studied. For older women, who are more likely to seek ART to conceive due to diminished ovarian reserve, they will experience longer cumulative exposure to these chemicals which may further reduce the chances of pregnancy. Future research needs to focus on the effects of multiple EDCs in combination, as this is more representative of real-life exposure.

Although the literature is growing, particularly examining BPA exposure, for other chemicals such as POPs the evidence is scarce and conflicting. Epidemiological studies have shown associations; however, future studies should be experimental in nature in order to prove causality and mechanisms of action.

Overall, moderate evidence supports associations between increasing EDC concentration, diminishing ovarian reserve and poor IVF outcomes. Women of a reproductive age should therefore minimise exposure to these chemicals where possible, with healthcare professionals actively advocating their avoidance where possible. Clinical practitioners working in reproductive medicine should be educated about the impacts of these endocrine disruptors and be trained to support patients in how to reduce and mitigate their effects. At a population level, social awareness campaigns are needed to advise on the use of consumer products, particularly personal care products, and healthy diets. Arguably, the most important action needed to reduce their impact, however, is urgent worldwide government regulation of EDC use. These measures will help to achieve the ultimate goal of ART treatment for as many patients as possible: the birth of a healthy baby.

8.3   Impact of EDCs on Pregnancy Outcomes

Outside of fertility treatments, there is increasing evidence to suggest EDCs can also impact pregnancies conceived naturally. Maternal exposure to EDCs has been identified as a risk factor for many complications of pregnancy, including recurrent miscarriage, intrauterine growth restriction, gestational hypertensive disorders and pre-term birth [70]. The specific mechanisms by which EDCs cause these adverse effects are unknown; however, it is thought EDCs are able to either accumulate within or act upon the placenta, in order to regulate signalling pathways within trophoblasts resulting in altered cell viability and invasiveness [71]. It is also possible that they trigger an inappropriate immune response in the mother, causing inflammation and oxidative stress [70, 72]. Overall, resultant abnormal placentation can lead to gestational disorders including pre-eclampsia and intrauterine growth restriction, [71, 73] significantly impacting both maternal and foetal health. EDCs can also cause chromosomal abnormalities within oocytes and affect embryo development and implantation, resulting in early pregnancy loss [74].

8.3.1   Miscarriage

Miscarriage—the loss of a pregnancy before 20 weeks gestation—is the most common complication of pregnancy, [75] occurring in 15–25% of all clinically recognised pregnancies [76]. Whilst the majority of sporadic losses are due to chromosomal abnormalities in the foetus, other factors including maternal anatomy, endocrine system and psychological status are also known to influence viability. Environmental exposure is another important factor; it is thought that exposure to air pollutants, including carbon monoxide and particulate matter, is associated with an increased risk of both miscarriage and stillbirth [77]. A similar pattern of evidence is emerging regarding exposure to EDCs. One of the earliest studies in this area found that maternal exposure to BPA was associated with recurrent miscarriage [78]. Laboratory models have confirmed this, with showing that BPA increases the risk of miscarriage at the level of both the endometrium and the oocyte, [75] by affecting the differentiation of uterine stromal to decidual cells [79]. Furthermore, certain phthalate metabolites have also been associated with higher rates of miscarriage, [80, 81] as have the PFAS perfluorooctanoic acid (PFOA) and perfluoroheptane sulfonate (PFHpS) [82]. In contrast, other EDCs including PCBs have not been associated with pregnancy loss, in patients with a history of recurrent miscarriage [83]. However, it is important to recognise that the evidence for all EDCs under consideration is poor; largely comprising of small studies with weak statistical power. Larger epidemiological studies are needed to confirm the effects seen, alongside mechanistic studies in order to understand the mechanism behind the EDCs, this may then allow for the development of therapeutics to mitigate their effects, enabling mothers to have a better chance of a viable pregnancy.

8.3.2   Preeclampsia

Hypertensive disorders of pregnancy, including preeclampsia, pose a significant risk to the health of both mother and foetus; an estimated 10–15% of all maternal deaths are associated with preeclampsia and eclampsia [84]. This complex disease is associated with improper placentation and remodelling of the uterine spiral arteries [85], although the pathophysiology is not completely understood. Evidence is emerging regarding exposure to environmental toxicants as risk factors for the development of gestational hypertensive disorders. Strong evidence already exists to show significant associations between maternal exposure to chemicals including cadmium and lead and the development of preeclampsia [84, 86, 87]. In contrast, the evidence in regard to endocrine disruptors is scarce and conflicting. Six PFAS subtypes have shown positive associations with hypertensive disorder development in several studies, but not all [88]. Inconclusive results have also been found for BPA and phthalates [75, 87, 89]. However, mechanistic studies have shown EDCs such as BPA and PFAS are able to cause abnormal trophoblast invasion, impaired placental function and placental ischaemia [71, 90]. Given what is understood about the aetiology of preeclampsia, it would fit that EDCs may be able to mediate increased incidence of this disease; it is therefore imperative that future research into this area is prioritised.

8.3.3   Infant Birth Weight

Birth size acts as a significant predictor for future health. A low birth weight is associated with increased development of obesity, cardiovascular disease, and non-insulin-dependent diabetes in later life, [91, 92] and these infants are also more likely to need immediate healthcare in the days following birth. Conflicting evidence exists regarding the impact endocrine disruptors might have on infant birth weight. For example, maternal urinary concentration of several different phenols has been associated with decreased birth weight in male infants, with no such association seen for phthalate metabolites [93, 94]. In contrast, BPA alone has been associated with increased head circumference in the same population [93]. In female infants, antenatal exposure to PFAS has been identified as the EDC class to have the strongest association with reduced birth weight, [91] though several other studies have failed to find significant associations for other EDCs [95, 96]. A recent systematic review examining the role of BPA and foetal growth velocity found mounting evidence to suggest that BPA exposure leads to foetal growth restriction and reduced birth size, particularly when exposure occurs in the first half of pregnancy [97]. Additionally, exposure to EDCs including phthalates and PFOS have been associated with a significantly increased chance of premature delivery [98, 99], which is likely to lead to reduced birth weight and is an independent cause of neonatal mortality. Despite the contradictory evidence, as with the previous outcomes discussed, it should be recommended that exposure to endocrine disruptors is as minimal as possible during pregnancy. This reduction in exposure and hopeful subsequent reduction in incidence of low birth weight would have a significant public health impact, given its downstream consequences.

8.3.4   Cholestasis

There is very little evidence pertaining to associations between EDCs and intra-hepatic cholestasis of pregnancy (ICP). ICP causes reversible liver dysfunction from the third trimester of pregnancy until delivery in between 0.5% and 1.8% of pregnancies in Europe [100]. It is characterised by elevated serum bile acid and transaminase levels, with resultant maternal pruritis, particularly on the palms and soles. There is a little evidence to suggest that maternal exposure to EDCs acts as a risk factor for the development of ICP. For example, a recent cohort study found an association between higher urine levels of certain phthalates in the first trimester and increased incidence of ICP [101]. Further research is clearly needed to determine the impact exposure to EDCs may have on the development of ICP, and by what mechanism.

8.4   Conclusion

It is clear that endocrine-disrupting chemicals pose significant risk to the reproductive health of both males and females, with potential downstream effects lasting for years to come. Although evidence is frequently conflicting, it is undeniable that public health is significantly threatened by these offending chemicals. Identifying the risk of a single EDC is challenging and relatively meaningless, as we are exposed to many hundreds of these substances from conception. Further research is therefore needed to examine the impact of different mixtures of EDCs in combination and to understand their mechanisms of action. In the meantime, a strategy is urgently needed to prevent the exposure to these chemicals, especially in women of child-bearing age, to mitigate their impact on the outcomes of pregnancy and assisted reproduction. This will not only improve public health but ensure as many parents as possible achieve reproductive success with the birth of healthy babies.

References

1. Vander Borght M, Wyns C. Fertility and infertility: definition and epidemiology. Clin Biochem. 2018;62:2–10.
2. Conforti A, Mascia M, Cioffi G, De Angelis C, Coppola G, De Rosa P, et al. Air pollution and female fertility: a systematic review of literature. Reprod Biol Endocrinol. 2018;16(1):117.
3. Yuan M, Bai M-Z, Huang X-F, Zhang Y, Liu J, Hu M-H, et al. Preimplantation exposure to Bisphenol A and Triclosan may Lead to implantation failure in humans. Biomed Res Int. 2015;2015:1.
4. Ravitsky V, Kimmins S. The forgotten men: rising rates of male infertility urgently require new approaches for its prevention, diagnosis and treatment. Biol Reprod. 2019;101(5):872–4.
5. Canipari R, De Santis L, Cecconi S. Female fertility and environmental pollution. Int J Environ Res Public Health. 2020;17(23):8802.
6. Punab M, Poolamets O, Paju P, Vihljajev V, Pomm K, Ladva R, et al. Causes of male infertility: a 9-year prospective monocentre study on 1737 patients with reduced total sperm counts. Hum Reprod. 2017;32(1):18–31.
7. Li Q, Zheng D, Wang Y, Li R, Wu H, Xu S, et al. Association between exposure to airborne particulate matter less than 2.5 μm and human fecundity in China. Environ Int. 2021;146:106231.
8. Altmäe S, Haller K, Peters M, Hovatta O, Stavreus-Evers A, Karro H, et al. Allelic estrogen receptor 1 (ESR1) gene variants predict the outcome of ovarian stimulation in in vitro fertilization. Mol Hum Reprod. 2007;13(8):521–6.
9. Talmor A, Dunphy B. Female obesity and infertility. Best Pract Res Clin Obstet Gynaecol. 2015;29(4):498–506.
10. Diamanti-Kandarakis E, Bourguignon J-P, Giudice LC, Hauser R, Prins GS, Soto AM, et al. Endocrine-disrupting chemicals: an Endocrine Society scientific statement. Endocr Rev. 2009;30(4):293–342.
11. Green MP, Harvey AJ, Finger BJ, Tarulli GA. Endocrine disrupting chemicals: impacts on human fertility and fecundity during the peri-conception period. Environ Res. 2021;194:110694.
12. Gore AC, Chappell VA, Fenton SE, Flaws JA, Nadal A, Prins GS, et al. EDC-2: the Endocrine Society's second scientific statement on endocrine-disrupting chemicals. Endocr Rev. 2015;36(6):E1–E150.
13. Yilmaz B, Terekeci H, Sandal S, Kelestimur F. Endocrine disrupting chemicals: exposure, effects on human health, mechanism of action, models for testing and strategies for prevention. Rev Endocr Metab Disord. 2020;21(1):127–47.
14. Björvang RD, Damdimopoulou P. Persistent environmental endocrine-disrupting chemicals in ovarian follicular fluid and in vitro fertilization treatment outcome in women. Ups J Med Sci. 2020;125(2):85–94.
15. Mínguez-Alarcón L, Gaskins AJ. Female exposure to endocrine disrupting chemicals and fecundity: a review. Curr Opin Obstet Gynecol. 2017;29(4):202–11.
16. Dominguez F. Phthalates and other endocrine-disrupting chemicals: the 21st century's plague for reproductive health. Fertil Steril. 2019;111(5):885–6.
17. Neel BA, Sargis RM. The paradox of Progress: environmental disruption of metabolism and the diabetes epidemic. Diabetes. 2011;60(7):1838.
18. Xu X, Yang M. Effects of environmental EDCs on oocyte quality, embryo development, and the outcome in human IVF process. Adv Exp Med Biol. 2021;1300:181–202.

19. Sharma A, Mollier J, Brocklesby RWK, Caves C, Jayasena CN, Minhas S. Endocrine-disrupting chemicals and male reproductive health. Reprod Med Biol. 2020;19(3):243–53.
20. Eskew AM, Jungheim ES. A history of developments to improve in vitro fertilization. Mo Med. 2017;114(3):156–9.
21. Stephen EH, Chandra A, King RB. Supply of and demand for assisted reproductive technologies in the United States: clinic- and population-based data, 1995-2010. Fertil Steril. 2016;105(2):451–8.
22. Human Fertilisation and Embryology Authority Fertility treatment 2019: trends and figures. 2021. hfea.gov.uk.
23. Chiang C, Mahalingam S, Flaws JA. Environmental contaminants affecting fertility and somatic health. Semin Reprod Med. 2017;35(3):241–9.
24. Karwacka A, Zamkowska D, Radwan M, Jurewicz J. Exposure to modern, widespread environmental endocrine disrupting chemicals and their effect on the reproductive potential of women: an overview of current epidemiological evidence. Hum Fertil (Camb). 2019;22(1):2–25.
25. Mok-Lin E, Ehrlich S, Williams PL, Petrozza J, Wright DL, Calafat AM, et al. Urinary bisphenol A concentrations and ovarian response among women undergoing IVF. Int J Androl. 2010;33(2):385–93.
26. Ziv-Gal A, Flaws JA. Evidence for bisphenol A-induced female infertility: a review (2007–2016). Fertil Steril. 2016;106(4):827–56.
27. Peretz J, Vrooman L, Ricke WA, Hunt PA, Ehrlich S, Hauser R, et al. Bisphenol A and reproductive health: update of experimental and human evidence, 2007–2013. Environ Health Perspect. 2014;122(8):775–86.
28. Sifakis S, Androutsopoulos VP, Tsatsakis AM, Spandidos DA. Human exposure to endocrine disrupting chemicals: effects on the male and female reproductive systems. Environ Toxicol Pharmacol. 2017;51:56–70.
29. Ehrlich S, Williams PL, Missmer SA, Flaws JA, Ye X, Calafat AM, et al. Urinary bisphenol a concentrations and early reproductive health outcomes among women undergoing IVF. Hum Reprod. 2012;27(12):3583–92.
30. Ehrlich S, Williams PL, Missmer SA, Flaws JA, Berry KF, Calafat AM, et al. Urinary bisphenol A concentrations and implantation failure among women undergoing in vitro fertilization. Environ Health Perspect. 2012;120(7):978–83.
31. Radwan P, Wielgomas B, Radwan M, Krasiński R, Klimowska A, Kaleta D, et al. Urinary bisphenol A concentrations and in vitro fertilization outcomes among women from a fertility clinic. Reprod Toxicol. 2020;96:216–20.
32. Machtinger R, Orvieto R, Bisphenol A. Oocyte maturation, implantation, and IVF outcome: review of animal and human data. Reprod BioMed Online. 2014;29(4):404–10.
33. Cao Y, Qu X, Ming Z, Yao Y, Zhang Y. The correlation between exposure to BPA and the decrease of the ovarian reserve. Int J Clin Exp Pathol. 2018;11(7):3375–82.
34. Zhou W, Liu J, Liao L, Han S, Liu J. Effect of bisphenol A on steroid hormone production in rat ovarian theca-interstitial and granulosa cells. Mol Cell Endocrinol. 2008;283(1):12–8.
35. Alur S, Wang H, Hoeger K, Swan SH, Sathyanarayana S, Redmon BJ, et al. Urinary phthalate metabolite concentrations in relation to history of infertility and use of assisted reproductive technology. Fertil Steril. 2015;104(5):1227–35.
36. Du Y, Guo N, Wang Y, Teng X, Hua X, Deng T, et al. Follicular fluid concentrations of phthalate metabolites are associated with altered intrafollicular reproductive hormones in women undergoing in vitro fertilization. Fertil Steril. 2019;111(5):953–61.
37. Wu H, Ashcraft L, Whitcomb BW, Rahil T, Tougias E, Sites CK, et al. Parental contributions to early embryo development: influences of urinary phthalate and phthalate alternatives among couples undergoing IVF treatment. Hum Reprod. 2017;32(1):65–75.
38. Meeker JD, Ferguson KK. Urinary phthalate metabolites are associated with decreased serum testosterone in men, women, and children from NHANES 2011–2012. J Clin Endocrinol Metab. 2014;99(11):4346–52.

39. Paoli D, Pallotti F, Dima AP, Albani E, Alviggi C, Causio F, et al. Phthalates and Bisphenol A: presence in blood serum and follicular fluid of Italian women undergoing assisted reproduction techniques. Toxics. 2020;8(4):91.
40. Pallotti F, Pelloni M, Gianfrilli D, Lenzi A, Lombardo F, Paoli D. Mechanisms of testicular disruption from exposure to Bisphenol a and Phtalates. J Clin Med. 2020;9(2):471.
41. Serrano SE, Braun J, Trasande L, Dills R, Sathyanarayana S. Phthalates and diet: a review of the food monitoring and epidemiology data. Environ Health. 2014;13(1):43.
42. Rudel RA, Gray JM, Engel CL, Rawsthorne TW, Dodson RE, Ackerman JM, et al. Food packaging and bisphenol a and bis(2-ethyhexyl) phthalate exposure: findings from a dietary intervention. Environ Health Perspect. 2011;119(7):914–20.
43. Chen M, Tang R, Fu G, Xu B, Zhu P, Qiao S, et al. Association of exposure to phenols and idiopathic male infertility. J Hazard Mater. 2013;250–251:115–21.
44. Yueh M-F, Tukey RH. Triclosan: a widespread environmental toxicant with many biological effects. Annu Rev Pharmacol Toxicol. 2016;56(1):251–72.
45. Wang X, Ouyang F, Feng L, Liu Z, Zhang J. Maternal urinary Triclosan concentration in relation to maternal and neonatal thyroid hormone levels: a prospective study. Environ Health Perspect. 2017;125(6):067017.
46. Radwan P, Wielgomas B, Radwan M, Krasiński R, Klimowska A, Zajdel R, et al. Triclosan exposure and in vitro fertilization treatment outcomes in women undergoing in vitro fertilization. Environ Sci Pollut Res. 2021;28(10):12993–9.
47. Hua R, Zhou Y, Wu B, Huang Z, Zhu Y, Song Y, et al. Urinary triclosan concentrations and early outcomes of in vitro fertilization-embryo transfer. Reproduction. 2017;153(3):319–25.
48. Lange A, Carignan CC, Minguez-Alarcon L, Williams P, Calafat AM, Toth TL, et al. Triclosan exposure and treatment outcomes in women undergoing in vitro fertilization. Fertil Steril. 2015;104(3):e86.
49. Braun JM, Just AC, Williams PL, Smith KW, Calafat AM, Hauser R. Personal care product use and urinary phthalate metabolite and paraben concentrations during pregnancy among women from a fertility clinic. J Expo Sci Environ Epidemiol. 2014;24(5):459–66.
50. Tavares RS, Martins FC, Oliveira PJ, Ramalho-Santos J, Peixoto FP. Parabens in male infertility—is there a mitochondrial connection? Reprod Toxicol. 2009;27(1):1–7.
51. Smith KW, Braun JM, Williams PL, Ehrlich S, Correia KF, Calafat AM, et al. Predictors and variability of urinary paraben concentrations in men and women, including before and during pregnancy. Environ Health Perspect. 2012;120(11):1538–43.
52. Cabaleiro N, de la Calle I, Bendicho C, Lavilla I. An overview of sample preparation for the determination of parabens in cosmetics. TrAC Trends Anal Chem. 2014;57:34–46.
53. Gomez E, Pillon A, Fenet H, Rosain D, Duchesne MJ, Nicolas JC, et al. Estrogenic activity of cosmetic components in reporter cell lines: parabens, UV screens, and musks. J Toxicol Environ Health A. 2005;68(4):239–51.
54. Nowak K, Ratajczak-Wrona W, Górska M, Jabłońska E. Parabens and their effects on the endocrine system. Mol Cell Endocrinol. 2018;474:238–51.
55. Smith KW, Souter I, Dimitriadis I, Ehrlich S, Williams PL, Calafat AM, et al. Urinary paraben concentrations and ovarian aging among women from a fertility center. Environ Health Perspect. 2013;121(11–12):1299–305.
56. Sabatini ME, Smith KW, Ford J, Ehrlich SR, Toth TL, Hauser R. Urinary paraben concentrations and in vitro fertilization (IVF) outcomes. Fertil Steril. 2011;96(3):S154.
57. Mínguez-Alarcón L, Chiu YH, Messerlian C, Williams PL, Sabatini ME, Toth TL, et al. Urinary paraben concentrations and in vitro fertilization outcomes among women from a fertility clinic. Fertil Steril. 2016;105(3):714–21.
58. Ma X, Cui L, Chen L, Zhang J, Zhang X, Kang Q, et al. Parental plasma concentrations of perfluoroalkyl substances and in vitro fertilization outcomes. Environ Pollut. 2021;269:116159.
59. Olsen GW, Burris JM, Ehresman DJ, Froehlich JW, Seacat AM, Butenhoff JL, et al. Half-life of serum elimination of perfluorooctanesulfonate,perfluorohexanesulfonate, and per-

fluorooctanoate in retired fluorochemical production workers. Environ Health Perspect. 2007;115(9):1298–305.

60. McCoy JA, Bangma JT, Reiner JL, Bowden JA, Schnorr J, Slowey M, et al. Associations between perfluorinated alkyl acids in blood and ovarian follicular fluid and ovarian function in women undergoing assisted reproductive treatment. Sci Total Environ. 2017;605–606:9–17.

61. Younglai EV, Foster WG, Hughes EG, Trim K, Jarrell JF. Levels of environmental contaminants in human follicular fluid, serum, and seminal plasma of couples undergoing in vitro fertilization. Arch Environ Contam Toxicol. 2002;43(1):121–6.

62. Petro EM, Leroy JL, Covaci A, Fransen E, De Neubourg D, Dirtu AC, et al. Endocrine-disrupting chemicals in human follicular fluid impair in vitro oocyte developmental competence. Hum Reprod. 2012;27(4):1025–33.

63. Bloom MS, Fujimoto VY, Storm R, Zhang L, Butts CD, Sollohub D, et al. Persistent organic pollutants (POPs) in human follicular fluid and in vitro fertilization outcomes, a pilot study. Reprod Toxicol. 2017;67:165–73.

64. Al-Hussaini TK, Abdelaleem AA, Elnashar I, Shabaan OM, Mostafa R, El-Baz MAH, et al. The effect of follicullar fluid pesticides and polychlorinated biphenyls concentrations on intracytoplasmic sperm injection (ICSI) embryological and clinical outcome. Eur J Obstet Gynecol Reprod Biol. 2018;220:39–43.

65. Meeker John D, Maity A, Missmer Stacey A, Williams Paige L, Mahalingaiah S, Ehrlich S, et al. Serum concentrations of polychlorinated biphenyls in relation to in vitro fertilization outcomes. Environ Health Perspect. 2011;119(7):1010–6.

66. Khanjani N, Sim MR. Maternal contamination with PCBs and reproductive outcomes in an Australian population. J Expo Sci Environ Epidemiol. 2007;17(2):191–5.

67. Jirsová S, Masata J, Jech L, Zvárová J. Effect of polychlorinated biphenyls (PCBs) and 1,1,1-trichloro-2,2,-bis (4-chlorophenyl)-ethane (DDT) in follicular fluid on the results of in vitro fertilization-embryo transfer (IVF-ET) programs. Fertil Steril. 2010;93(6):1831–6.

68. Müller MHB, Polder A, Brynildsrud OB, Grønnestad R, Karimi M, Lie E, et al. Prenatal exposure to persistent organic pollutants in Northern Tanzania and their distribution between breast milk, maternal blood, placenta and cord blood. Environ Res. 2019;170:433–42.

69. Pocar P, Brevini TA, Fischer B, Gandolfi F. The impact of endocrine disruptors on oocyte competence. Reproduction. 2003;125(3):313–25.

70. Schjenken JE, Green ES, Overduin TS, Mah CY, Russell DL, Robertson SA. Endocrine disruptor compounds-a cause of impaired immune tolerance driving inflammatory disorders of pregnancy? Front Endocrinol. 2021;12:607539.

71. Yang C, Song G, Lim W. A mechanism for the effect of endocrine disrupting chemicals on placentation. Chemosphere. 2019;231:326–36.

72. Rogers JA, Metz L, Yong VW. Review: endocrine disrupting chemicals and immune responses: a focus on bisphenol-A and its potential mechanisms. Mol Immunol. 2013;53(4):421–30.

73. Ferguson KK, McElrath TF, Cantonwine DE, Mukherjee B, Meeker JD. Phthalate metabolites and bisphenol-a in association with circulating angiogenic biomarkers across pregnancy. Placenta. 2015;36(6):699–703.

74. Liu Q. Effects of environmental endocrine-disrupting chemicals on female reproductive health. Adv Exp Med Biol. 2021;1300:205–29.

75. Krieg SA, Shahine LK, Lathi RB. Environmental exposure to endocrine-disrupting chemicals and miscarriage. Fertil Steril. 2016;106(4):941–7.

76. The Practice Committee Of The American Society For Reproductive Medicine. Evaluation and treatment of recurrent pregnancy loss: a committee opinion. Fertil Steril. 2012;98(5):1103–11.

77. Grippo A, Zhang J, Chu L, Guo Y, Qiao L, Myneni AA, et al. Air pollution exposure during pregnancy and spontaneous abortion and stillbirth. Rev Environ Health. 2018;33(3):247–64.

78. Sugiura-Ogasawara M, Ozaki Y, Sonta S, Makino T, Suzumori K. Exposure to bisphenol A is associated with recurrent miscarriage. Hum Reprod. 2005;20(8):2325–9.

79. Li Q, Davila J, Bagchi MK, Bagchi IC. Chronic exposure to bisphenol A impairs progesterone receptor-mediated signaling in the uterus during early pregnancy. Receptors. Clin Investig. 2016;3(3):e1369.
80. Mu D, Gao F, Fan Z, Shen H, Peng H, Hu J. Levels of phthalate metabolites in urine of pregnant women and risk of clinical pregnancy loss. Environ Sci Technol. 2015;49(17):10651–7.
81. Toft G, Jönsson BAG, Lindh CH, Jensen TK, Hjollund NH, Vested A, et al. Association between pregnancy loss and urinary phthalate levels around the time of conception. Environ Health Perspect. 2012;120(3):458–63.
82. Liew Z, Luo J, Nohr EA, Bech BH, Bossi R, Arah OA, et al. Maternal plasma Perfluoroalkyl substances and miscarriage: a nested case-control study in the Danish National Birth Cohort. Environ Health Perspect. 2020;128(4):47007.
83. Sugiura-Ogasawara M, Ozaki Y, Sonta S, Makino T, Suzumori K. PCBs, hexachlorobenzene and DDE are not associated with recurrent miscarriage. Am J Reprod Immunol. 2003;50(6):485–9.
84. Rosen EM, Muñoz MI, McElrath T, Cantonwine DE, Ferguson KK. Environmental contaminants and preeclampsia: a systematic literature review. J Toxicol Environ Health B Crit Rev. 2018;21(5):291–319.
85. Pijnenborg R, Vercruysse L, Hanssens M. The uterine spiral arteries in human pregnancy: facts and controversies. Placenta. 2006;27(9–10):939–58.
86. Poropat AE, Laidlaw MAS, Lanphear B, Ball A, Mielke HW. Blood lead and preeclampsia: a meta-analysis and review of implications. Environ Res. 2018;160:12–9.
87. Gómez-Roig MD, Pascal R, Cahuana MJ, García-Algar O, Sebastiani G, Andreu-Fernández V, et al. Environmental exposure during pregnancy: influence on prenatal development and early life: a comprehensive review. Fetal Diagn Ther. 2021;48(4):245–57.
88. Erinc A, Davis MB, Padmanabhan V, Langen E, Goodrich JM. Considering environmental exposures to per- and polyfluoroalkyl substances (PFAS) as risk factors for hypertensive disorders of pregnancy. Environ Res. 2021;197:111113.
89. Pergialiotis V, Kotrogianni P, Christopoulos-Timogiannakis E, Koutaki D, Daskalakis G, Papantoniou N. Bisphenol A and adverse pregnancy outcomes: a systematic review of the literature. J Matern Fetal Neonatal Med. 2018;31(24):3320–7.
90. Marinello WP, Mohseni ZS, Cunningham SJ, Crute C, Huang R, Zhang JJ, et al. Perfluorobutane sulfonate exposure disrupted human placental cytotrophoblast cell proliferation and invasion involving in dysregulating preeclampsia related genes. FASEB J. 2020;34(11):14182–99.
91. Marks KJ, Howards PP, Smarr MM, Flanders WD, Northstone K, Daniel JH, et al. Prenatal exposure to mixtures of persistent endocrine-disrupting chemicals and birth size in a population-based cohort of British girls. Epidemiology. 2021;32(4):573–82.
92. Barker DJ. The developmental origins of chronic adult disease. Acta Paediatr Suppl. 2004;93(446):26–33.
93. Philippat C, Mortamais M, Chevrier C, Petit C, Calafat AM, Ye X, et al. Exposure to phthalates and phenols during pregnancy and offspring size at birth. Environ Health Perspect. 2012;120(3):464–70.
94. Wolff MS, Engel SM, Berkowitz GS, Ye X, Silva MJ, Zhu C, et al. Prenatal phenol and phthalate exposures and birth outcomes. Environ Health Perspect. 2008;116(8):1092–7.
95. Whitworth KW, Haug LS, Baird DD, Becher G, Hoppin JA, Skjaerven R, et al. Perfluorinated compounds in relation to birth weight in the Norwegian mother and child cohort study. Am J Epidemiol. 2012;175(12):1209–16.
96. Longnecker MP, Klebanoff MA, Brock JW, Guo X. Maternal levels of polychlorinated biphenyls in relation to preterm and small-for-gestational-age birth. Epidemiology. 2005;16(5):641–7.
97. Vrachnis N, Loukas N, Vrachnis D, Antonakopoulos N, Zygouris D, Kolialexi A, et al. A systematic review of Bisphenol A from dietary and non-dietary sources during pregnancy and its possible connection with fetal growth restriction: investigating its potential effects and the window of fetal vulnerability. Nutrients. 2021;13(7):2426.

98. Ferguson KK, McElrath TF, Meeker JD. Environmental phthalate exposure and preterm birth. JAMA Pediatr. 2014;168(1):61–7.

99. Deji Z, Liu P, Wang X, Zhang X, Luo Y, Huang Z. Association between maternal exposure to perfluoroalkyl and polyfluoroalkyl substances and risks of adverse pregnancy outcomes: a systematic review and meta-analysis. Sci Total Environ. 2021;783:146984.

100. Floreani A, Caroli D, Lazzari R, Memmo A, Vidali E, Colavito D, et al. Intrahepatic cholestasis of pregnancy: new insights into its pathogenesis. J Matern Fetal Neonatal Med. 2013;26(14):1410–5.

101. Wang JQ, Gao H, Sheng J, Tao XY, Huang K, Zhang YW, et al. Urinary concentrations of phthalate metabolites during gestation and intrahepatic cholestasis of pregnancy: a population-based birth cohort study. Environ Sci Pollut Res Int. 2020;27(11):11714–23.

Chapter 9
Endocrine-Disrupting Chemicals and the Offsprings: Prenatal Exposure

Maria Laura Solerte ⓘ **and Erich Cosmi** ⓘ

9.1 Introduction

The role of environmental impact on humans is extensively studied and evaluated throughout the world, given that numerous correlations between harmful environmental factors and many pathologies had been extrapolated. Following several clinical evidences, numerous research were carried out by study groups; since the late 1980s to early 1990s, the need to monitor the major long-term consequences on human health of exposure to chemical factors capable of altering the endocrine system during life has been underlined. In 1991, during the World Health Organization (hereafter **WHO**) seminar, topics regarding the impact of the toxic environmental factors on reproductive health (hereafter **RH**) were extensively discussed by scientists coming from all parts of the world; particularly, they focused on the correlation between harmful environment, pollutants, sperm quantitative/qualitative alterations [1], and reproductive rate/outcome, also with a view to fill the absence of essential notions.

Therefore, the modalities of **RH** study and surveillance were indicated, also retracing the dramatic events of Seseo 1976, Bhopal 1984, and Chernobyl 1986,

M. L. Solerte
The Residency School of Specialization in Obstetrics and Gynaecology, University of Padua, Padua, Italy

E. Cosmi (✉)
The Residency School of Specialization in Obstetrics and Gynaecology, University of Padua, Padua, Italy

Department of Woman and Child Health, University of Padua, Padua, Italy

The Maternal and Fetal Medicine Unit for High-Risk Pregnancies, Padua Hospital, University of Padua, Padua, Italy

Midwifery School, Faculty of Medicine, University of Padua, Padua, Italy
e-mail: erich.cosmi@unipd.it

and their harmful effects; research and intervention activities were here promoted to define the nature of harmful substances [1]. This era marked the elevation of the issues linked to the environmental modifications, which several countries had to face up in light of the increasing diseases derived from this extremely negative correlation. A great attention had begun to actively focus on global topics such as climate change, alteration of biological diversity, stratospheric depletion of ozone, air transport pollution, acid deposition, and toxic waste. In particular, attention was focused on the nature and mechanism of action of the various environmental factors as the cause of negative influence on the male and female reproductive system, for example, the sperm quantity and quality anomalies, infertility, cell/chromosome pathology, intrauterine growth retardation, prenatal or postnatal death, offspring defects, neurological issues, and early aging, which arose later [1]. Even as early as 1962, a well-documented articulate text, defined as "changing the world," was introduced as a debated and intricate scientific concept of the interaction between dichlorodiphenyltrichloroethane (hereafter **DDT** or **p,p'-DDT**, which is the isomer that has the greatest insecticidal capacity), a powerful lasting chemical pesticide, and health, significantly active on human sexual development and reproduction, which greatly affected public opinion [2]. The uncontrolled use of synthetic pesticides was emphasized; the book was so significant that it inspired the environmental movements that took place in the following years against ambient poisons. Even then, although the evidences of severe, negative, and cancerous pesticide effects were highlighted, they became a controversial and debatable topic. Hence, natural or man-made disaster and/or chemical pollutants began to be blatantly indicated as the main culprits for the negative consequences on humans. Moreover, the research needs for methodologies and surveillance data were explained, especially in order to improve the international works for similar approaches and database, to get alert systems, and to assess the toxic levels of the chemicals indicated. In addition, the need to introduce multidisciplinary studies has emerged, for evaluations of both the environment and cellular processes negatively affected by biological, chemical, and physical factors and pesticide exposure: those damages result in problems related to fertility, spontaneous abortion, womb growth anomalies of the fetus, and various pathologies found at birth, due to the indirect exposure in pregnancy or early stages of extrauterine life [1, 3]. Chemical pollutions, radiation, and various forms of stress could be the main elements worthy of further study in order to increase the knowledge of their interaction with the physiological function of the reproductive system and the early prenatal organogenesis development [3] of the subsequent offsprings and long-term effects [3]. Also, excesses of vitamin A, radiation caused by Chernobyl disaster, and methylmercury, for specific neurological defects, were cited [1]. Some harmful chemical substances, including industrial ones, released in the ambient, starting from the World War II [1–3], were also listed, for example, pesticides (later also biocides), fungicides, metals, insecticides, and nematocides, provided with the ability to bind to the receptors for steroid hormones situated in specific target cells and tissues, therefore simulating their action. With increasing relevance for the control of the conditions necessary for general health, the well-documented concept was formed about the negative interference of polychlorinated biphenyls (henceforth **PCBs**), **DDT**, other pesticides, and their metabolites. With the knowledge gained in that historical period, some studies had emerged, to spread greater long-term awareness as well as health consequences associated with the exposure to chemicals that disrupt the endocrine system during the breastfeeding (for the bioaccumulation of toxic substances) and the first years of life and, therefore, to assess the related lifelong health risks. Some evidence has been exposed that pollutants found in rainwater, well water, lakes, and oceans, as well as freshwater and marine food products, can interfere with the endocrine development in wild and laboratory animals or humans. Furthermore, hundreds of synthetic elements, some with the agonist, antagonist, and synergist abilities as endocrine disruptors, disseminated in the ambient, were identified: their capacity to negatively affect development functionality of both the immune and endocrine systems was pointed out [3]. The production of oocytes, pregnancy, and lactation time are phases of female life that can cause the mobilization of toxic environmental molecules accumulated in the fat tissue and therefore can give rise to the phenomenon of transgenerational exposure [3]. Starting from the consolidated concept that, in mammals and other vertebrates, the connections between cells are indispensable for their physiological development, the embryology science had begun to orient itself on studies of substances produced by a group of cells that synthesize a product, which then influences the course of development determining the regular function of other cell units. Groups of hormones have been identified, derived from mother's ovaries, adrenal glands, placenta, fetal gonads, and adrenal glands, with a primary regulation role in many tissues' development. Moreover, the type of organs in the offsprings of subjects exposed to toxic substances had therefore been identified, who could have been more affected by the adverse consequences; the presence of receptors, also in the fetal life, for ovarian, testicular, placenta, and adrenal steroid hormones made the reproductive system one of the main targets [3]. Indeed, it is known that at the end of the second month of pregnancy, highly sensitive stages of growth are identified, regulated by steroid hormones indicated above [3, 4]. Prior to 1980s–1990s, over the past 30–50 years, some authors have argued about the correlations between abnormalities of the male genital tract, decrease of semen volume and sperm counts, and mothers' intake of synthetic estrogens during their gestation: the association with diethylstilbestrol (hereafter **DES**) was also mentioned, which is a synthetic estrogen discovered in 1938 and used from 1945 to 1971 [3, 4] to prevent spontaneous miscarriage [3], and the major increase in the incidence of hypospadias, cryptorchidism, and testicular cancer [5]. This theory was born from the concept that the latter three pathologies, mentioned above, could have the same etiological factors involved in the male fetal life.

In addition, **DES** was named as a "mother substance," due to its powerful estrogenic capacity as well as its effects; **DES** exposure was followed as an estrogen agonist model to analyze the impact of substances with equal potential for estrogenic endocrine disruption [3]. Moreover, in order to understand the route and mechanism of access of the estrogens (endogenous/exogenous) and therefore of their harmful action, it was reported that the physiological hormone, by an enterohepatic recirculation, was metabolized and reabsorbed by the bowel, more easily when the bowel contained low amounts of fiber. This last aspect, formulated as hypothesis, could explain the lower incidence of breast cancer, estrogen linked, in women on a high-fiber diet, which conferred some sort of protection right against estrogen [5]. This above assumption has allowed to expand the research and knowledge also on subsequent synthesis molecules and on their overexposure effects on the development of reproductive traits and related disorders [5]. Previously, other authors also published the adverse effects of **DES** [6], affecting the male and female reproductive tract; the pathological forms had been well listened: defects of vagina-cervix-uterus-fallopian tubes, adenosis, clear cell adenocarcinoma, breast cancer for female subjects, and testicle hypoplasia, cryptorchidism, microphallus, and ependymal cysts for the male ones; later, these anomalies led to dysfunction in reproductive age and pregnancy time. As known, **DES** was banned as drug for pregnant women, by the Food and Drug Administration (hereafter **FDA**), in 1971, and for meat production in 1979 and in 1978/1981 in Europe [3, 5]. Moreover, the effects of cigarette smoking were addressed; not least in importance but only for narrative scheme, already before 1976, the issue of smoking in pregnancy had been examined, revealing that fetal growth retardation, linked to birth weight under 2.500 g, exhibited in two somatic types: the first with a low ponderal index and the second one with short crown-heel length for fetal age [7]. The reduction of the average body length for dates of full-term births from smoking mothers had speculated that the use of cigarettes was associated with the second type of growth retardation [7]. Given the evident correlations, continuously published, between causes and effects, several policies of careful monitoring were initiated, extended, continuously used, and updated, with the aim of identifying and regulating the numerous chemicals present in the environment responsible for the endocrine alteration effects, also with long-term problems. Indeed, even that time, the major danger for the physiological functions of the reproductive system seemed to be chemical pollutants [8]. Furthermore, epidemiology science, based on precise and systematic data takeover, could already identify diseases and their related risk factors and try to develop adequate prevention strategies [8]. To better understand the various deleterious components that negatively affect important physiological body functions, it might be interesting to review some definitions [8].

9.2   Identification Overview

9.2.1   *Endocrine Glands and System*

The univocal definition of the endocrine system is, for a broad view, the union of the balanced interdependent body glands (Fig. 9.1), including hypothalamus, pituitary, pineal, thyroid, pancreas, adrenal, gonads, and fat tissue, distant for topographical and physical connections, with internal secretion.

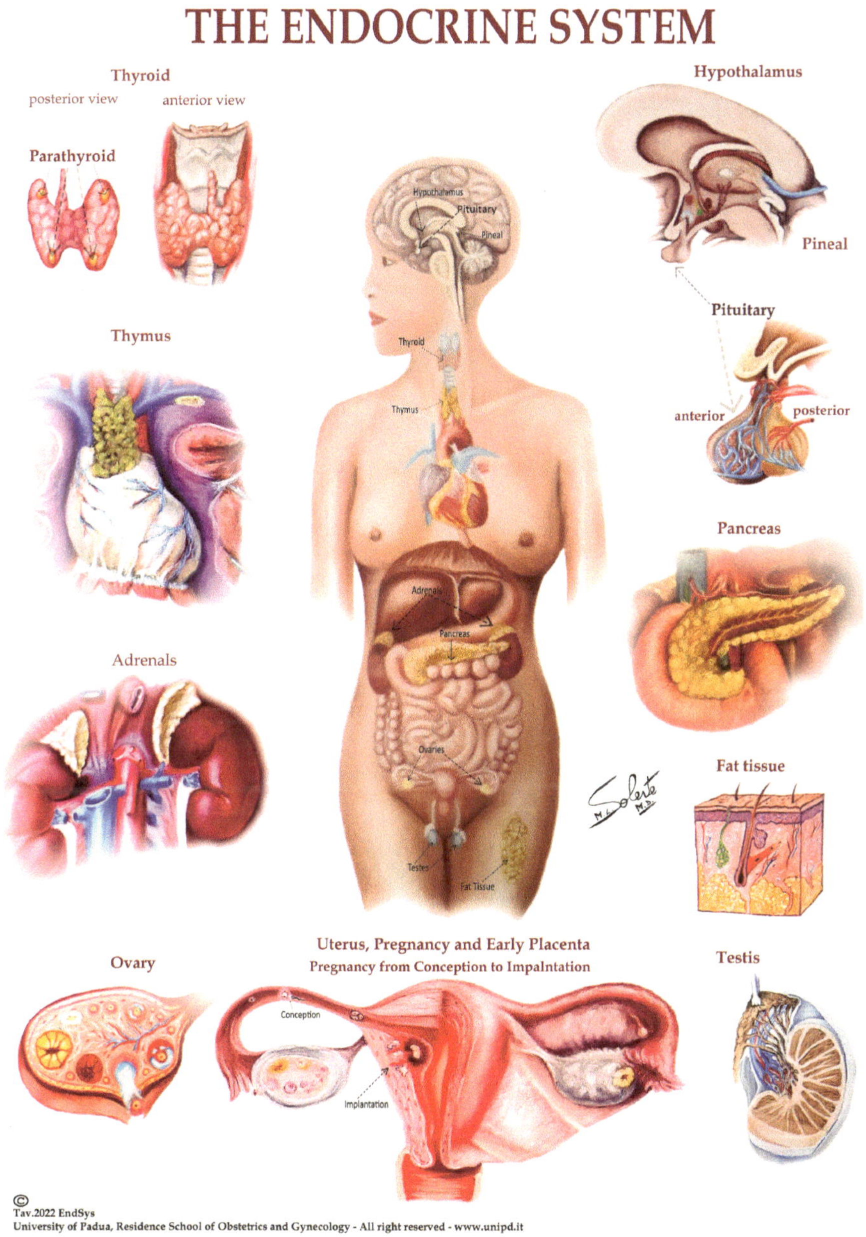

THE ENDOCRINE SYSTEM

Fig. 9.1 Schematic illustration of the main organs, with some anatomical details, composing the endocrine system. The Endocrine System Table—by Maria Laura Solerte: copyright and royalties for University of Padua, The Residency School of Obstetrics and Gynaecology Specialization—based on (1) The Endocrine System, Anatomical Chart Company 2002, permission by Wolters Kluwer Health, (2) youtube.com/c/HumanAnatomyLessons, permission by Geetha Hari MD

This system is basilar for the control, regulation, and coordination of several human usual tasks as development, sleep, homeostasis, immunity, metabolism, growth, response to stress or injury stimulations, behavior, sexual functions, and reproductive processes; it has very specific functions and complex interactions with the physiology of the human body and its hormonal targets. The endocrine system interacts, in a complementary and bidirectional sense, with the nervous system which, however, differs from the first one for the greater speed of signal transmissions [9, 10]. The endocrine structure organs are essential to maintain organic homeostasis, also when external conditions vary; the system is provided with dense vascularization (Fig. 9.2a) [9], extremely sensitive to changes by aging and pathologies [9].

Fig. 9.2 (**a**) Vascular niche functions in the endocrine system. In the testis, ECs release various endocrine signals to maintain SSCs and spermatogenesis. OSC maintenance is supported by pericytes. During follicular and luteal stages of the cycle, growth factors regulate periodic growth and regression of ovarian vasculature that is needed for follicular and luteal development. In the thyroid, angiogenic signals from TSCs and pericytes regulate angiogenesis and endothelial fenestrae formation that is important for thyrocyte function. Pituitary ECs and pericytes promote maintenance and function of neurosecretory cells in the neurohypophysis and pituitary stem cells in the adenohypophysis. Angiocrine signals also regulate endocrine function of the adrenal cortex that, in turn, promotes angiogenesis via the endocrine gland-specific growth factor EG-VEGF. In the pancreas, reciprocal interaction between ECs and b-cells is required for angiogenesis and insulin secretion. *EC* endothelial cell, *SSC* spermatogonial stem cell, *FGF2* fibroblast growth factor 2, *GDNF* glial cell line-derived neurotrophic factor, *CSF-1* colony-stimulating factor 1, *OSC* ovarian stem cell, *PDGF* platelet-derived growth factor, *VEGF* vascular endothelial growth factor, *ANG1* angiopoietin 1, *MMP* matrix metalloproteinase, *TSC* thyroid stem cell, *TSH* thyrotropin-releasing hormone, *BMP* bone morphogenetic protein, *bFGF* basic fibroblast growth factor, *NGF* nerve growth factor, *EGF* epidermal growth factor, *EG VEGF* endocrine gland-derived vascular endothelial growth factor, *NO* nitric oxide, *HGF* hepatocyte growth factor, *IGF* insulin-like growth factor, *TSP-1* thrombospondin-1, *TGF-b1* transforming growth factor b1. Original images and text: full original version, published under Frontiers Copyright Statement by CC-BY-4.0 license and permission [9]. (**b**) Vascular niche function in the endocrine system during aging. Young ECs secrete angiocrine signals to promote proliferation of endocrine cells and support endocrine function. In the young endocrine system, ECs produce low ROS levels that support Leydig cell proliferation in the testis and promote ovulation OSC maturation in the ovaries. Angiogenic growth factors from pituitary endocrine cells and others promote angiogenesis. Upon aging, endothelial ROS production increases, impairing sperm cell motility, quality, and quantity of follicular cells in the ovary and proliferation and hormone production of various endocrine cells, including pancreatic b-cells and endocrine cells and neurosecretory axon terminals in the pituitary gland. In contrast, elevated ROS levels increase the production of inflammatory mediators such as ICAM-1 in the thyroid and increase the release of glucocorticoids from the adrenal cortex, promoting the stress response. *EC* endothelial cell, *ROS* reactive oxygen species, *T3* triiodothyronine, *T4* thyroxine, *VEGF* vascular endothelial growth factor, *BCAM* basal cell adhesion molecule, *OXT* oxytocin, *BMP* bone morphogenetic protein, *HGF* hepatocyte growth factor, *AVP* arginine vasopressin, *GH* growth hormone, *TSH* thyroid-stimulating hormone, *ICAM-1* intercellular adhesion molecule 1, *MMP* matrix metalloproteinase. Original images and text: full original version, published under Frontiers Copyright Statement by CC-BY-4.0 license and permission [9]

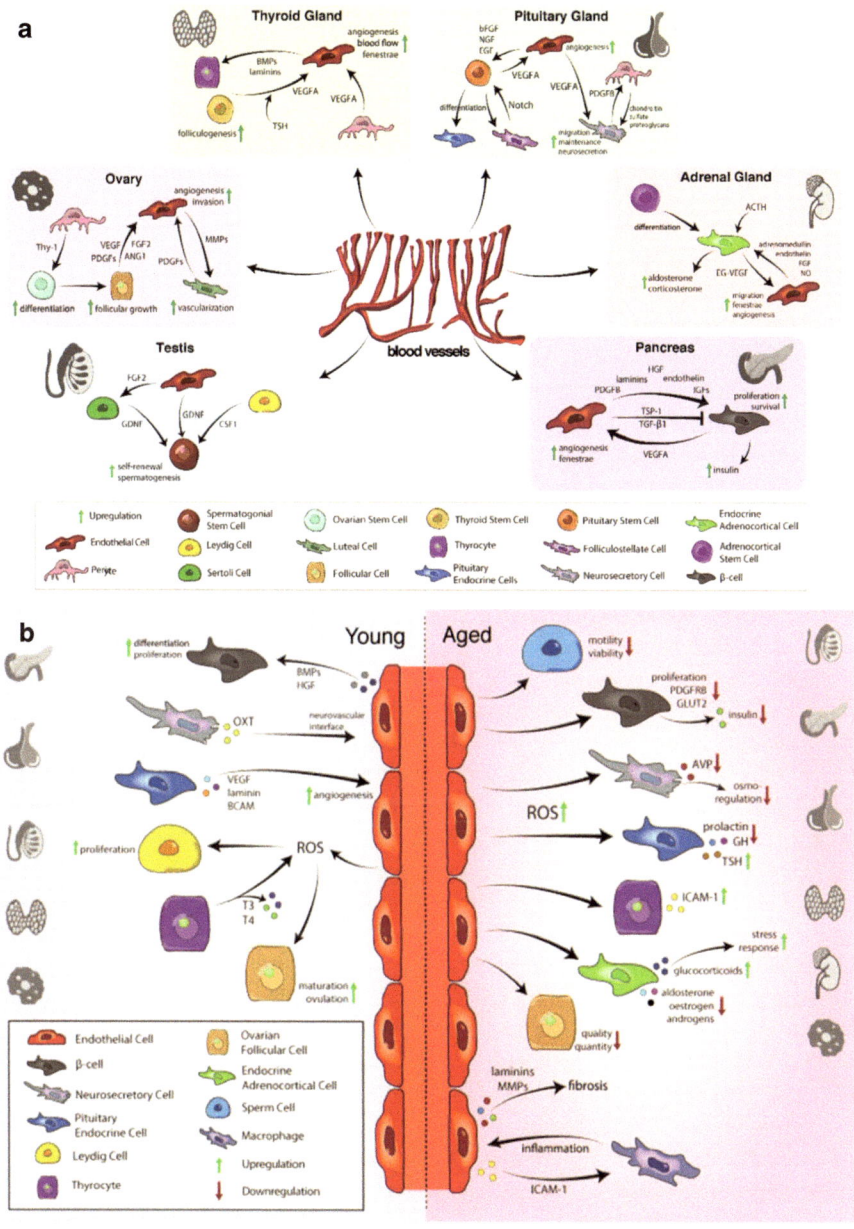

Therefore, endocrine unit has been identified as a refined and, at the same time, complicated and effective network, also equipped with vascular niches, of connection between their three main components: endocrine glands, hormones, and receptors. This elementary scheme is a prerequisite for understanding the multiple mechanisms of the endocrine system, in constant balance, which is responsible for organizing, influencing, and controlling several human functions, mainly guided by the hypothalamus-pituitary axis, which is a neuroendocrine organ essential for regulating growth and development [9]. The endocrine system can be affected by disorders that can alter the normal balance of human body functions and cause adverse health effects; individuals are constantly exposed to a wide range of substances, which cause a negative impact on endocrine products, in several contexts such as work, drug or consumer product use, and even natural resources. Furthermore, the aging effects on the endothelial cells of the vascular niches have been reported in detail, identifying it as a vascular microenvironment that governs the function of cells and subtypes (Fig. 9.2b) [9].

9.2.1.1   Glands

The endocrine glands (hereafter **EGs**), called also "specialized ductless glands" due to their characteristic functional capacity, are highly vascularized (Fig. 9.2a) [9] tissues or organs, equipped with internal secretion mechanism and cellular/ molecular cues, that signal each other in pulsatile, sequence, and feedback patterns. Their dense vascular systems are in turn a "microenvironment" [9] and are formed by endothelial cells (hereafter **ECs**) that also play an endocrine role, by changing the vascular diameters, and vascular endothelial growth factor (hereafter **VEGF**), which plays roles in angio-vasculo-morphogenesis: **ECs**, microenvironment vessels, and their roles are shown in Fig. 9.2a, b [9], with the related sophisticated circuits linked to the main endocrine organs. The **EGs** are located in various anatomic districts of the human body and have peculiar capacities to synthesize chemical messengers, called hormones, released directly into the blood circulatory system, unlike the exocrine glands, which hence carry their products up to the target organs all over the body. Thanks to the rich vascular systems, there are rapid interactions between productive cells and the endothelium and its **ECs**. The **EGs** have different anatomic, histological, morphological, and functional characteristics; as the main pattern, they are vascular and commonly have intracellular vacuoles or granules that store their hormonal products [10].

As mentioned above, the endocrine system, and therefore the **EGs**, also suffers from the effects of aging, as shown, for instance, by the development of the reactive oxygen species (hereafter **ROS**), from altered mitochondrial work (Fig. 9.2b) [9]. The entire endocrine system is made up of an elaborate network of **EGs** and axes, provided with feedback, negative or positive, working by circadian rhythm or pulsatile model, and, precisely, comprises the following main elements (Fig. 9.1) [9, 10]:

1. *Hypothalamus*: It is a smart essential control coordinating center, located deep in the center of the brain; through its paraventricular nucleus neuroendocrine, it can control homeostasis, pituitary gland, blood pressure, energy, water balance, mood, appetite, reproductive behaviors, temperature, and stress. It produces corticotropin-releasing hormone (hereafter **CRH**), thyrotropin-releasing hormone (hereafter **TRH**), growth hormone-releasing hormone (hereafter **GH-RH**), somatostatin-releasing hormone (hereafter **SRIH**), gonadotropin-releasing hormone (**Gn-RH**), luteinizing hormone-releasing hormone (hereafter **LH-RH**), prolactin-releasing peptide (hereafter **PRH**), and prolactin-releasing/inhibitor factor (hereafter **PRF/PIF**). Hypothalamus is the main component of the neuroendocrine system with the interaction of (x) hypothalamus-(anterior) pituitary-adrenal axis (hereafter **HPA**), (xx) hypothalamus-(anterior) pituitary-thyroid axis (hereafter **HPT**), (xxx) hypothalamus-(anterior) pituitary-gonadal axis (hereafter **HPG**), and axis-relative feedback reactions.

2. *Pituitary gland*: Named also hypophysis, it has anterior/adeno and posterior/neuro lobes located in the brain, at the base of the hypothalamus, and is "the most highly vascularized mammalian tissue" [9]; it is an essential control center, via its anterior lobe endocrine cell hormones, of the other endocrine glands and several body functions as metabolism, blood pressure, stress, growth reproduction, labor, lactation, and water balance. Hypophysis produces thyroid-stimulating hormone (hereafter **TSH**), adrenocorticotropic hormone (hereafter **ACTH**), follicle-stimulating hormone (hereafter **FSH**), and luteinizing hormone (hereafter **LH**). This gland is an integral part of the **HPA, HPG,** and **HPT** axes.

3. *Pineal gland*: Named also epiphysis cerebri, it is located in the epithalamus brain zone, behind the hypophysis; it produces a hormone that regulates sleep, puberty, circadian rhythms, and other functions; in particular, it obtains retina information on the light-dark cycle from the external ambient and uses these data to rhythmically synthesize and secrete, by pinealocytes, its hormone called melatonin (tryptophan-serotonin derived).

4. *Thyroid gland*: With its left and right lobes, it is located on the anterior side of the neck, in front of the trachea and larynx; it mainly regulates the metabolism through its hormones and the feedback of the **HPT** axis. Thyroid gland produces two hormones: thyroxine (hereafter **T4**) and triiodothyronine (hereafter **T3**).

5. *Parathyroid glands*: These are four small glands located on both sides of the thyroid, posteriorly, but their activities are unrelated; they control, through the production of parathyroid hormone parathormone (hereafter **PTH**), calcium level regulation, bone structures, and the body's calcium balance; moreover, **PTH** involves the kidney and small intestine activities.

6. *Thymus*: With its two small lobes, it is a primary organ of the lymphatic system, located on the upper front side of the chest, between the lungs, behind the sternum, below the sternum manubrium, in action before birth until puberty to produce T-lymphocytes; this gland goes to a progressive involution during the human aging. Thymus produces an array of hormones, like thymosin, thymopoietin, thymic humoral factor, and thymulin, regulating the nervous-endocrine circuits and the immune cell production. Thymus is a hub between

the endocrine-immune-nervous interdependent systems, also equipped with a feedback mechanism (thymus as regulator of **HPG** axis).

7. *Adrenal glands*: With their subzones (cortex and medulla), they are located on the top of both kidneys, in the retroperitoneum; they regulate blood pressure, electrolyte balance, immune response, stress reaction, metabolism, and salt and water balance; some of their functions/steroidogenesis also regulate **HPA** axis, of which it is a component. Adrenal cortex and medulla produce (a) the corticosteroids cortisol and aldosterone, (b) the catecholamines adrenaline/ epinephrine and noradrenaline/norepinephrine, (c) corticotropin-releasing hormone (hereafter **CRH**), (d) dehydroepiandrosterone (precursor for male/ female sex hormones, hereafter **DHEA**), and (e) antidiuretic hormone named vasopressin (hereafter **ADH**).

8. *Pancreas*: It is located across the back of the abdomen: its endocrine cells and subtypes (almost 1–2%, of the entire organ, clustered in islets) regulate blood glucose levels by insulin and glucagon hormone production and secretion.

9. *Placenta*: It is a fetal annex located into the maternal womb, derived from the early fusion of chorion and allantoid in the precocious stage of gestation; at the same development time, in the very early time of pregnancy, the syncytio/ cytotrophoblast cells synthetize (a) the chorionic gonadotropin glycoprotein hormone (hereafter **hCG**), (b) progesterone, (c) a group of estrogens (estrone, 17β-estradiol, estriol, esterol (hereafter, respectively, E_1, E_2, E_3, and E_4), (d) human placental lactogen (hereafter **hPL**), and (e) human placental growth hormone (hereafter **hPGH**), which are fundamental for gestation. It is an essential endocrine gland, thanks to its hormones, during pregnancy for itself, the mother, and the fetus; other hormones are linked to the placental functions [11].

10. *Ovaries*: With their three zones, they are located on both sides of the uterus; they are the female reproductive unit by production of follicles, oocytes, and corpus luteum cells via their main steroid hormones, estrogen, progesterone, and androgens. These glands are an integral part of the **HPG** axis, necessary for fertility and breast growth.

11. *Testes*: With their two zones, they are located behind the penis; they are the male reproductive unit by sperm and testosterone hormone production. Also, the endothelium has endocrine function by its **ECs**, which release vasoactive signals for vasodilatation or vasoconstriction. Through the feedback, they influ- ence the hypothalamic-pituitary axis.

Other organs own secondary endocrine functions, including bones, kidneys, fat tissue, liver, and heart. Thanks to the hormones produced, **EGs** can exert control of most of the bodily functions: mood, emotions, sexual function, reproduction, sleep, metabolism, growth, and others; their functionality is interconnected with the link of microvasculature and **ECs** [9, 10]. To define the origin of the functional anomalies, which also occur when a mechanism step of these glands does not work as it should, and therefore to treat them to restore their interface with the reproductive mechanisms, the pathophysiology and the endocrine clinic of human reproduction have emerged as dedicated medical branches. Moreover, it has also become possible to understand the importance of the physiological and necessary correct functioning of the endocrine glands and the metabolic control of adipose tissue, liver tissue, and other units, in turn regulated by adrenaline, noradrenaline, and insulin; the renal functions checked by angiotensin and renin; and the sex differentiation and growth regulated by sex hormones. These precious and essential functional units can be altered by endocrine disruptors, which are numerically relevant environmental substances, which can potentially cause serious damage to the health, especially in particular crucial and sensitive phases of the human life.

9.2.1.2 Hormones

Hormones are chemical substances produced, in several types, by switching on several genes responsible for their synthesization; the word hormone was derived from "hormao," a Greek term that means "put in motion" and later "that sets in motion"; they are active molecules, with chemical features of proteins derived also from steroids, equipped with high functionality and specificity. In order to perform their tasks, they are able to carry information and instructions from units of cells to another one, stimulating several cellular activities by specific receptors recognized. They act as "messengers" and are flowed out, by endocrine glands, directly into the glandular interstitial spaces (and not in ducts as occurs for the products of the exocrine glands) where they are then absorbed into the blood to be then distributed by the circulatory stream; hence, an adequate blood supply is then necessary to transport them to all the target body sites of these sophisticated molecules. After reaching the programmed destination, they bind to their target receptors, triggering a cascade of intracellular signals that induces that related cell's actions. To ensure bodily functions, certain processes must be carried out correctly; therefore, hormones have to be produced in the right quantity; if an abnormal quantity of hormones is produced, common endocrine disorders develop as well as hormonal imbalance. Hormones are classified into five main categories and then subclassified into other much more specific groups: (1) by effects: metabolic (for instance insulin), kinetic (pineal), and morphogenetic (for instance thyroxine); (2) by chemical nature being water or lipid soluble: steroid, amine, peptide, protein, glycoprotein, and eicosanoid; (3) by the stimulation of endocrine glands: tropic or no tropic; (4) by the action mechanism: group I binds to intercellular receptors and group II binds to cell surface receptors, using then a second messenger; and (5) by the nature of function: local or general; furthermore, a direct classification sees the distinction in amine/peptide (that fit the cell membrane receptors) and steroid (that fit the intracellular receptors) hormones [12]. The sex steroids testosterone, dihydrotestosterone, and estradiol are linked, through the bloodstream, in an inactive form, to the sex hormone-binding globulin (hereafter **SHBG**), a globulin produced by the liver, probably involved even in metabolic disorders. To determine the effect of a hormone, its dose-response activity is necessary: therefore, for example, the concept of "non-monotonic dose-response curve" was also introduced and enhanced, defined as "a complex relationship between the dose of a substance and its effect, such that instead of a certain response simply increasing or decreasing with dose, the curve may be for example 'U' shaped" [13].

The main hormones, synthesized by the relative glands, are shown in Fig. 9.1 through a schematic illustration of the main organs, with some anatomical details, composing the endocrine system.

9.2.1.3 Receptors

Hormone receptors are specific sites that are already created in the fetal life, also known as "*docking*" molecules [10], placed on the surface membrane (hereafter **MemRs**) or inside the cell, like cytoplasmatic receptors (hereafter **CRs**) or nuclear receptors (hereafter **NRs**) as a sort of lock-and-key model (Fig. 9.3) [14]; for example, the estrogen receptors (hereafter **ERs**: ER_α and ER_β), that act as transcription factors, are intracellular proteins of the superfamily of **NRs** and are both **CRs** and **NRs**; **NRs** include also the androgen receptors (hereafter **ARs**), glucocorticoid receptors (hereafter **GRs**), progesterone receptors (hereafter **PRs**), and mineralocorticoid receptors (hereafter **MRs**) [12]. They can interact and bind "with and to" the endocrine hormones through different times and modalities: for example, protein-structured hormones react with the receptors located on the cell surface, and therefore the resulting events are more rapid; in contrast, fat-soluble products have the ability to diffuse through the cytoplasmatic membrane and the nuclear envelope, as steroids, estradiol, testosterone, progesterone, cortisone, aldosterone, thyroid hormones, and retinoids, and typically have interactions with receptor sites in the intracellular area, effectively causing protein synthesis which requires a relatively slower functioning. The link between the hormones and a specific receptor causes a distinct and precise physiological effect in the targeted cells. The bodily functions listed in the above section originate from the binding between hormones and receptors, which triggers a cascade of different events according to the type of hormone and the way of link; they need to be in adequate amounts in the target tissues and/or cells, which must be able to give the right reply to the hormonal signal; furthermore, if there is no correspondence among a receptor site and a hormone, no physiological reaction occurs. The connection between receptor functions and mechanism of action of environmental interferers has been extensively studied to obtain information as detailed as possible [14]; **MemRs**, **CRs**, and **NRs** can be stimulated, via several alterations, to increased, decreased, or reverse activities, based on the nature of the ambient substances and their biological potential.

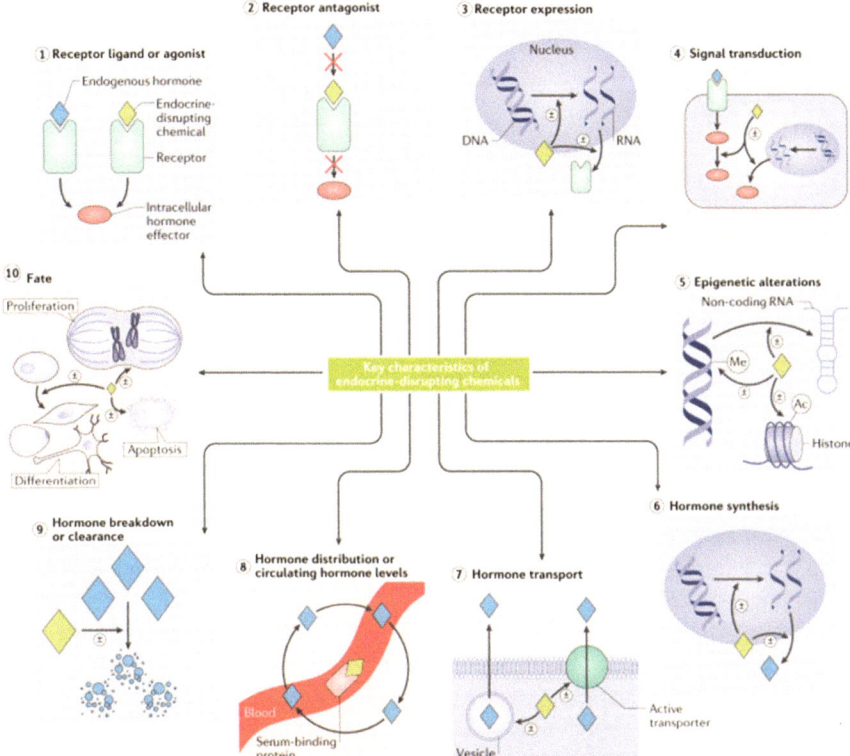

Fig. 9.3 The key characteristics of endocrine-disrupting chemicals. Arrows identify the ten specific key characteristics (KCs) of endocrine-disrupting chemicals (EDCs). The ± symbol indicates that an EDC can increase or decrease processes and effects. KC1 states that an EDC can interact with or activate hormone receptors. KC2 states that an EDC can antagonize hormone receptors. KC3 states that an EDC can alter hormone receptor expression. KC4 states that an EDC can alter signal transduction (including changes in protein or RNA expression, posttranslational modifications, and/or ion flux) in hormone-responsive cells. KC5 states that an EDC can induce epigenetic modifications in hormone-producing or hormone-responsive cells. KC6 states that an EDC can alter hormone synthesis. KC7 states that an EDC can alter hormone transport across cell membranes. KC8 states that an EDC can alter hormone distribution or circulating hormone levels. KC9 states that an EDC can alter hormone metabolism or clearance. KC10 states that an EDC can alter the fate of hormone-producing or hormone-responsive cells. Depicted EDC actions include amplification and attenuation of effects. *Ac* acetyl group, *Me* methyl group. Original image and text: full version, published under CC-BY-4.0 license and permission of the author Michele A. La Merrill, PhD, MPH Associate Professor, Chancellor's Fellow, Department of Environmental Toxicology Genome, Environmental Health, Comprehensive Cancer, and Perinatal Origins of Disparities Center, University of California at Davis [14]

9.2.2 Reproductive Health

Reproductive health (hereafter **RH**) had already been identified and addressed in its relationship with the likely adverse effects related to the environment [1, 8]. **RH** is a situation *"in which the human reproductive activity is expressed in full physical, mental, and social well-being"* [8] and occurs despite the presence of diseases also of the reproductive system [8, 15]; expressing a series issue for both health and environment, **RH** has become a thermometer of infertility-related disorders that can often be traced back to environmental etiology [8]. **RH** is constantly threatened by polluting and disrupting environmental factors such as consumer products, chemical substances, radiation, and stress [1, 7, 8, 14]. On the other hand, based on global assessments, socio-economic-cultural status, malnutrition, and infections must also be considered; it includes wide-ranging issues related to ambient and health [8].

Furthermore, adverse effects on human well-being and development of **RH** were reported, due to dibromochloropropane (hereafter **DBCP**), **PCBs**, methyl mercury, arsenic, and lead, having more data from studies on animals [8, 14]. The protection of **RH** starts from the purely technical concept that indicates the endocrine system as the organ responsible for controlling the reproductive system, its development, and its function; through the **RH** monitoring and its reproduction rate, the results in several areas with the identification of risk factors can be assessed and analyzed also to acquire new forecast and prevention strategies [8]. The factors capable of negatively influencing the reproductive health, in each of its phases, had and have gradually become the subject of important international studies and policies, aimed at epidemiological control and at development of preventive measures. Factors involved in negative **RH** interference, such as smoking, nutritional habits, infection disease, and stress, should also be mentioned for the sake of completeness; in fact, it has been widely reported in the literature how the use of smoking in pregnancy [8, 15] has been associated with an increase in maternal and fetal pathologies such as placenta previa, placental abruption and premature or low birth weight of the newborn at birth, and other serious obstetrical conditions due to "the numerous environmental toxins present in tobacco smoke" [15].

9.2.3 Environment

Ambient is a word derived from the Latin "ambiens," interpreted as "everything to go around and to surround something"; it plays the predominant role in determining and regulating life cycles. In order to have a unique language and therefore frame the context examined below, the need to define the term environment thus emerges: it is the physical space, with physical and biological elements that interact, as well as the system of external situations, with biological features, surrounded and influenced by its conditions, where multiple forms of life are inserted [16]; it can be natural or artificial, and marine or mountain, and must be under sets of measures, regulations, and laws to protect it from any kind of pollution or alteration; the ambient issue, as it concerns health, has become central in every state and nation, globally. In terms of etiological agent, the same definition of environment, provided by the **WHO**, denotes the importance of the issue: "environment is a major determinant of health, estimated to account for almost 20% of all deaths in the **WHO** European Region"; in 2010, the focus was also on the need to open a new action era, based on the past world indications, for our health and environment [17]. The Italian Constitution protects the ambient through the fundamental principles and states: "The Republic promotes the development of culture and scientific-technical research; it protects the landscape and the artistic heritage of the nation. It protects the environment, biodiversity and ecosystems also in the interest of future generations. State law governs the methods and forms of animal protection" [18]. The known dependence between living beings and nature has therefore increased worldwide awareness of the negative effects of the ambient on the health; stratospheric depletion of ozone, long-range air pollution, climate change, toxic waste export, acid deposition, and loss of biological diversity are some of the global issues. It has been known for decades that the endocrine system and the reproductive functions, regulated by the first one, are greatly influenced by unfavorable environmental factors. Moreover, in the world, there are negative socio-economic conditions and cultural and/or pollution influences that compromise the reproductive health, which are constantly increasing; multiple chemicals are present at the same time and overlap with other ambient problems, making it difficult to assess the effect of a single harmful element. Epidemiological studies and data collected, under precise and systematic surveillance, are basic sources for the acquisition of new and updated information and are areas of fundamental importance for weighing the gravity and the risk of harmful factors and for developing approaches and prevention strategies. The determination of the concept environment, in ecology, includes two closely related aspects: the physical and biological contexts (for example: humidity, climate) that surround an organism, a human population, or a biotic community; from the biological point of view, it means what influences life and development; therefore, the entire biosphere is a set of several environments on the Earth, expressing varying levels of global environmental contamination by several xenobiotic compounds, also used for metabolic functions of microorganisms capable of catabolizing them by processes such as degradation and enzyme or gene pathways. Understanding of the real environmental conditions of a given population can lead to the study of useful performances for the protection of the public health, and therefore also in the field of reproductive health. Many environmental factors have the capacity to influence the latter aspect of a particular community: for example: ethnic, health, chemicals, radiations, infections, malnutrition, stress, social status, and cultural influences. Absurdity situation as in some countries and major technological and economic developments have shown a negative impact on the population, while, in countries developing with many life difficulties, the lack of accessibility to reproductive health care is the main issue of people. Furthermore, several scientific evidences have demonstrated that the negative effects of various environmental components, as in consumer products, water, and food, as smoke, polycyclic aromatic hydrocarbons, air pollution, pesticides, and others, are much more incisive on minority populations, fetus in utero, children, and adolescents. Thus, it gave rise to the need to carry out also targeted studies on individual exposure to the environmental chemicals, as well as on the ecological action as an etiological factor.

9.2.4 Endocrine Disruptor Chemicals

Following the above introductive overview, it can be understood that a malfunction caused by human diseases of some endocrine unit components may, less or more severely, influence a lot of matters regarding human physiological homeostasis and/ or functions. Indeed, in mammals, endocrine messages, coordinated by **MemRs**, **CRs**, and **NRs** [14], are responsible for **RH**, placenta-embryo-fetal growth, energy management, electrolyte balance, and other fundamental bodily functions and systems [19].

Several evidences had demonstrated that various ambient negative conditions are able to falsify human endogenous hormone activities, compromising their regular tasks; many natural (for instance, phytoestrogens in soya) or synthetic hormones and other man-made chemical molecules (for instance, pesticides or air pollutions or plastic substances for industry) may mimic, interact, or interfere with the physiological hormonal functions [1, 3, 8, 14, 20]. Various hypotheses had been formulated of dangerous disturbance capacity for almost 2000 substances, concerning, at least, one of the three main hormonal pathways, that is, estrogens, androgens, and thyroid [3]. The U.S. Environmental Protection Agency (hereafter **EPA**), studying both ecological and human health effects and working on recommendations for future research, had given those molecules, which have a disturbing action on the endocrine system, a definition as "an exogenous substance that interferes with synthesis, secretion, transportation, metabolism, binding action, or elimination of natural blood-borne hormones that are present in the body responsible for homeostasis, reproduction, and development" [21]. Moreover, there was the WHO 2002 **EDC** formulation as "an exogenous substance or mixture that alters functions of the endocrine system and consequently causes adverse health effects in an intact organism, or its progeny, or (sub)populations" [20]. Therefore, data collection studies had begun, also according to the guidelines developed to date by the **EPA**, first on the consequences of the aforementioned molecules on human and animal health, rather than the mechanisms of action and the involved organs [21]. Those dangerous materials were also identified as heterogeneous and deriving from various sources present in the ambient; therefore, the **WHO** and the United Nations Environment Programme (hereafter **UNEP**) had elaborated another indicative and explanatory term for those compounds, "Endocrine **D**isrupting Chemicals" (hereafter **EDCs**), that are artificial molecules (exogenous/xenobiotics substances) with a heavy negative impact on health, being able to interrupt, as well as selectively modify, the hormonal axes regulating the reproductive systems and those responsible for body growth; moreover, the same **WHO/UNEP** Programme had identified "800 environmental chemicals that are known or suspected to be capable of interfering with hormone receptors, hormone synthesis, or hormone conversion" [22, 23]. A similar definition for **EDCs** has been elaborated by the Endocrine Society: "an exogenous (not natural) chemical that interferes with any aspect of hormone action" [24]. **EDCs**, due to their adverse aftermath, are in continuous analysis and monitoring by the **WHO**, the European Commission (hereafter **EC**), the **UNEP**, the United States Food and Drug Administration, the National Institute of Environmental Health Sciences (hereafter **NIEHS**), the **EPA**, the Endocrine Society, the International Federation of Gynecology and Obstetrics (hereafter **FIGO**), the Chemical Agency, the European Food Safety Authority (hereafter **EFSA**), the European Commission's Joint Research Center, the Centers for Disease Control and Prevention (hereafter **CDC**), the American Academy of Pediatrics, the Organisation for Economic Co-operation and Development (hereafter **OECD**), and international societies and groups as **EDC** experts, active in updates and debates on the evaluation of risks, epidemiology, and prevention of the negative exposure effects on the public heath [25]. In fact, through the **WHO/UNEP** 2012 document, the **UNEP** and the **WHO** updated the 2002 International Chemical Safety report, which highlighted the significant increase in endocrine abnormalities and disease in humans, likely due to **EDC** exposure [23].

Moreover, critical "windows" of greater sensitivity in human life, as puberty or embryo-fetal growth, were identified as highly at risk for the development of irreversible effects after exposure to **EDCs**, which act with a mechanism of action both at the tissue and cellular levels. The **WHO** and the **EC** have screened more than 100 molecules as potential **EDCs**, in order to perform strategies for the sustainability of an ambient of better quality [22, 25, 26]. Furthermore, following a precise rationale and starting from the model of carcinogens, ten **EDC** key characteristics have also been developed (Fig. 9.3), based on their effects and hormonal action; to perform this evaluation, models such as (1) diethylstilbestrol, (2) bisphenol A, and (3) perchlorate were used. In fact, this interesting, schematic, and effective approach to this global issue offers a uniform evaluation of the key characteristics to identify the **EDCs** with research methods aimed mainly at their ability to negatively interact with the endocrine systems [14]. The three models used can be summarized as follows: (1) already abovementioned drug used to avoid miscarriage or preterm labor with controversial results; (2) a drug used in the first half of the 1900s and currently found, for example, in plastics also for medical and sports artifacts, in materials to transport food, or in dental issues and paints; (3) those found in water, vegetation, and soil and associated with rocket propellants, pyrotechnic article explosives, and missile motors [14]. The **EDC** was even described as "an exogenous agent that interferes with synthesis, secretion, transport, metabolism, binding action, or elimination of natural blood-borne hormones that are present in the body and are responsible for homeostasis, reproduction, and developmental process" [22]. Compared to the initial studies, from which it seemed that they could only act on and through the nuclear hormone receptors (for example receptors of estrogen, androgen, progesterone, thyroid), recent evidence shows that their function is also directed on membrane hormone receptors, steroid and nonsteroid, and orphan receptors (with no ligand identified) [13, 27]. Besides, environmental factors were evaluated (also by numerous epidemiological and observational studies on animal models and few on human cohorts) during pregnancy for maternal-fetal and/or neonatal well-being without observing the preconception period: therefore, it was also recognized as a further sensitive window for both sexes. Epidemiological studies have found associations between **EDCs** and adverse outcomes on **RH**, pregnancy, male/female fertility, and early life; furthermore, biological samples were collected in a study of male and female cohorts in preconception, pregnancy, and various trimester times [28]. The access routes of the dangerous substances can be several: humans, during everyday life or work activity, come into contact with **EDCs** through food and beverages, pesticides, and household and cosmetic products; concretely, the contact may be through diet, air, skin, and water. The physiological endocrine functions are sensitive to even small changes caused by low hormone levels, determining the related significant biological effects; therefore, scientists and expert analysts have realized that even low doses of **EDCs** may be dangerous.

9.2.4.1   Endocrine Disruptor Chemicals and Action Overview

The negative consequences of **EDCs** in wildlife had already been reported since the 1950s [3]; since then, hundreds of thousand dangerous materials to the health of living beings, humans in particular, have been identified; epidemiological and biomonitoring studies were then carried out to control **EDCs** which, as already indicated, could interfere with hormonal functions. At first, there were evidenced issues regarding the amount of the sperm and cancers as prostate, testicular, or breast, which could have an endocrine correlation and ambient etiologies [5, 21]. Subsequently, over some decades and thanks to the acquisition of new scientific elements, researchers have refined the key characteristics such as receptor agonist/ antagonist or modifiers of the receptor expression, or directly the target tissues; particular mention must be made of the three main human endocrine axes, **EDCs'** potential targets: hypothalamus-pituitary, adrenal, and thyroid glands [23, 27, 29, 30]. Moreover, in 2013, the **EC** indicated three **EDC** action criteria as fundamental for their recognition: "(1) endocrine activity, (2) deleterious and/or pathologic endocrine-mediated activity, and (3) cause–effect relationship between substances and endocrine activity in exposed subjects" [31, 32]; the Endocrine Society also mentioned more **EDC** action mechanisms such as "interference with any aspect of hormone action" [24], DNA methylation, DNA acetylation, and histone alterations, defined as "epigenetic changes" [24, 32]; they can act like genomic to exert some biological effects practically [12]. Furthermore, a greater susceptibility and sensitivity to toxic molecules of the subjects were also highlighted monitored by various research studies in the ambient epidemiology area, based on the human age, such as fetal life and childhood [33, 34]. Those molecules are mostly from the sectors of industrial productivity, aimed at forming materials for various activities and jobs; their access routes are mainly through solid and/or liquid feeding, breathing, and contact; along with the most popular lists, the main distinction is based on the criteria of persistence, as **DDT**, or non-persistence, as phthalates (hereafter **PhTh**) [33–37]. On the other hand, for other opinions, there seems to be no international effective schedule of **EDCs**, not even based on the action mechanism [14]; a basic list is given below, in order to better outline the known data, obtained from studies on human and experimental models that concern global environmental contaminants.

- *Persistent Organic Pollutants*: Persistent organic pollutants (hereafter **POPs**), coming from air pollution, are carbon-based organic chemicals, also classified as a group of **EDCs** man-made for industrial utilization, then released into the ambient, characterized by stability and a long half-life, and therefore classified as "persistent"; they get the ability to be bioaccumulated in living organisms and also humans. Some **POPs**, even of high molecular weight, can also cross the human placenta and get to the fetus, consequently [38]. The **POPs'** effects on human/animal health involve multiple organ systems such as nervous, reproductive, and endocrine up to carcinogenic effects; their toxic potential on health is associated with several variables, such as the synergy with other substances, the capacity to be absorbed and accumulated, as well as the ability to interact with **MRs**, **CRs**, and **NRs** present on the target tissue of hormone products.

  - *Dichlorodiphenyltrichloroethane* (**DDT, DDTs, p,p'-o,p'-DDT, p,p'DDE**): This **POP** is an insecticide used in agriculture, banned in the United States since 1972 [39] and still in use in some countries, especially for the malaria control; **DDT** was paid attention due to its effects on the reproductive and sexual systems [2, 14, 25]; **DDTs** and its metabolites, as diphenyl dichloroethane (hereafter **DDE**), often present together, have different chemical structures and different capacity of estrogenic/antiandrogenic actions as **EDCs** [14, 40] and possess a long persistence in animal tissues and environment; therefore, it can be taken by humans in every vital epoch through food or by contact with contaminated products, which then accumulates in the adipose tissues and crosses the placental barrier in pregnant women (see also in dedicated section). There were many papers, also contradictory ones, that found obesogenic and ovarian **DDT** consequences, based on animals' experimental evidences. In order to highlight the concept of perinatal exposure, it was examined in the **C**hild **H**ealth and **D**evelopment **S**tudies (hereafter **CHDS**) cohort, starting from the serum samples collected from the first generation of patients, the grandmother ones, in the 1960s; this was the first human research to hypothesize the association of grandmothers' exposure to **o,p'DDT** with the outcomes in daughters and granddaughters (subjects belonging to three generations) regarding early menarche time, adiposity, and obesity that are mostly risk factors for breast cancer [40].
  - *Per- and Poly-fluoroalkyl Substances* (**PFASs/PFCs/PFOA/PFBA/PFHxA/ PFHpA**): **PFASs** are man-made fluorinated polymer/non-polymer

compounds; due to their chemical-physical characteristics (repellency, lipophilicity, thermal-chemical stability, hydrophobicity, and resistance and also the natural process of biotic degradation, photolysis, or hydrolysis), they are widely used in industrial sectors, such as tools for personal care, firefighting foams, nonstick pan, paper, textiles, coatings, and food storage. **PFASs** are released into the environment, are persistent and nonbiodegradable, continue to bioaccumulate without decomposing, and could contaminate water and foods, also by their packaging; new evidences indicate that **PFASs** are dispersed through the air over long distances: widespread exposure to **PFASs** has been detected in the US population. The human body comes into contact through food, ingestion, or inhalation of dust. **PFASs** could hold antiandrogenic potential and a link with male effects due to the interaction with androgen receptor activity and sex hormones; moreover, they can reach the fetus through the passage of the placental barrier [41, 42].

– *Polybrominated Diphenyl Ethers* (**PBDEs**): **PBDEs** are organohalogen substances used to make flame retardants for household products such as furniture foam and carpets. They can get bromine, fluorine, or chlorine atoms and from these characteristics derive the various names assigned to the different chemical compounds. **PBDEs** are lipophilic with androgenic and estrogenic abilities, thus being able to interfere with the development of the sexual sphere by postponing the male pubarche or anticipating the female menarche [43]; further data will be needed to support the evidence ascertained so far, also related to the **b**ody **m**ass **i**ndex (hereafter **BMI**) of the monitored subjects in some studies [43].

– *Polychlorinated Biphenyls* (**PCBs**): **PCBs** are man-made compounds used to make electrical materials like transformers and are also used in hydraulic fluids, heat transfer fluids, lubricants, and plasticizers; they are persistent and therefore difficult to dispose of; even though they have been banned in the United States since 1979, the exposure still occurs today, due to the presence of previous artifacts or mixtures and even 15 years of estimated half-life; several monitoring and epidemiology studies on humans and experiments on animals have reported the effects of **PCBs** on newborns' weight and on head circumference, due to prenatal exposure, through their passage from the placental barrier [44]. Moreover, **PCBs** have androgenic, estrogenic, and antiestrogenic effects [45–47], and, with their subtype, they turn on a thyroid receptor [14].

• *Bisphenol A* (*BPA*): **BPA** was first synthesized in 1891 and is used to make plastics and epoxy resins and found in many products (for instance, food storage containers); humans can be exposed through food mostly; it was never used as a drug. Of all **EDCs**, **BPA** is probably the substance endowed with greater estrogen-like capacity; many epidemiological research studies reported about the sexual maturation of girls exposed to this substance in utero, but the findings were controversial [43]. This theme is most in focus by experts in the sector, even if further investigations are needed to identify all the molecular routes involved in their capacity to act as endocrine disruptors [14, 43].

- *Dioxins* (2,3,7,8-tetrachlorodibenzo-p-dioxin, hereafter **TCDD**): These are industrial substances achieved as a byproduct in herbicide manufacturing and paper bleaching; their presence in the ambient is due to waste burning and wildfire; dioxin exposure appears to be associated with female cancers, impaired fertility/fecundity, endometriosis, and incorrect time of puberty age [19, 24, 42].
- *Perchlorate*: It is a substance produced by the pharmaceutical, aerospace, fireworks, and weapon industries. It can be found in drinking water. Its structure is in some respect similar to the iodide ion; therefore, it can interfere with the physiological function of the thyroid, causing, as an inhibitor, the blocking of the synthesis of thyroid hormones with its negative effects on body metabolism [48].
- *Phthalates* (**PhThs**: dibutyl phthalate (**DBP**), di-2-ethylhexyl phthalate (**DEHP**), benzyl butyl phthalate (**BzBP**), polyethylene terephthalate (**PET**), and others): These are diesters of phthalic acid with the characteristic of non-persistence; for example, **p**oly**v**inyl **c**hloride (**PVC**) is used to make plastics more flexible; they are also used in some food packaging, detergents, cosmetics, children's toys, and medical devices; they can come into contact with humans by food, water, skin, or breath. These **EDCs** may interfere with the physiological function of the hormonal axes that regulates the development of the male and female sex; several studies have evaluated phthalate metabolites in urine to obtain data without risk of external contamination [49]. Alterations in the masculinization of the male fetus were reported: experts showed, on a small cohort, a decreased anogenital distance in human male infants, exposed to phthalates, during their intrauterine life [49]. Furthermore, phthalates' association with earlier puberty in females and the hypothesis of an estrogenic role by **PET** of water packaging were highlighted [50]. However, other studies show different evidences; therefore, phthalates could exercise antiandrogenic and estrogenic skills, but more data is needed to confirm the aspects related to pubertal timing [43].
- *Phytoestrogens*: For decades, they have been included in the diet to meet nutritional needs, and also as a substitute for animal proteins, found in many food plants, like soya; they have hormone-like activity; and their presence has already been correlated to the increase in **SHBG** synthesis, probably interfering with the action of endogenous estrogens [5]. However, to date, other phytoestrogens, such as genistein and daidzein, are in soy products, like tofu or soy milk, and their biological effects are probably attributable not only to their environmental presence [5].
- *Parabens*: Parabens are chemical preservative substances, 4-hydroxybenzoic esters in fact, used, for over 50 years, in detergents, food, pharmaceutical, and cosmetics to fight fungi and several harmful matters; in vivo and in vitro evidences show that, among other types, butylparaben and propylparaben can negatively interfere with the endocrine system also due to their possible estrogen-like action [51]. In addition, parabens' presence in the adipose tissues of breast was observed, made possible by their characteristic of mild hydrophobicity;

however, to examine the connection with the breast cancer etiology, future observation of parabens' and other EDCs' exposure effects is needed [51].

- *Triclosan* (**TCS**; 5-chlorophenol): **TCS**, first used as a pesticide since 1969, is an antibacterial and antifungal chlorophenol and may be found in some antiseptic, disinfectant, and personal care products, like toothpaste, soaps, shampoo, body lotions, and creams; antibacterial products with triclosan were excluded by the **FDA** from sale in 2016–2017; it was included to **EDC** list [23]. People may be exposed to this endocrine disruptor through the antiseptic and cosmetics used, via skin and/or ingestion; it was first detected in human milk and then in plasma and urine [52]. In essence, **TCS** could get the ability to interfere with the endocrine functions and **RH**, related to its chemical structure (2-phenol); its action as estrogenic molecules has been highlighted, even if the available little epidemiological data are contradictory [52].

- *Metals*: Metals are substances with high electrical and thermal conductivity, and also high *ductility* and malleability among other properties; heavy metals are difficult to metabolize; therefore, aquatic environment/organisms can accumulate them; in fact, these metals are considered the main pollutants of aquatic reserves. There are experimental and epidemiological studies on both animals and humans, which have reported negative effects on organisms. Moreover, some of them can cross the placental barrier, reach the fetus having a teratogenic effect, and disrupt the hormones needed for the pregnancy, with the following increased risk for stillbirths or spontaneous abortion [12, 35]. According to the **EDCs'** definition [20–24], this group of contaminants have been included in the category of endocrine disruptors, and several analyses have been oriented on their unsafe action mechanism. Some heavy metals (copper, aluminum, cadmium, and lead, for example) have been identified as "metalloestrogens" for their interference, also as agonist, with the physiological function of estrogens, **ERs**, and the consequent elements of response to that hormone.

  - *Cadmium*: Cadmium (hereafter **Cd**) is a contaminant that can be found in significant quantities in ground and water collection basins due to its release from various polluting industries [12]; humans can come into contact with **Cd** through the intake of contaminated food or contact with relative adverse consequences on organs and body systems, as widely reported in literature also for the production of **GH**, **ACTH**, and **TSH** hormones in rats (even ovary progesterone) and human plasma [12]. Furthermore, experimental study had evidenced a decrease of placental progesterone production from human trophoblast cell cultures, by interfering in the cholesterol accumulation, necessary for this essential process; on the other hand, these data were contradictory according to other in vitro and in vivo studies [12]. On the other hand, prenatal exposure to Cd had been correlated to negative pregnancy outcome resulting from placental defects, and female early puberty time in the offsprings [12]. Due to **Cd** body storage organs, researchers also highlighted that **Cd** exposure can contribute to diabetes mellitus (hereafter **DM**) development and progression [53].

– *Mercury*: Mercury (hereafter **Hg**) is widely present in several natural environments, industrial artifacts, and food chain; frequently, **Hg** exposure occurs through the intake of contaminated food, especially fish. **Hg** exposure appeared to be related to the increase in female hormones, by testing, for instance, **Hg** levels in human blood and hair, therefore hypothesizing its role in the stimulation of this precise hormonal synthesis [12]. There were also evidences that correlated Hg levels with the thyroid endocrine system [12]. Methyl mercury, the natural/synthetic bioaccumulation form (hereafter **MeHg**), had already begun to be considered in relation to neurological alterations in exposed fetus and children [1]; in 2004, the **FDA** and the **EPA** set up a declaration to warn the population about the adverse effects of **MeHg** on brain development.

– *Arsenic*: Arsenic (hereafter **As**), like other heavy metals, is present in several sectors of industry and agriculture; human exposure occurs both for work activity and for contaminated food intake. Many experimental studies have shown that the **As'** dangerous consequences for human health are due to its role as **EDCs**, interfering with some types of hormone receptors and their expression [12, 53]. Moreover, due to **As** body storage organs, such as **Cd**, was also highlighted as **As** exposure can contribute to **DM** development and progression [53].

– *Lead*: The role of lead (hereafter **Pb**) exposure in pregnancy should be mentioned; several scientific evidences have hypothesized the correlation between lead and preeclampsia (a serious pregnancy disease characterized by hypertension and proteinuria); moreover, a meta-analysis showed a high association between the bodily presence of this metal and the development of gestational hypertension through the possible increase of vasoconstrictive substances and reduction of vasodilating ones, with the related vasoconstriction and placental ischemia, and also proteinuria due to a direct adverse action on the renal and endothelial physiology. In particular, the lead levels and their dose effect were also detected.

• *Tobacco Smoke*: The use of tobacco smoke is a matter of great impact: it is evaluated that for 2030, the deaths caused by its use will be more than 8 million [15]. Over any period of human life, tobacco smoke is a notoriously harmful habit for every aspect related to human health. Moreover, it had already been reported that the use of tobacco smoke, or its passive exposition, in the gestational period negatively affects the fetal development and is associated with respiratory disorders and defects of brain development, intrauterine growth, and respiratory disorders in the fetus and newborn [8, 15]. Cigarette smoke has more than 4000 chemical compounds, and many of them have the ability to interfere with brain processes; in particular, nicotine holds the detrimental, serious "neuro-teratogen power" by interfering with neuroanatomy, cell life, and subtypes of neuronal nicotinic acetylcholine receptors (hereafter **AChRs**), which are found in the fetus as early as the first trimester of pregnancy [15]. Besides, cigarette use is an important source of **Cd**, with the consequences, among others, mentioned in the dedicated section [53, 54]. Also, there is severe constellation of obstetric and neonatal complications such as premature delivery, placental disorders, and sudden infant death syndrome (hereafter **SIDS**) [15].

- *Microplastics* (hereafter **MPs**): **MPs**, which term had been coined by a marine biologist professor in 2004, are polymer chains made up of carbon and hydrogen atoms, are not biodegradable, are very small snippets of plastic, and measure less than 5 mm (0.2 in.) in length, according to the definition of the world's leading experts (United Nations Expert Panel: United Nations Environmental Programme, **UNEP**) [55]. **MPs** result from man-made matters or from the environmental degradation of various plastic products; chemical additives are also present in **MPs**, as phthalates, **PBDE**, and tetrabromobisphenol A (hereafter **TBBPA**). In recent years, the public opinion has led to greater scientific investigations to increasingly promote strategic operational interventions for monitoring and prevention. Through a prospective preclinical observational study, based on a plastic-free protocol, **MP** particles have also been identified, for the first time, in the human placenta (Fig. 9.4), and precisely both in the maternal-fetal sides and in the amniochorial membranes [56]. In this context, **MPs** seem to be realistically transported in the blood system by maternal respiratory or gastrointestinal organs, where they could also become carriers for environmental pollutants and additives which are elements with adverse effects, thus becoming **EDCs** with probable consequences first on embryo/fetal growth and maternal well-being and afterwards in the long-term life periods. Given the well-known fundamental role of the placenta as an interface between maternal environment and fetus, the above evidences would underline the need to investigate the entry routes and the consequences of **MPs** [56].

9.2.4.2 Endocrine Disruptor Chemical Effects on Human Body Systems, Reproductive Health, Prenatal Exposure, and Offsprings

To the Barker "fetal origins" hypothesis, which highlighted, in models in utero, deficiencies nutrition and metabolic syndrome/cardiovascular malfunctions, and which had been focused on the analysis of the development of the state of health and disease, the prospective role of environmental impact, by its chemicals elements, was also included, especially in the most sensitive "windows" life of the human growth, already from the oocyte stage that is *"in egg exposure"* [14, 22, 40, 46, 57–59]. From this hypothesis, "a new vision of an optimal early human development" is highlighted as a starting point for likely short- and long-term effects on the well-being of the infant and child, considering both birth weight and body load during infancy and beyond [57]. Importantly to emphasize that, during pregnancy and early life, the developmental phases of the fetal organs and systems, which then continue into the postnatal time, make, among the others, the second trimester of gestation particularly vulnerable to negative interference from external stimulation [3, 46]. Particularly, in womb, **EDC** exposure may interfere with the life of the fetus and the following generations, as well [25]. Afterwards, later in life, ambient substance

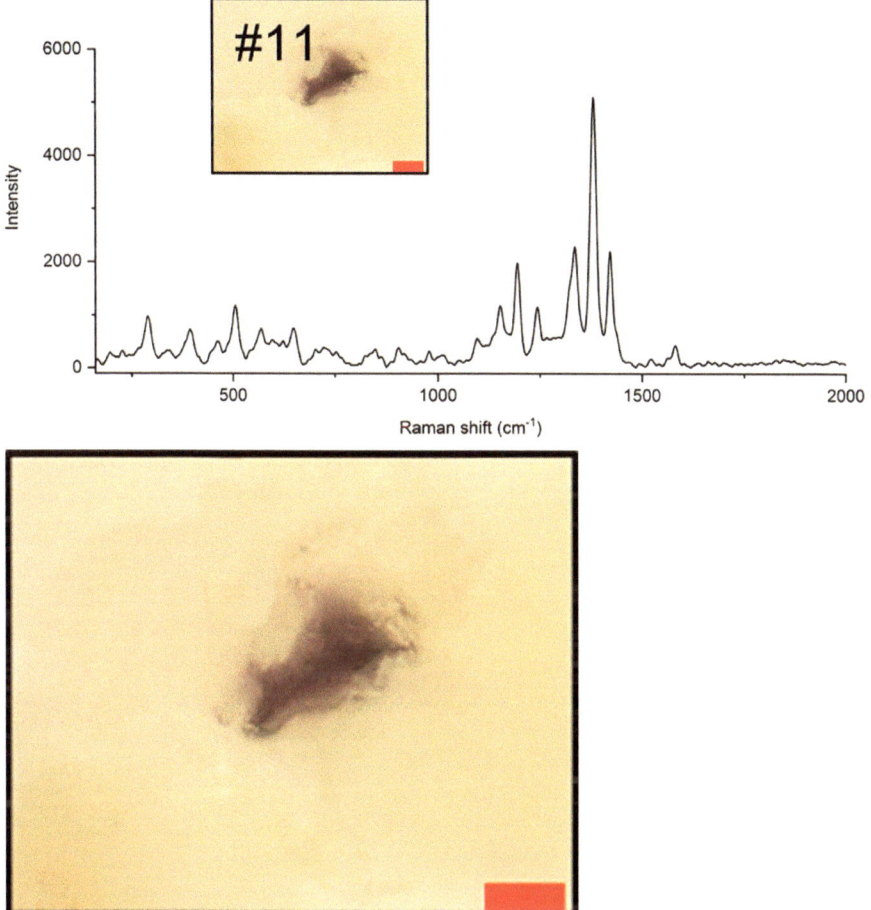

Fig. 9.4 Particle #11. Stained MP particles identified as polypropylene. Original image: full version, published under CC BY-NC-ND license and permission of the author Prof. Antonio Ragusa M.D. Ph.D. Director of Department of Obstetrics and Gynecology, San Giovanni Calibita Fatebenefratelli Hospital, Isola Tiberina, Rome, Italy [56]

interference could alter the physiological development of gonadal cells, resulting in fertility/implantation reduction, fetal chromosomal disorders, and dysregulation of fetal growth with consequences like small for gestational age (hereafter **SGA**) or intrauterine growth retardation (hereafter **IUGR**) and similar [8]. Epidemiological human studies, both in men and women, which is the fundamental discipline for public health/disease analysis [8], highlighted the correlation between environmental chemicals, even enhanced by nutritional factors, and adverse **RH** effects, on male and female fertility and on pregnancy outcomes and increased risk of childhood and adult diseases; moreover, clinical studies projected to assess the impact of the ambient and lifestyle components on fertility and pregnancy outcomes were

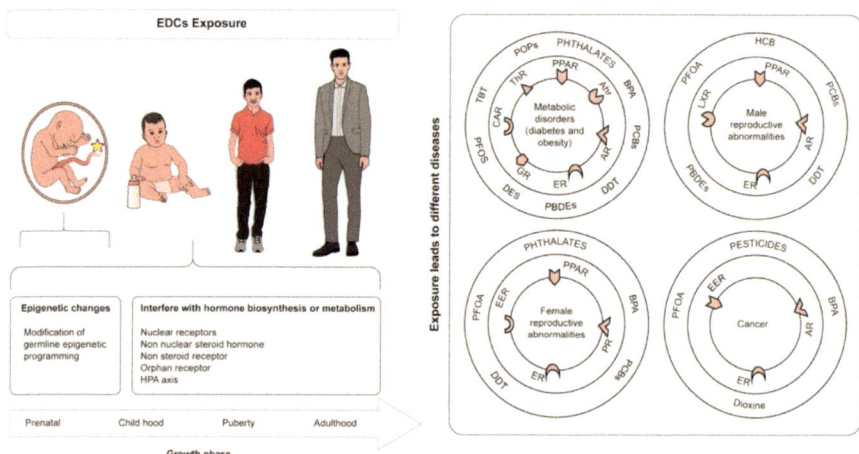

Fig. 9.5 Health effect of endocrine-disrupting chemicals. Tributyltin (TBT), perfluorooctanesulfonate (PFOS), perfluorinated compounds (PFCs), perfluorooctanoic acid (PFOA), bisphenol A (BPA), diethylstilbestrol (DES), hexachlorobenzene (HCB), dichlorodiphenyltrichloroethane (DDT), polybrominated diphenyl ethers (PBDEs), constitutive androstane receptor (CAR), thyroid hormone receptor (ThR), glucocorticoid receptor (GR), aryl hydrocarbon receptor (Ahr), androgen receptor (AR), peroxisome proliferator-activated receptor (PPAR), estrogen receptor (ER), liver X receptor (LXR). Original image and text: full version, with permission of the author Manoj Kumar, Scientist "C," ICMR-National Institute for Research in Environmental Health, Department of Health Research, Ministry of Health and Family Welfare, Govt. of India [36]. The unique modification is the addition of a symbol to connect Fig. 9.5 to Fig. 9.6 to include and show the human placenta, with the role as target, during its development phases

therefore created. An extremely important concept has been focused on the issue that the "susceptibility to the potential health impacts of toxic environmental chemicals can increase when exposure occurs during critical and sensitive developmental periods, such as during pregnancy, childhood, and adolescence" [22]. Therefore, it can understand the link with the Developmental Origins of Health and Disease (hereafter **DOHaD**) theory related to the lifestyle of each lifetime (Fig. 9.5) with both epigenetic and endocrine interferences. Hence, the identification of the critical and sensitive "windows" of the individual's life was possible, which represent the points of greatest danger during the presence of potential **EDCs** [22, 59]. The extent of endocrine interference caused by **EDCs** is also influenced by the age of the exposed subjects, types of molecules, their capacity for negative action as agonists or as antagonists, and dose-response dynamics [3]. Moreover, in the body context, the placenta, which is a transitory organ essential for reproduction, is studied in particular for its vulnerability given by the presence of hormone receptors for steroids on its tissues [46]; the **EDC** exposure of this functional unit is being evaluated, thanks to studies that will, in any case, need to be standardized, even taking into account the product of conception sex [46].

The complex and fascinating process of placental development (Fig. 9.6) could be compromised at various times by many **EDC** molecules and their congeners,

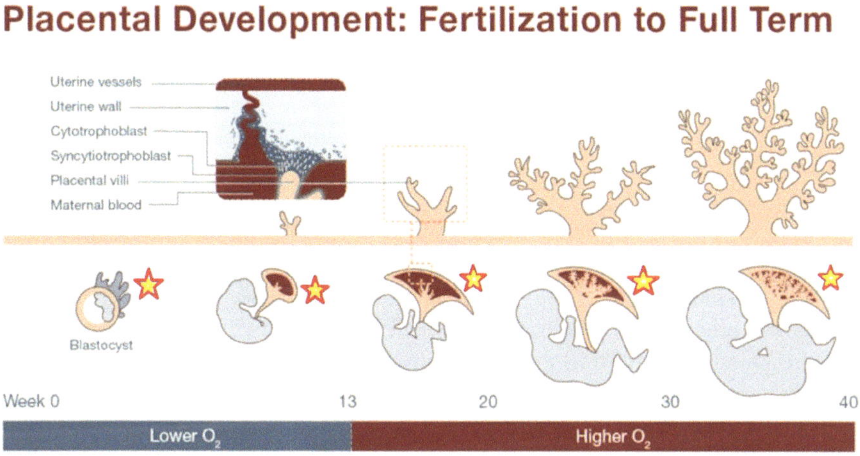

Placental Development: Fertilization to Full Term

Fig. 9.6 The fetus and placenta: development: from fertilization to full term; with permission from Eunice Kennedy Shriver, National Institute of Child Health and Human Development (NICHD), NIH, HHS; https://www.nichd.nih.gov/. Original image; the only modification is the addition of symbol to connect Fig. 9.5 to Fig. 9.6 in order to include and show the human placenta role during its development phases

thus interfering with the multiple placental functions [60]. At present, the data revealing the adverse effects of **EDCs** derive from research studies on animals, mainly; evidence of plausible links between **EDCs** and multiple human pathological pictures includes adverse outcome on metabolism and **RH** and increase in risk factors for cancer [36]; it is evident that diseases affecting various organs or systems can directly or indirectly extend to the **RH**, which is physiologically subject to changes in other body compartments. In addition, exposures to low levels of toxic substances must also be considered, especially in vulnerable life periods, with the probable transgenerational and/or epigenetic injurious impacts; these are some of the criteria that differentiate EDCs from other toxic substances, including the body response to non-monotonic doses [36].

9.2.4.3 EDCs' Effects on Female and Male Reproductive Health

RH should depend on "*cultural, ethnic, social, political, economic, and health factors and strategies to ensure that RH protects the community from the environment and develops a positive interaction between humans and their environment, taking into consideration that the environment affects the persons, and the person affects the environment*" [8]. **EDCs'** adverse effects on the global human reproductive systems were already reported in relation to, for instance, Bhopal and Chernobyl events [8], as well as to the **DDT** [2, 25] and **DES** [14, 25] interference, also due to prenatal exposure; through only partially known mechanism of action, these harmful compounds can interfere with **RH** [36]. Moreover, it had already been reported that the prenatal/antenatal exposure events to chemical environmental substance could have serious adverse effects on **RH**, because of malfunction of the relative interconnected systems [8]. Like any other aspect related to bodily health, **RH** shows its first signs of development and differentiation in the intrauterine life and then continues in well-coded postnatal stages; during each week of gestation, pathological processes of various types can be established that have a decisive influence on the life of the individual, both in terms of endocrine pathologies and in terms of impaired reproductive functions due to **RH** alterations. Furthermore, few epidemiologic data point out how environmental chemicals, more than lifestyle and nutrition, should be, at the same time, factors evaluated as a whole on the **RH** impact during the preconception period for both women and men [8, 59]. Through an ongoing monitoring, during male and female preconception time and pregnancy time, a prospective study showed, for example, higher urinary concentrations of phthalate metabolites in association with low egg production, a decreased odds of embryo implantation, and an increased risk of pregnancy loss in patients treated for infertility (drug-induced pregnancy) [28]. Moreover, in a dose-dependent way, a decrease of semen quality and higher monobutyl phthalate levels were found to be correlated; more than 40 biomarkers of environmental chemical exposure were tested for a broad overview of the association between substances, dietary elements, and negative effects [28].

1. *Female.* In the etiopathogenesis of female pathologies, the responsibility of adverse environmental factors, as xenoestrogens, or other **EDCs** must be carefully considered; those chemical substances exercise their function by mimicking the endogenous estrogens. The presence of pesticides, as **PCBs**, dioxins, **DDT**, and others, in fat tissue and breast milk, had already been reported [3]. Factors related to poor access to medical and health care, stress, and socio-economic status could potentiate the effects of the chemical exposures on female/maternal safety. In general, many alterations of the female reproductive mechanisms have been observed, probably also due to **EDC** exposure [19]; the main disfunctions are in the following two classes: (1) short-term issues, i.e., pregnancy outcome with **SGA/IUGR**, embryo implantation, and fertility rate, and (2) long-term issues, i.e., puberty timing, fibroids, polycystic ovary syndrome, endometriosis, and cancer; the latter three alterations are more frequently associated with **EDCs** [24, 42].

   The puberty, precocious-central or early, and menarche age have been studied in relation to the effects of endocrine disruptors that could interfere with hypothalamic function by epigenetic alterations. Animal model studies have found epigenetic transgenerational impacts from **DDT** and **DDE** exposure, resulting in polycystic ovary syndrome and primary ovarian insufficiency with granulosa cell alterations [40]. Also in the first human cohort study, the effects of **POPs** were analyzed on three generations: maternal, in utero, and *in egg* exposure, with the evidence of early menarche *in egg* exposure generation; obesity and premature menarche, known as risk factors for breast cancer, could alter the regular development of the studied generations [40]. On the other hand,

EDCs' role is still debated; however, other evidences have showed earlier breast development and earlier menarche time in female offspring, exposed in utero to **DDT/DDE** [43]; furthermore, a reduced fertility and a higher risk of breast and female genital tract cancers, in females exposed in utero to **DES**, during their embryo-fetal stages of life, had been widely reported, as also already mentioned in the introductory section [5, 42]. Likely, the main implicated compounds are bisphenol A, phytoestrogen, and dioxin [19, 24, 42].

2. *Male.* It has been shown that during in utero life, the differentiation of the male genital tract occurs, through the Müllerian duct involution and the testes' descent into the scrotal apparatus; hormonal management was found to be essential for these processes [5]. Over the decades, alterations in the physiological functioning of the male reproductive tract have been reported, but the role of exposure to **EDCs** has been debated [20]. It had already been suggested that sperm amount and quality could be compromised by milieu pollution [8]; furthermore, some glycol ethers and **Pb** may begin to be related to malfunction of male fertility [8]. Moreover, the concept of interference on Sertoli cells by estrogens, and related consequences on the fetal pituitary circuit in FSH production and later in sperm amount, was introduced [5]. Later, the acquired knowledge identified more alterations, such as the decrease in sperm count and motility, that could be attributed to the presence of pollutants with estrogen-like activity in the fetal life of affected patients; "phthalate syndrome" in animals was also reported, related to the testicular dysgenesis in humans exposed prenatally, probably due to those antiandrogen **EDCs'** impact on programming and development of fetal gonads [49]. Furthermore, there were results indicating male sex-dependent pharmacodynamics and clearance of **PFAS** with much higher exposure; through the passage of the placental barrier, the male fetus could suffer from their antiandrogenic potential [41]. The main male effects are in the following issues: anomalies of the pubertal period, poor semen quality, cryptorchidism (undescended testis), hypospadias, low serum testosterone level, and testicular cancer [36].

3. *Maternal Health, Placenta, and Pregnancy Outcomes.* As already highlighted in the previous sections, the **EDCs'** exposure can be detrimental and dangerous, associated with the development of pathologies affecting both the maternal and the fetal compartments, during the gestational (Fig. 9.6) and postnatal "critical" and "sensitive" periods (Figs. 9.5 and 9.6) [22, 24, 57, 59–61].

Generally, long-term consequences on the reproductive systems, from prenatal and perinatal exposure to toxic agents, have already been suspected and identified [8, 24]. Hence, starting to delineate the dimension of "**Pregnant Utero Biosphere**" (hereafter **PUB**, Fig. 9.7), the pregnancy time must be particularly attentive as the developing fetus and placenta could undergo alterations and consequent long-term diseases [22, 24, 59, 62]; it is a well-established notion that fetus is extremely more susceptible to external agents also due to the immaturity of both its detoxification mechanism and its immune system [38]. Moreover, many papers concerning pregnancy and lactation have already identified the effects on both the maternal and

Fig 9.7 "Pregnant uterus biosphere" (PUB). The concept of PUB protection through total environmental defense and security. Handmade pencil drawings and digital processing by Maria Laura Solerte: copyright and royalties for University of Padua, The Residency School of Obstetrics and Gynaecology Specialization

fetal organism and their offspring, introducing the concept of transgenerational inheritance, where an alteration, **EDC** induced, can manifest itself in the future generations [30, 34]. Among a lot, there were evidences of neurotoxicity evaluated on animals' and humans' prenatal exposures to **MeHg**, **Pb**, and pesticides [58, 59]; **MeHg** had even been found at higher levels in the fetal cord blood than in the mother [59, 63]. The placental development, with the relative co-mixture of the maternal-fetal sides, known as "placentation," is undertaken by the embryo which literally attaches itself to the uterine wall with the invasion of its trophoblastic cells; they gradually enter deeper and deeper uterine body in a stage called invasion [64]. Any events/substances with adverse effects/disruption on placental development phases and function, with their natural repercussions on the pregnancy outcome, must be scientifically and inevitably a basic concept for the protection of **RH**, which must be known and managed globally; several "environmental sphere" and **PUB** (Fig. 9.7) components can be adverse factors and targets with varying severe degrees of prenatal impacts such as risk of miscarriage, stillbirth, premature delivery, placental alterations, **SGA**, **IUGR**, low birth weight, and congenital anomalies and/ or **SIDS** [35]. Furthermore, it is also necessary to mention again one of the nicotine effects, which are possible due to its action on the **AChRs**; it has been shown that maternal tobacco smoking, and some of its components, during pregnancy, has detrimental effects on the placental physiological functions and on the fetus because of their capacity to cross the placental barrier and to modify the brain cell proliferation and differentiation by the **AChRs**; the increased risk of cognitive and/ or auditory alterations would therefore derive from the cell loss and neuronal deficiencies [15, 65].

Also exposure to inhalational anesthetic exposure or to low levels of **Pb** could be associated with an increased risk of infertility and miscarriage, respectively; moreover, exposure to radiation, pollutants, pesticides, and organic solvent can be associated with the obstetrical complications listed above; in addition, forms of stress, even physical and occupational, can influence gestational outcome even up to an increase in the incidence of preeclampsia probably linked to excessive release of catecholamines [11]. Even if the **PFAS** interference on **hCG** (Fig. 9.5) levels is documented, further studies will be needed to evaluate the negative effects of any **PFAS** exposures on female well-being and gestation [42]. Moreover, a **PFOA** inhibitory capacity on the rodent placenta and, consequently on pregnancy has even been identified [42].

From epidemiological evidences in also in vivo and ex vivo models, the placental presence of **PCBs** had been related to the decrease in the placental size [45, 46]; precisely, an alteration of syncytiotrophoblast volume and of placental growth factor (hereafter **PIGF**), with a compromised remodeling of the spiral artery and a likely placental disruption, has been documented in a little cohort of normal pregnancies [45]; already in the past, the impact of **PCBs'** prenatal exposure on newborns' weight and head circumference, due to their placental transition, was documented in pregnancy outcome of women who had taken contaminated lake fish [44]; however, since **PCBs** are present in a mixture, any specific placenta effects of the various components are complicated to select [45, 46]. The role of exposure to **EDCs** in relation to premature delivery must also be considered, in particular, the presence of **PhTh** metabolites [66].

In the general framework increased by the **EDC** interference, such as neurodevelopment and metabolic and cancer diseases [24], it is useful to schematize the following systemic adverse effects:

– *Metabolic alterations*: resulting from the **EDCs'** chemical interference [67, 68]; some **POPs**, **BPA**, and **PhThs** could interfere with the physiological processes of development, already in in utero life of adolescence, which would also depend on the time of exposure [14]. Obesity, in subjects up to 20 years of age, had been associated with prenatal and perinatal/infancy exposure to **p,p'DDE**, in a meta-analysis of prospective studies, which therefore confirmed previous similar evidence on individuals exposed to **o,p'-DDT**; later, also experimental analyses on animals highlighted this association [40].

- *Pituitary gland*: may be impaired in its development and all aforementioned endocrine axis functions, with neuroendocrine control (initially neuronal and later endocrine), by those **EDCs** that have the ability to interfere with neurotransmitter receptors [24, 32, 69]; the consequences of these alteration vary, according to the endocrine system affected, with the central role of the hypothalamus [69]. For instance, a disruption in puberty central time and/or in the circadian rhythm can occur [32].
- *Thyroid gland*: This involves possible compromises assuming the consolidated notion that indicates iodine as a basic component of thyroid hormones; the **EDCs** may interfere with thyrocyte activities and, especially by modifying the necessary channel to transport iodine in those cells, leads to a consequent reduction in the hormonal activity of the gland, with related hypothyroidism [32].
- For example, a **PFOA** adverse outcome on the thyroid has been identified in rodents [42]. Further research is needed to evaluate the dose-effect relationship responsible for the effects of **EDC** interference and thyroid functionality [32, 48] and correlated **Hg** levels with the thyroid endocrine system [12].
- *Adrenal gland*: Mostly for xenoestrogens, it is a preferential **EDC** target, due to its lipophilic structure on the cell membranes, the peculiar enzymatic activity, and the presence of a dense vascular network [32]. The main disrupted adrenal activity is correlated with **HPA** alterations also with the enzymatic function responsible for the steroid hormone production and every phase of steroidogenesis [24, 32, 70]. Adrenal interferences on its fetal time development are recognized congenitally also in adrenogenital disease occurring in the postbirth periods.
- *Brain development and behavior*: In addition to what has already been reported, modifications evaluated, for instance, by one longitudinal research of groups of subjects at birth have been systematically studied for the possible association between neurodevelopment and **EDCs**, which could exert their endocrine interference on the individual already from his in utero *life* [47]; however, these systematic analyses are difficult due to the presence of incomplete side-by-side hardly comparable data, also following the review evaluation of over 100 papers where hypotheses on the consequences of **PCBs** pre-postnatal exposures have been investigated [47]. In addition, an evaluation, by magnetic resonance imaging of **BPA** effects on children brain white matters in utero exposed, gave evidence of a probable causal link; on the other hand, the results of these evaluations report consequences on childhood behavior, but extreme caution must be exercised in reading these data, even if specific questionnaires were validated [71, 72]. However, the study analyzed made it possible to list a sort of guideline to be applied when evaluating the cause-effect relationship between **BPA** pre-postnatal exposure and behavioral defects [72].
- *Cancers*: The tendency to get sick with certain types of tumors could be related to noxious events during extremely sensitive periods of human life, such as embryo-fetal development and infancy [5, 19]. Moreover, the increased risk factors due to **EDC** exposure for hormone-dependent cancers are, for instance,

around 90% of breast cancers could be related to the environment, with known probability [51], and prenatal exposure to exogenous estrogens had been documented as an element for the increased risk of breast and genital cancer [42].

A case-control retrospective analysis must be mentioned, in which Herbst and his group, in 1971, published a cluster of seven cases, affected by vaginal adenocarcinoma (clear and hobnail cells or endometrial type), in exposed young women to diethylstilbestrol (**DES**), during their intrauterine life; the association between the cancer and therapy was observed, with oral estrogens of their mother, prescribed for high-risk pregnancy, also underlined by the absence of pathology in the daughters of untreated women [73]. In this context, the concept of "diethylstilbestrol syndrome" was therefore also introduced to highlight the effects of each estrogenic chemical with antagonist capacity [12].

Generally, correlation data had and have been identified between exposure to **EDCs** and onset of tumor pathologies [14] also affecting prostate (**As**, **Cd**, pesticides, **PCBs**), testicles (also between **As**, **Cd**, **PCBs**, **DDT**, **DDE**, and **PDBE** with testicular dysgenesis syndrome (**TDS**)), thyroid (**PCBs**, biocides, pesticides, **TCDD**), and breast (**PCBs**, phytoestrogens, **DES**, **Cd**, dioxin) [24, 30, 32]; on the other hand, concrete association between thyroid and testicular tumors and **EDC** exposure is not possible, given the modest case studies [32].

9.3   States of EDC Science

It was the mid-1930s when a British medical researcher firstly, accidentally identified the estrogenicity of **BPA**, already synthesized in 1891, while attempting to find a synthetic estrogen agent, and earlier the endocrine-disrupting capacity of **DDT** and phenanthrene derivatives was discovered [74, 75]. Up to that historical moment, no chemically estrogen-like active substance, without the phenanthrene nucleus, had been identified [75]. On the other hand, subsequently, after some research years, the powerful estrogenic mother substance **DES** was obtained with particular methods, proving other evidence on ovariectomized rats [75]; by vaginal cornification test, **DES** had been classified as an estrogen-like substance, harder than **BPA** [75]. Afterwards, the first epidemiological research on female offspring patients, born between 1946 and 1951, was conducted for their vaginal cancers with anamnestic history of first-trimester in utero exposure to **DES**, administered to mothers, following the indications of high-risk pregnancy. This was the first association of cancer induced by prenatal contact with drugs endowed with the estrogenic capacity, as the current **EDCs** [73]. The potentially harmful substances, on animals and probably also on humans, were then evaluated, up to the formation of groups of interferers, to assess the scientific basis of risk from the ambient exposure to them. Continuing with some highlights, in 1997, the 50th World Health Assembly 50.13 included, among the strategic points to be developed, the following: "take the necessary action to strengthen **WHO** leadership in risk-taking evaluation as a basis for addressing emerging high-priority problems, and in promoting and coordinating correlated research, for example, on potential health related to the endocrine system effects to exposure to chemicals." Thereafter, several aspects linked to the human and wildlife exposure to several compounds, single or mixed, were taken into consideration to evaluate their effects on several physiological body systems [20]. At that point, the definition of **EDCs**, since its basic formulation [3], has been well established by the **WHO**, since 2002, as "an exogenous substance or mixture that alters functions of the endocrine system and consequently causes adverse health effects in an intact organism, or its progeny, or (sub) populations" [20], as well as "an exogenous chemical, or mixture of chemicals, that can interfere with any aspect of hormone action," also indicated in the **EC** programmatic four options [76, 77]. As already mentioned above, the 2012 **UNEP** and the **WHO** became a milestone, publishing an update of the 2002 report of the International Programme on Chemical Safety (**IPCS**) [76, 78], noting the increase of endocrinal multicausal disorders also associated with exposure to **EDCs**, and recognizing the main sensitive human window—life (fetal, childhood, and puberty) of exposure [16, 23, 26, 54, 59].

In our time, different states which concern **EDC** science are at the global attention; nearly 86,000 toxic compounds [79] have been reported by the **EPA** in the Toxic Substances Control Act (**TSCA**) [46, 79]; the **TSCA**, which is a US law since 1976, had already defined the chemical substance "as any organic or inorganic substance of a particular molecular identity, including any combination of these substances occurring in whole or in part as a result of a chemical reaction or occurring in nature, and any element or uncombined radical," and had already inventoried 62,000 compounds in 1982 [79]. The **EPA** has also formulated, and constantly updated and shared, the Endocrine Disruption Screening Program (**EDSP**), starting in 1996, based on research projects aimed at identifying guidelines for the main **EDCs** [24, 32, 80, 81]. Furthermore, the **EU** Commitment has therefore gradually intensified, under the Community Strategy for **EDCs**, with important advances in highlighting the mechanism of action of these molecules [26]. Currently, a systematically updated **EDC** list by the **NIEHS** is available which supports studies on the **EDC** mechanism of action that negatively affects human health, summarizing them as follows: (1) "decrease or increase normal hormone levels," (2) "mimic the body's natural hormones," and (3) "alter natural hormone production" [29]. Moreover, the **NIEHS** has taken part in the Consensus on the **EDCs'** specific key characteristics and coordinates projects in different research areas [14, 29]. On the other hand, extreme caution is required to evaluate in vitro/in vivo data, to be used as experimental evidences that can be translated to the human situation, also with the issue of analysis of the **EDCs'** dose-response/duration of exposure and mechanism of action as interference with hormonal and enzymatic activity; for instance, some hypotheses published regarding the last aspect need to be deepened [46, 52]. Thanks to the research on animals, from which derive the most substantial data [36], and to human epidemiological results, which have made it possible to acquire the technical information available today, common guidelines can be defined to prevent **EDCs'** effects, with the safe use of chemical compounds through shared monitoring systems [82]: with permission of the Endocrine Society. Hence, from the aforementioned epidemiological plus animal model studies, and the further known **EDC** definition as "an exogenous chemical, or mixture of chemicals, that can interfere with any aspect of hormone action" [82], a relationship emerges between common noncommunicable disease and **EDC** levels, as well as doses, in the relative milieu [36, 82]. The scientific community does not yet report an unambiguous result on the **EDCs'** non-monotonic dose-response/low-dose effects and their "safety threshold" [26]. However, thanks to the acquired knowledge, some of the mechanisms of action of **EDCs** have been detected on hormonal receptors of different types, on epigenetic modifications, and on transgenerational effects at different doses of endocrine-disrupting substances, underlining the foundations of scientific research on **EDCs** [14, 36].

Definition of a systematic scientific work on **EDCs**, the use of their **KCs**, and their transgenerational-epigenetic effects could be the starting point for weighing the risks associated with the exposure to that type of substance [14, 36].

Besides, the need to develop several appropriate and uniform testing methods is highlighted [26]. From the evidence that emerges, relating to a milieu of **EDCs**, there are possible associations between several adverse health outcomes, as well as interferences on sex hormones/receptors, reproductive mechanisms (as implantation/placentation) and thyroid [83, 84], and **EDC** exposure. Moreover, it was identified that several biological effects of **EDCs** are also mediated through the gene expression alterations, already mentioned as epigenetic modification effects in their three systems of action [36]. Experimental human research shows that pre- and postnatal exposure to **EDCs**, and/or a mixture of **EDCs**, may interfere with hormone axis with relatively negative effects on the male endocrine reproductive system; however, there are elements that make these studies limited and worthy of review on the EDC-level measuring methods, in relation to the different periods of the individual life [83]. The probable worrying and growing threat to environmental and human health, globally, could be determined by agricultural and industrial processes; therefore, other biomonitoring accurate systems will be needed [36]. To date, for example, starting from the lack of knowledge of the **EDCs'** precise mechanism of interference with mammalian endometrium, a 2021 systematic review summarized the available studies, in vivo, in literature on the **EDCs'** effects on mouse-model blastocyst-endometrial implant: following **BPA** and phthalate exposures, the detrimental alteration of implantation sites, receptivity endometrial markers, pregnancy hormone receptors, and pregnancy rates were highlighted [84]. Moreover, the same alteration was recorded in the adult age of the subjects who were in utero exposed to **BPA** [84]. On the other hand, the negative pregnancy issues, linked to the **EDCs**, seem to be related to both immature-mature and maternal-fetal side placental dysfunctions with alterations of the trophoblast cells' metabolic capacity [62]. Recently, to deepen the study of the mechanisms that lead to these changes, the concept of "trophoblast organoid," as an entity to be examined, has been introduced [62].

Already for years, there have been scientific evidences that would lead to the identification of urgent measures to reduce or avoid the exposure of **EDCs**, both temporally and quantitatively, also through the systematic monitoring of the potency of those substances, which could have an even more impact on human health, considering, moreover, the absence of a "safe dose" [59]. With regard to waste and harmful chemicals, in anticipation of the objectives until 2030, the Organisation for Economic Co-operation and Development (**OECD**) defines a "red light" matter that needs urgent application measures [59, 85].

Therefore, a common strategy was formulated, understood as the correlation between environment and public health, involving "*global, regional, national, and local environmental factors, including external physical, chemical, and biological factors,*" because the global objective is "*a healthy environment vital to ensure healthy lives and promote well-being for all at all ages*" [86, 87].

It will be appropriate to examine and investigate every aspect relating to the effects of any molecular endocrine disruptors, using standardized methods indicated by international guidelines, also for drinking water [36, 88].

Conflict of Interest The authors declare that the chapter was conducted in the absence of any commercial or financial relationships that could be construed as a potential, previous and/or actual, conflict of interest.

References

1. WHO; Department of Growth and Reproduction, Copenhagen University. Impact of the environment on reproductive health. Prog Hum Reprod Res. 1991;20:1–11.
2. Carson R. Silent spring. Boston: Houghton Mifflin Company; 1962.
3. Colborn T, Vom Saal FS, Soto AM. Developmental effects of endocrine-disrupting Chemicals in Wildlife and Humans. Environ Health Perspect. 1993;101(5):379–84.
4. Moore KL. The developing human: clinically oriented embryology. Philadelphia, PA: W.B. Saunders; 1982.
5. Sharpe RM, Skakkebaek NE. Are oestrogens involved in falling sperm counts and disorders of the male reproductive tract? Lancet. 1993;341:1392–5.
6. Stillman RJ. In utero exposure to diethylstilbestrol: adverse effects on the reproductive tract and reproductive performance and male and female offsprings. Am J Obstet Gynecol. 1982;142(7):905–21.
7. Miller HC, Hassanein K, Hensleigh PA. Fetal growth retardation in relation to maternal smoking and weight gain in pregnancy. Am J Obstet Gynecol. 1976;125(1):55–60.
8. Michal F, Grigor KM, Negro-Vilar A, Skakkebaek NE. Impact on the environment on reproductive health: executive summary. Environ Health Perspect. 1993;101(Suppl 2):159–67.
9. Stucker S, De Angelis J, Kusumbe AP. Heterogeneity and dynamics of vasculature in the endocrine system during aging and disease. Front Physiol. 2021;12:624928.
10. Hiller-Sturmhöfel S, Bartke A. The endocrine system-an overview. Alcohol Health Res World. 1998;22(3):153–64.
11. Costa MA. The endocrine function of human placenta: an overview. Reprod BioMed Online. 2016;32:14–43. https://doi.org/10.1016/j.rbmo.2015.10.005.
12. Iavicoli I, Fontana L, Bergamaschi A. The effects of metals as endocrine disruptors. J Toxicol Environ Health. 2009;12:206–23.
13. European Food Safety Authority. 2022. www.efsa.europa.eu. Accessed May 2022.
14. La Merrill MA, Vandenberg LN, Smith MT, Goodson W, Browne P, Patisaul HB, Guyton KZ, Kortenkamp A, Cogliano VJ, Woodruff TJ, Rieswijk L, Sone H, Korach KS, Gore AC, Zeise L,

Zoeller RT. Consensus on the key characteristics of endocrine- disrupting chemicals as a basis for hazard identification. Nat Rev Endocrinol. 2020;16:45–57.
15. Smith AM, Dwoskin LP, Pauly JR. Early exposure to nicotine during critical periods of brain development: mechanisms and consequences. J Pediatr Biochem. 2010;1(2):125–41.
16. World Health Organization. Regional Office for Europe. https://apps.who.int/iris/handle/10665/107985.
17. Parma Declaration on Environment and Health, Parma, Italy, 10-12 March 2010.
18. Italia Constitution. https://www.senato.it/istituzione/la-costituzione/principi-fondamentali/articolo-9. Accessed May 2022.
19. Caserta D, Maranghi L, Mantovani A, Marci R, Maranghi F, Moscarini M. Impact of endocrine disruptor chemicals in gynaecology. Hum Reprod Update. 2008;14(1):59–72.
20. Damstra T, Barlow S, Bergman A, Kavlock R, Van Der Kraak G. World Health Organization. In: Global assessment of the state-of-the-science of endocrine disruptors. Geneva: WHO; 2002.
21. Kavlock RJ, Daston GP, DeRosa C, Fenner-Crisp P, Gray LE, Kaattari S, Lucier G, Luster M, Mac MJ, Maczka C, Miller R, Moore J, Rolland R, Scott G, Sheehan DM, Sinks T, Tilson HA. Research needs for the risk assessment of health and environmental effects of endocrine disruptors: a report of the U.S. EPA-sponsored workshop. Environ Health Perspect. 1996;104(4):715–40.
22. Diamanti-Kandarakis E, Bourguignon JP, Giudice LC, Hauser R, Prins GS, Soto AM, Zoeller RT, Gore AC. Endocrine-disrupting chemicals: an endocrine society scientific statement. Endocr Rev. 2009;30(4):293–342.
23. World Health Organization. In: Bergman A, Heindel JJ, Jobling S, Kidd KA, Zoeller RT, editors. United Nations environment Programme: state of the science of endocrine disrupting chemicals–2012. Geneva: WHO; 2013.
24. Gore AC, Chappell VA, Fenton SE, Flaws JA, Nadal A, Prins GS, Toppari J, Zoeller RT. The Endocrine Society's second scientific statement on endocrine-disrupting chemicals. Endocr Rev. 2015;36(6):593–602.
25. Kahn LG, Philippat C, Nakayama SF, Slama R, Trasande L. Endocrine-disrupting chemicals: implications for human health. Lancet Diabetes Endocrinol. 2020;8:703–18.
26. European commission. Commission staff working document fitness check on endocrine disruptors. 2020. Brussels SWD 251 final.
27. European Food Safety Authority. Guidance for the identification of endocrine disruptors in the context of Regulations (EU) No 528/2012 and (EC) No 1107/2009. EFSA J. 2018;16(6):5311.
28. Messerlian C, Williams PL, Ford JB, Chavarro JE, Mínguez-Alarcón L, Dadd R, Joseph M, Braun JM, Gaskins AJ, Meeker JD, James-Todd T, Chiu YH, Nassan FL, Souter I, Petrozza J, Keller M, Toth TL, Calafat AM, Hauser R. The environment and reproductive health (EARTH) study: a prospective preconception cohort. Hum Reprod Open. 2018;2:1–11.
29. NIEHS. Endocrine Disruptors. 2021. www.niehs.nih.gov/health/topics/agents/endocrine/index.cfm. Accessed 12 Jun 2022.
30. Egalini F, Marinelli L, Rossi M, Mott G, Prencipe N, Giaccherino RR, Pagano L, Grottoli S, Giordano R. Endocrine disrupting chemicals: effects on pituitary, thyroid and adrenal glands. Endocrine. 2022;78(3):395–405. https://doi.org/10.1007/s12020-022-03076-x.
31. Committee ES. Scientific opinion on the hazard assessment of endocrine disruptors: scientific criteria for identification of endocrine disruptors and appropriateness of existing test methods for assessing effects mediated by these substances on human health and the environment. EFSA J. 2013;11:3132.
32. Lauretta R, Sansone A, Sansone M, Romanelli F, Appetecchia M. Endocrine disrupting chemicals: effects on endocrine glands glands. Front Endocrinol. 2019;10:178.
33. Perera FP, Rauh V, Tsai WY, Kinney P, Camann D, Barr D, Bernert T, Garfinkel R, Tu YH, Diaz D, Dietrich J, Whyatt RM. Effects of Transplacental exposure to environmental pollutants on birth outcomes in a multiethnic population. Environ Health Perspect. 2003;111(2):201–5.
34. RJ S, Binková B, Dejmek J, Bobak M. Ambient air pollution and pregnancy outcomes: a review of the literature. Environ Health Perspect. 2005;113:375–82.

35. Hossain N, Westerlund-Triche E. Environmental factors implicated in the causation of adverse pregnancy outcome. Semin Perinatol. 2007;31(4):240–2.
36. Kumar M, Sarma DK, Shubham S, Kumawat M, Verma V, Prakash A, Tiwari R. Environmental endocrine-disrupting chemical exposure: role in non-communicable diseases. Front Public Health. 2020;8:553850. https://doi.org/10.3389/fpubh.2020.553850.
37. Centre for Disease Control and Prevention. National biomonitoring program. U.S. Department of Health & Human Services last reviewed. 2017. https://www.cdc.gov/biomonitoring/chemical_factsheets.html.
38. Vizcaino E, Grimalt JO, Fernández-Somoano A, Tardon A. Transport of persistent organic pollutants across the human placenta. Environ Int. 2014;65:107–15. https://doi.org/10.1016/j.envint.2014.01.004.
39. U.S. Environmental Protection Agency. DDT, a review of scientific and economic aspects of the decision to ban its use as a pesticide. Washington, D.C.: United States Environmental Protection Agency; 7. Report nr EPA-540/1-75-022. 1975.
40. Cirillo PM, La Merrill MA, Krigbauma NY, Cohna BA. Grandmaternal perinatal serum DDT in relation to granddaughter early menarche and adult obesity: three generations in the Child Health and Development Studies cohort. Cancer Epidemiol Biomark Prev. 2021;8:1480–8. https://doi.org/10.1158/1055-9965.EPI-20-1456.
41. Foresta C, Tescari S, Di Nisio A. Impact of perfluorochemicals on human health and reproduction: a male's perspective. J Endocrinol Investig. 2018;41(6):639–45.
42. Rickard BP, Rizvi I, Fenton SE. Per- and poly-fluoroalkyl substances (PFAS) and female reproductive outcomes: PFAS elimination, endocrine-mediated effects, and disease. Toxicology. 2022;465:153031.
43. Papadimitriou A, Papadimitriou DT. Endocrine-disrupting chemicals and early puberty in girls. Children. 2021;8:492. https://doi.org/10.3390/children8060492.
44. Fein GG, Jacobson JL, Jacobson SW, Schwartz PM, Dowler JK. Prenatal exposure to polychlorinated biphenyls: effects on birth size and gestational age. J Pediatr. 1984;105(2):315–20. https://doi.org/10.1016/s0022-3476(84)80139-0.
45. Tsuji M, Aiko Y, Kawamotoa T, Hachisuga T, Kooriyama C, Myoga M, Tomonaga C, Matsumura F, Anan A, Tanaka M, Yu HS, Fujisawa Y, Suga R, Shibata E. Polychlorinated biphenyls (PCBs) decrease the placental syncytiotrophoblast volume and increase placental growth factor (PlGF) in the placenta of normal pregnancy. Placenta. 2013;34:619–23.
46. Gingrich J, Ticiani E, Veiga-Lopez A. Placenta disrupted: endocrine disrupting chemicals and pregnancy. Trends Endocrinol Metab. 2020;31(7):508–24.
47. Goodman M, Li J, Flanders WD, Mahooda D, Anthony LG, Zhang Q, LaKind JS. Epidemiology of PCBs and neurodevelopment: systematic assessment of multiplicity and completeness of reporting. Global Epidemiol. 2020;2:100040. https://doi.org/10.1016/j.gloepi.2020.100040.
48. Greer MA, Goodman G, Pleus RC, Greer SE. Health effects assessment for environmental perchlorate contamination: the dose response for inhibition of thyroidal radioiodine uptake in humans. Environ Health Perspect. 2002;110(9):927–37. https://doi.org/10.1289/ehp.02110927.
49. Swan SH, Main KM, Liu F, Stewart SL, Kruse RL, Calafat AM, Mao CS, Redmon JB, Ternand CL, Sullivan S, Teague JL. Decrease in anogenital distance among male infants with prenatal phthalate exposure. Environ Health Perspect. 2005;113(8):1056–61. https://doi.org/10.1289/ehp.8100.
50. Wagner M, Oehlmann J. Endocrine disruptors in bottled mineral water: total estrogenic burden and migration from plastic bottles. Environ Sci Pollut Res. 2009;16:278–86. https://doi.org/10.1007/s11356-009-0107-7.
51. Hager E, Chen J, Zhao L. Minireview: parabens exposure and breast cancer. Int J Environ Res Public Health. 2022;19:1873. https://doi.org/10.3390/ijerph19031873.
52. Wang CF, Tian Y. Reproductive endocrine-disrupting effects of triclosan: population exposure, present evidence and potential mechanisms. Environ Pollut. 2015;206:195–201.
53. Sabir S, Akasha MSH, Fiayyaza F, Saleemb U, Mehmoodb MH, Rehmand K. Role of cadmium and arsenic as endocrine disruptors in the metabolism of carbohydrates: inserting the asso-

ciation into perspectives. Biomed Pharmacother. 2019;114:10880. https://doi.org/10.1016/j.
biopha.2019.108802.

54. Tweed JO, Hsia ST, Lutfy K, Friedman TC. The endocrine effects of nicotine and cigarette smoke. Trends Endocrinol Metab. 2012;23(7):334–42. https://doi.org/10.1016/j.
tem.2012.03.006.

55. UNEP. Microplastics. https://www.unep.org/resources/report/microplastics. Accessed
Jun 2022.

56. Ragusa A, Svelato A, Santacroce C, Catalano P, Notarstefano V, Carnevali O, Papa F,
Rongioletti MCA, Baiocco F, Draghi S, D'Amore E, Rinaldo D, Matta M, Giorgini
E. Plasticenta: first evidence of microplastics in human placenta. Environ Int.
2021;146:106274. https://doi.org/10.1016/j.envint.2020.106274.

57. Barker DJP. The developmental origins of adult disease. J Am Coll Nutr.
2004;23(6):588S–95S. https://doi.org/10.1080/07315724.2004.10719428.

58. Zama AM, Uzumcu M. Epigenetic effects of endocrine-disrupting chemicals on female
reproduction: an ovarian perspective. Front Neuroendocrinol. 2010;31(4):420–39. https://doi.
org/10.1016/j.yfrne.2010.06.003.

59. Di Renzo GC, Conry JA, Blake J, De Francesco MS, De Nicola N, Martin JN Jr, McCueKA
RD, Shah A, Sutton P, Woodruff TJ, van der Poel SZ, Giudice LC. International Federation
of Gynecology and Obstetrics (FIGO) opinion on reproductive health impacts of exposure
to toxic environmental chemicals. Int J Gynaecol Obstet. 2015;131(3):219–25. https://doi.
org/10.1016/j.ijgo.2015.09.002.

60. Yang C, Song G, Lim W. A mechanism for the effect of endocrine disrupting chemicals on placentation. Chemosphere. 2019;231:326–36. https://doi.org/10.1016/j.
chemosphere.2019.05.133.

61. Street ME, Bernasconi S. Endocrine-disrupting Chemicals in Human Fetal Growth Int. J Mol
Sci. 2020;2020(21):1430. https://doi.org/10.3390/ijms21041430.

62. Rice D, Barone S. Critical periods of vulnerability for the developing nervous system: evidence
from humans and animal models. Environ Health Perspect. 2000;108(3):511–33. https://doi.
org/10.1289/ehp.00108s3511.

63. Stern AH, Smith AE. An assessment of the cord blood: maternal blood methylmercury ratio:
implications for risk assessment. Environ Health Perspect. 2003;111(12):1465–70. https://doi.
org/10.1289/ehp.6187.

64. Pascular F. Trophoblast organoids: a new tool for studying placental development. Environm
Health. Perspective. 2022;130(5):054003-1/2. https://doi.org/10.1289/EHP11351.

65. Centers for Disease Control and Prevention. Substance use during pregnancy. http://www.cdc.
gov/reproductivehealth/TobaccoUsePregnancy.

66. Welch BM, Keil AP, Buckley JP, Calafat AM, Christenbury KE, Engel SM, O'Brien KM,
Rosen EM, Todd TJ, Zota AR, Ferguson KK, Pooled Phthalate Exposure and Preterm Birth
Study Group. Associations between prenatal urinary biomarkers of phthalate exposure and
preterm birth. A pooled study of 16 US cohorts. JAMA Pediatr. 2022;176(9):895–905. https://
doi.org/10.1001/jamapediatrics.2022.2252.

67. Sheng JA, Bales NJ, Myers SA, Bautista AI, Roueinfar M, Hale TM, Handa RJ. The hypo-
thalamic-pituitary-adrenal Axis: development, programming actions of hormones, and
maternal-fetal interactions. Front Behav Neurosci. 2021;14:601939. https://doi.org/10.3389/
fnbeh.2020.601939.

68. Heindel JJ, Blumbergb B, Cavec M, Machtingerd R, Mantovani A, Mendezf MA, Nadalg
A, Palanzah P, Panzicai G, Sargisj R, Vandenbergk LN, vom Saal F. Metabolism disrupting
chemicals and metabolic disorders. Reprod Toxicol. 2017;68:3–33. https://doi.org/10.1016/j.
reprotox.2016.10.001.

69. Gore AC. Neuroendocrine targets of endocrine disruptors. Hormones. 2010;9:16–27. https://
doi.org/10.14310/horm.2002.1249.

70. Harvey PW. Adrenocortical endocrine disruption. J Steroid Biochem Mol Biol. 2016;155(Pt.
B):199–206. https://doi.org/10.1016/j.jsbmb.2014.10.009.

71. Grohs MN, Reynolds JE, Liu J, Martin JW, Pollock T, Lebel C, Dewey D, APrON Study Team. Prenatal maternal and childhood bisphenol A exposure and brain structure and behavior of young children. Environ Health. 2019;18(1):85. https://doi.org/10.1186/s12940-019-0528-9.
72. Mustieles V, Fernández MF. Bisphenol A shapes children's brain and behavior: towards an integrated neurotoxicity assessment including human data. Environ Health. 2020;19:66. https://doi.org/10.1186/s12940-020-00620-y.
73. Herbst AL, Ulfelder H, Poskanzer DC. Adenocarcinoma of the vagina. Association of maternal stilbestrol therapy with tumor appearance in young women. N Engl J Med. 1971;284:878–81.
74. Vandenberg LN, Maffini MV, Sonnenschein C, Rubin BS, Soto AM. Bisphenol-A and the great divide: a review of controversies in the field of endocrine disruption. Endocr Rev. 2009;30(1):75–95. https://doi.org/10.1210/er.2008-0021.
75. Dodds E, Lawson W. Synthetic estrogenic agents without the phenanthrene nucleus. Nature. 1936;137:996.
76. WHO, International Programme on Chemical Safety. 2002. Global assessment of the state-of-the-science of endocrine disruptors. 2002. https://apps.who.int/iris/handle/10665/67357
77. Slama R, Bourguignon JP, Demeneix B, Ivell R, Panzica G, Kortenkamp A, Zoeller RT. Scientific issues relevant to setting regulatory criteria to identify endocrine disrupting substances in the European Union. Environ Health Perspect. 2016;124:1497–503. https://doi.org/10.1289/EHP217.
78. UNEP. 2012 annual report. 2021. https://www.unep.org/resources/annual-report/unep-2012-annual-report.
79. United States Environment protection Agency. About the TSCA chemical substance inventory. 2022. https://www.epa.gov/tsca-inventory/about-tsca-chemical-substance-inventory
80. Zoeller RT, Brown TR, Doan LL, Gore AC, Skakkebaek NE, Soto AM, Woodruff TJ, Vom Saal FS. Endocrine-disrupting chemicals and public health protection: a statement of principles from the Endocrine Society. Endocrinology. 2012;153:4097–110. https://doi.org/10.1210/en.2012-1422.
81. Marty S. Introduction to "screening for endocrine activity-experiences with the US EPA's endocrine disruptor screening program and future considerations". Birth Defects Res B Dev Reprod Toxicol. 2014;101:1–2. https://doi.org/10.1002/bdrb.21100.
82. Endocrine Society. endocrine disrupting chemicals. an endocrine society position statement. 2018. https://www.endocrine.org/advocacy/position-statements/endocrine-disrupting-chemicals. Accessed 12 Sept 2022.
83. Rodprasert W, Toppari J, Virtanen EH. Endocrine disrupting chemicals and reproductive health in boys and men. Front Endocrinol. 2021;12:706532.
84. Caserta D, Costanzi F, De Marco MP, Di Benedetto L, Matteucci E, Assorgi C, Pacilli MC, Besharat AR, Bellati F, Ruscito I. Effects of endocrine-disrupting chemicals on endometrial receptivity and embryo implantation: a systematic review of 34 mouse model studies. Int J Environ Res Public Health. 2021;18:6840.
85. OECD. OECD environmental outlook to 2030. Paris: Organization for Economic Co-operation and Development; 2008.
86. PAHO/WHO. Environmental Determinants of Health. www.paho.org. Accessed 17 Jun 2022.
87. United Nation. The 2030 agenda for sustainable development.
88. Maffini MV, Vandenberg LN. Failure to launch: the endocrine disruptor screening program at the U.S. Environmental Protection Agency. Front Toxicol. 2022;4:908439. https://doi.org/10.3389/ftox.2022.908439.

Chapter 10
Policy Implication and Community Interventions to Reduce EDCs Exposure

Luigi Montano and Antonino Guglielmino

10.1 Introduction

The group of molecules identified as endocrine disruptors (EDCs) is highly heterogeneous and includes synthetic chemicals used as industrial solvents/lubricants and their by-products, such as plastic compounds, plasticizers, pesticides, pharmaceutical agents, heavy metals, phthalates, bisphenol A, flame retardants, alkyl phenols, dioxins and polycyclic aromatic hydrocarbons have been identified as endocrine disruptors [1]. Currently, an endocrine disruptor (ED) assessment list is available at the European Chemicals Agency (ECHA) and includes chemicals undergoing an ED assessment under Registration, Evaluation, Authorization and Restriction of Chemicals (REACH) or the Biocidal Products Regulation. These substances are listed on the ECHA website updated on 29 April 2022 for discussion to ECHA's ED expert group [2].

Endocrine disruptor, term first introduced in 1991 [3], is defined as "an exogenous chemical, or mixture of chemicals, that can interfere with any aspect of endogenous hormone action" [4, 5]. These chemicals can bind to the body's endocrine receptors to activate, block or alter natural hormone synthesis and degradation, which occur

---

Luigi Montano and Antonino Guglielmino contributed equally with all other contributors.

L. Montano (✉)
Andrology Unit and Service of Lifestyle Medicine in UroAndrology, Local Health Authority (ASL) Salerno, Oliveto Citra Hospital, Salerno, Italy

PhD Program in Evolutionary Biology and Ecology, University of Rome Tor Vergata, Rome, Italy
e-mail: l.montano@aslsalerno.it

A. Guglielmino (✉)
Reproductive Medicine Unit, Centro HERA, Catania, Italy

© The Author(s) 2023
R. Marci (ed.), *Environment Impact on Reproductive Health*,
https://doi.org/10.1007/978-3-031-36494-5_10

through a plethora of mechanisms resulting in "false" lack or abnormal hormonal signals that can increase or inhibit normal endocrine function [6]. These chemicals are also classified as emerging pollutants because they can be detected in the order of nanograms to micrograms (ng/L and µg/L) using gas chromatography with mass spectroscopy and high-performance liquid chromatography with mass spectroscopy; other methods, like enzyme-linked immune-sorbent assay, emergent new biological methods through biosensors, are also currently widely used [7]. EDCs group can be derivate from anthropogenic activities (synthetic endocrine disruptors) and from natural origins (natural endocrine disruptors [7]. An example of this classification of some EDCs is shown in Fig. 10.1.

In April 2016, the German Federal Institute for Risk Assessment (BfR) reached a consensus about the development of EDCs criteria in relevant EU legislation during a meeting [8]. Data from ecological studies, animal models, clinical observations in humans and epidemiological studies agree to consider endocrine-disrupting chemicals (EDCs) as a significant for wildlife and human health [9, 10].

EDCs are widespread in the environment, and the increase of non-communicable diseases (NCDs), such as cancer, diabetes, obesity, cognition deficit and neurodegenerative diseases, endometriosis, polycystic ovarian syndrome, early

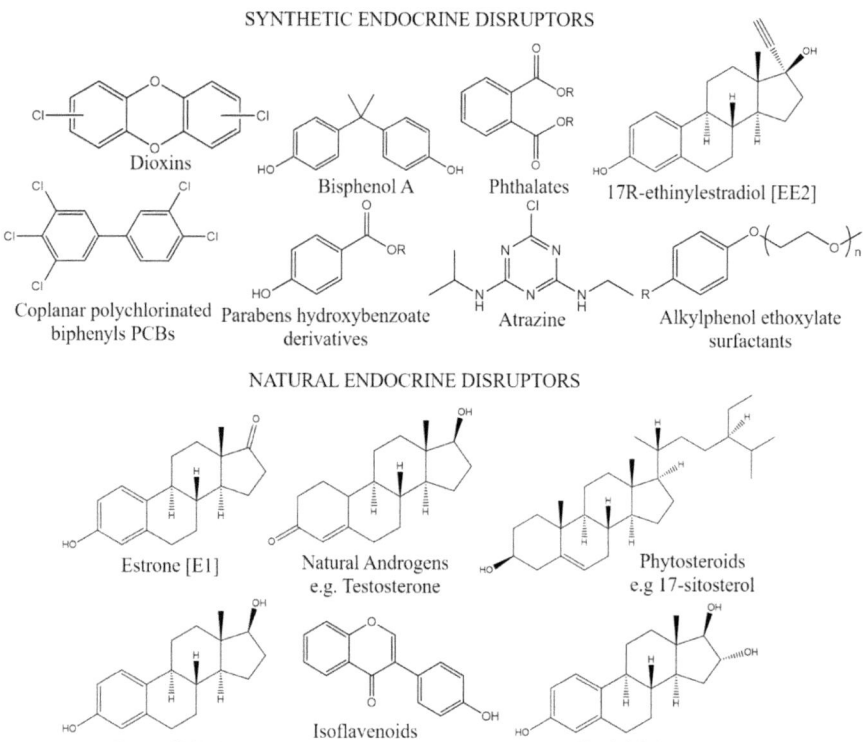

Fig. 10.1 Structure of common synthetic and natural endocrine disruptors

puberty, thyroid dysfunction, heart diseases and infertility, has all been linked to these substances exposure with costs in the hundreds of billions of Euros per year [11]. In particular, the exponential increase of cancer and metabolic disease, as well as obesity and diabetes worldwide, correlates with the widespread use of these substances and the costs in relation to morbidity and mortality are enormous [12–23].

The scientific research in the last three decades has solidified the knowledge of these chemicals and have been known the transgenerational effects through uterine exposure, for disease [24, 25].

However, the perturbative effects of EDCs on endogenous hormones, historically, have been focused on the reproductive system. In fact, already for some pesticides (thiocarbamates, chlororganics, imidazoles, triazoles and triazines), which determine an antiandrogenic action highlighted by the macroscopic sexual changes found in aquatic animals (particularly because of exposure to herbicides and fungicides) such as the demascolinization in rats and fish [26] and the production of estrogens and hermaphroditism in frogs [27] and other developmental disorder of the male gonad in alligators [28]. Certainly, the largest group of these substances accumulates in tissues and in the environment [29–32].

These substances cause an antiandrogenic effect in humans too, but they also mimic the estrogenic action, confirmed by both experiments in vivo and in vitro [33–41].

The great attention to the reproductive system underlines how it can represent a sentinel organ to environmental stresses, and the epidemiological and clinical data available today, in particular on male infertility, seem to confirm this sensitivity.

There is evidence that semen quality has declined in the last decades, and the incidence of male infertility has increased steadily in many countries [42–44]. An important decreasing trend has already been described for sperm concentration from 113×10^6/mL in 1940 to 66×10^6 in 1990 [45] and the same for testosterone levels [46, 47]. According to Levine, total sperm count had fallen by 59.3%/escalate between 1971 and 2011 in Europe, North America, Australia and New Zealand and sperm concentration/mL fell by 52.4% [48]. Other sperm decline is has reported also in China from 2001 to 2015 [49] in Africa, India, Brasil and Iran [50]. Changes in sperm production, initially thought to be due to maternal exposure to environmental oestrogens, corresponds precisely to the introduction of chemicals especially after 1940 [51], but with the most knowledge on experimental data, these effects seem to be due to different types of endocrine-disrupting chemicals (EDCs) [52, 53].

EDCs are now ubiquitous in the environment and their effects do not end with the exposed individual but are transmitted to future generation trough epigenetic changes to the germline, as reported in several studies [54–63].

However, if changes in behavioural factors and lifestyles, including the introduction and rapid growth of cell phones' use, the large increase in the consumption of opiates and marijuana, the increase in the consumption of cigarettes and increasing physical inactivity, may have potentially induced alterations in seminal parameters and thus reduce male fertility [64], environmental and chemical contaminants in the workplace are recognized as major risk factors especially for male infertility in both epidemiological and experimental studies [65–70].

The incidence of genitourinary tract malformations and reduced sperm quality is indeed higher in people living in areas with a high rate of pollution or in individuals exposed to EDCs for professional reasons. More strikingly, especially in industrialized countries, the reduction of semen quality and/or semen count present differences in areas in the same country or even in the same region, supporting the idea that environmental factors, present in some areas but not in others, may be responsible for the decline in semen quality and sperm count [71–84]. Furthermore, different studies have reported that in high environmental pressure areas there is an increase of infertility, urogenital malformation and chronic disease (cancer, diabetes, etc.) [11, 77–88].

These epidemiological data are important to understand the shared biological mechanisms mediated by contaminants. There are therefore several evidences that show how ubiquitous presence of chemicals in the environment and in food is actually the root cause of increased health reproductive problems, especially for the reduction of semen quality, and the increased incidence in recent years, even of testicular dyskinesias, induce to believe that harmful environmental factors can have a much more important role than people think.

However, for ethical reasons, it is difficult to establish a causal relationship on human beings. Clinically, the most common manifestation of contaminants is a reduced sperm concentration, while its most severe form can include an increased risk of testicular cancer [89].

Associating both environmental data and chemical factors of exposure to the data found in the body, as well as verifying the consistency and the determinism or order of passage from the environment to the organism, is a crucial step for a better understanding of the environment-health relationship. In light of this, the male reproductive system is sensitive to a broad variety of environmental pollutants; therefore, it represents an optimal model for the study of environmental health. Spermatogenesis from puberty onward is continuously exposed to insults at the stages of continuous replication; as a consequence the male germline accumulates mutations [90, 91]. Sperm cells are more susceptible than eggs to the effects of oxidative damage, because they lack significant antioxidant protection because of reductive cytoplasmic space for an appropriate armoury of defensive enzymes and significant amounts of polyunsaturated fatty acids [92]. Simultaneously, in semen it is possible to measure environmental contaminants and in vivo effects on sperm cells, which are readily available, with features sensitive to environmental pollutants such as motility, morphology and the integrity of the DNA strand. If human semen seems an earlier and sensitive source of biomarkers than blood in monitoring high environmental pressure on human health, therefore it can be considered a reliable environmental sentinel [77, 78, 93]. More evidence in literature indicates the human semen as an important health marker. In fact the spermatogenesis cycle is extremely complex and vulnerable to endogenous and exogenous stress, so it is not surprising that it can be an important indicator of the state of well-being of the organism.

Recent studies have demonstrated the association between semen quality and state of health [94–96], correlating the former with chronic degenerative diseases.

As a matter of fact, male infertility is becoming a public health priority, and it's also related to an increased risk of later onset adult diseases, especially cancer [11, 97–101] not only testicular cancer [102–104], medical comorbidity [105], shorter life expectancy [106] and trans-generational effects [107, 108].

In this prospective, fertility assessment, sperm may be an indicator of overall health and the attention on maximum fertility age (18–35 years) can be important for chronic diseases prevention. In addition to the potential preventive and predictive role of reproductive biomarkers for chronic adult degenerative diseases, the growing interest on the transgenerational effects induced by pollution and lifestyles through epigenetic modifications on gametes shifts the interest of prevention as far as preconception; therefore, the interest for the reproductive system and biomarkers assumes a greater significance for safeguarding the health of future communities [107, 109].

However, given that a healthy environment and the mother's lifestyle are crucial for the offspring's health, and the utero window represents a field of study of Developmental Origins of Health and Disease (DOHaD) [110, 111], the Paternal Origins of Health and Disease paradigm (POHaD) should be taken into serious consideration [112–118]. Nevertheless, in spite of having few epidemiological studies on humans, the perspective opening the systematic study of reproductive biomarkers in environmental impact assessment and early and predictive health risk assessment is enormous [107].

Considering both the great impact of EDCs on the environment and health system and the need to protect and reduce their impact, policy implication and community interventions are mandatory. In this sense, the following chapters will point out the most important plan directions of principal public institutions in the frame of more recent knowledge on these contaminants.

10.2 Europe Police Priorities on EDCs: REACH Regulation, European Plant Protection Products Regulation (PPPR), Biocidal Products Regulation (BPR)

In Europe, general future priorities for protecting humans and wildlife from adverse effects of EDCs were reported at the Weybridge meeting in 1996 [119] and later in 1999 [120] a community strategy for EDCs was adopted. This was a fundamental step for addressing a regulatory basis on health and environmental effects caused by EDCs. European policy embraces the precautionary principle aimed at limiting exposure to agents harmful to humans, animals and the environment even in the absence of scientific certainties. The precautionary principle in the rules of the UE gives particular attention in the comparison of the danger of exposure of the fetus in utero and in its subsequent development. The EDC strategy, indeed, was characterized by a series of actions for monitoring programs and estimating exposure and effects of EDCs, to define and check testing methods and other actions for research on EDCs and consequently to develop regulatory actions. All instruments adopted for long-term actions under ED research group of European Union (EU), for example the 7th Environment Action Programs (EAPs) [121] were meant to protect all living species in EU. In 2006, REACH, officially Regulation No. 1907/2006, a regulation of the European Union, dated 18 December 2006, concerning the registration, validation, authorization and restriction of chemical substances [122], defines EDCs as substances of very high concern (SVHCs) for both health and environment, in order to reduce their use and replace them with other safer substances. Before 2007, EDCs had been considered responsible for the development of reproductive problems and cancer lesions, and therefore regulated. Currently, the EDCs risk assessment is specifically applied in the context of chemical classes in use. REACH requires companies to register substances and give data for ensuring safe handling; if the chemical is identified as an SVHC, it is included in a list of restricted chemicals under consideration of REACH for possible authorization. The European Chemicals Agency (ECHA), indeed, evaluates in first instance the chemicals included in the Authorization List for allowing their entry into the market after evaluation on the basis of article 57 of REACH regulation, which refers to the toxicity, carcinogenic bio accumulative, environment persistence properties and possibility to be replaced with safer alternatives.

The European Plant Protection Products Regulation (PPPR) on EDCs [123], although approved after REACH regulation, it was the first (EU) to take into account health effects and non-target organisms, evaluating the substances at mutagenic, carcinogenetic or toxic level on the basis of the regulatory system of classification, labelling and packaging (CLP). Nevertheless, the PPPR does not contain indications to define a substance with endocrine-interfering properties, consequently it referred with amendments to the regulation based on the WHO IPCS regarding the definition of "endocrine disorder" and "adverse health" (Table 10.1) [124]. Two derogations to the PPPR on EDCs were applied, but no agreement has been reached on the handling of these derogations until now *(Derogations: 1. the necessity for an active substance to control a serious danger to plant health (Article 4, paragraph 7). 2: negligible exposure towards an active substance, safener or synergist (Annex II, point 3.6.5)* [125]

Table 10.1 WHO IPCS definitions

WHO IPCS Definition of adversity (2009):
'A change in the morphology, physiology, growth, development, reproduction or life span of an organism, system or (sub)population that results in an impairment of functional capacity, an impairment of the capacity to compensate for additional stress or an increase in susceptibility to other influences'
WHO IPCS Definition of endocrine disruptors (2002):
'An ED is an exogenous substance or mixture that alters function(s) of the endocrine system and consequently causes adverse health effects in an intact organism, or its progeny, or (sub) populations'

In 2009, Plant Protection Products Regulation banned EDCs in pesticides and in the 2018 the European Food Safety Authority and European Chemical Agency produce a guideline for identifying EDCs present in pesticides.

The Biocidal Products Regulation (Regulation (EU) No 528/2012 of the European Parliament and of the Council of 22 May 2012 concerning the making available on the market and use of biocidal products) [126], has been the second European Regulation on substances used in biocidal products and has been applicable since 4 June 2018 (European Commission, Commission Delegated Regulation (EU) 2017/2100 of 4 September 2017 setting out scientific criteria for the determination of endocrine-disrupting properties pursuant to Regulation (EU) No 528/2012 of the European Parliament and Council, 2018) [127]. The most important difference compared to PPPR is that, in case of a substance having endocrine-disrupting properties, the application is approved for 5 years only and a product containing them it cannot be authorized for public use.

For medical and in vitro diagnostics devices, an EU regulation consolidated data on 24 April 2020 (Consolidated text: Regulation (EU) 2017/745 of the European Parliament and of the Council of 5 April 2017 on medical devices, amending Directive 2001/83/EC, Regulation (EC) No 178/2002 and Regulation (EC) No 1223/2009 and repealing Council Directives 90/385/EEC and 93/42/EEC (Text with EEA relevance) [128] and individualized only substances that in contact with the body pass to higher concentrations 0.1% weight by weight and have a justified endocrine interference property. These findings have been reported and discarded. A consolidated text of EU regulation for in vitro diagnostic devices on 28 February 2022 (Consolidated text: Regulation (EU) 2017/746 of the European Parliament and of the Council of 5 April 2017 on in vitro diagnostic medical devices and repealing Directive 98/79/EC and Commission Decision 2010/227/EU (Text with EEA relevance) [129], labelling the presence of endocrine interfering properties must be mandatory. For cosmetics, in the 2009 regulation, there are no reported endocrine interfering properties, although the problem was later addressed in 2018 leading to a commitment of the ED commission to define these ED properties and possibly limit or in any case avoid their use. In the light of this, REACH regulation should also address the environmental matter.

In order to release these substances in contact with food, although Regulation (EC) No. 1935/2004 does not generally allow it the release, much is handed over postponed to national laws; however, greater attention is paid to the release of plastic material, which is absolutely not allowed to be used in children's food; the same level of Bisphenol A release was reduced from 0.6 mg kg^{-1} food to 0.05 mg kg^{-1} food (Commission Regulation (EU) No 10/2011 of 14 January 2011, on plastic materials and articles intended to come into contact with food). Bundesinstitut fu¨r Risikobewertung (BfR), Database BfR Recommendations on Food Contact Materials) [130].

The Water Framework Directive (WFD) 2000/60/EC regulate known or suspected EDCs in water. In fact there are substances that are detected and that have already been banned from the market for some time. In 2020, 205 substances were included in the substances of very high concern (SVHC) list, 16 of these substances were included due to their endocrine-disrupting characteristics, currently these are included in a list of 45 chemical substances. Another list of substances including pharmaceuticals are monitored and under evaluation for health and environmental risk substances.

10.3   Other EDCs Regulations

EDCs have been also identified as an emerging policy issue by the UN Environment Programme (UNEP), which oversees global policy through strategic alliance for International Chemical Management. In 2015, the alliance welcomed the 2012 WHO and UNEP State of the Science report on EDCs [131]. In this report, tests exclusively focus on the estrogen, androgen and thyroid pathways [132] and do not take into account neither other receptors nor many other receptors mechanisms of action [133]. In 2017, UNEP identified 28 Policy actions, worldwide characterized by the variability to regulate the use of hazardous EDCs [134].

Currently, the various existing regulations share an agreement to limit a subset of persistent organic pollutants, including many EDCs. At the same time, the USA do not take into account the agreement and continue to market chemicals that the agreement has banned.

10.4   Economic Burden

The Food and Drug Administration has identified more than 1800 chemicals; today, medical societies and governmental agencies are experiencing an increase in health problems created by the action of EDCs and their effect seems to be transgenerational, occurring over at least two or three generations.

Considering nutrition is the main route of exposure to EDCs in human beings, we are currently witnessing a global and massive process of orientation and categorization of food consumption, through customs and lifestyles. This widespread exposure to EDCs and their consequent pathologies are diversified not only by gender characteristics but also by different age groups as well as economic income.

As previously stated, the most common way one comes into contact to EDCs is through food; however, people who work using these chemicals are exposed to them daily. EDCs can have long term and severe adverse effects on unborn babies' hormone system too, as shown over the last generation. The next generation will experience the same, or worse, if no future regulatory strategy for prevention is adopted. Indeed, the presence of EDCs in the environment and people's daily habits can expose the unborn child, from the embryonic stage onward, to these chemicals through feta–placental contact with the mother. The severity of the pathologies created range from pathologies of the male and female reproductive systems, to diabetes, cancer and diseases that affect the cognitive system by conditioning the level of IQ [135–138].

The Global Burden of Disease project uses an approach that calculates disability-adjusted life-year (DALY) [139], thanks to which the costs [140] of intellectual diseases and disabilities are also assessed. Nevertheless, due to its complexity, the DALY system remains insufficient to evaluate the intellectual damage to the human being [141].

Recently, several assessments have been carried out on the costs of diseases and damage caused by exposure to EDCs, such as neurobehavioural defects, disorder of the male and female reproductive systems, obesity and diabetes [135–138]. The economic burden results in 163 billion euro for EU, and 340 billion dollars for the USA annually, which means that not only this evaluation has been underestimated but also that only a few EDCs and their effects on health have been investigated [142–144].

It is therefore clear that a regulation aimed at identifying EDCs facilitates their replacement and prevents exposure by informing the population on how to avoid EDCs.

10.5   Limiting Exposure to EDCs

To give a clear indication for the hoped regulations is to promote the EDCs definition given by Endocrine Society as "any chemical or mixture of chemicals that interferes with any aspect of hormone action" [4]. This definition has the possibility to be applied anywhere globally: in all sectors of the economy and jurisdictions of the world. It also has the clear potential to deeply analyze what the damage is to the general population and to the specific exposed workers. It has the ability to calculate what the costs of treating the diseases are and the inconveniences caused to humankind, surrounding nature and the environment.

Differently, the definition given by WHO specifically requires an adverse effect to be documented [6, 145, 146]. Pointing out an adverse effect to define an EDC is problematic because regulatory agencies often disagree on which outcomes are adverse [147].

Perhaps, the time has come to consider EDCs among global burden diseases; this assessment becomes more evident if we consider the health costs paid by people exposed to EDCs and the costs that are paid by communities to treat the diseases [148].

We are realizing day by day that the population's exposure to EDCs creates pathologies involving the female and male reproductive system, which reduce their abilities and also, in some minority groups of the population [149], create disabilities, including neurocognitive ones [138].

We should avoid searching with great difficulty the critical levels of tolerability as a risk-based approach assumes [150]. Furthermore, many EDCs can act at low concentrations and often present non-linear dose–responses; these properties represent a regulatory challenge.

Indeed, for several EDCs, it's not possible to evaluate a safe threshold for the toxicity, especially for neuro developmental deficits. In this regard, in a Danish report, the aim of obtaining an additional and pragmatic choice for the evaluation of safety was discussed [151].

Differently, in a hazard-based approach, once identified the hazardous properties of a chemical, it becomes sufficient to prohibit market use regardless of the exposure and economic cost. It is difficult to imagine that the regulation of an EDC must wait for full-blown pathological effect instead of preventing it. If we consider pathologies such as cancer, diabetes, or all malfunctions of the male and female reproductive apparatus and neurocognitive problems, it becomes very complicated to accept a logic that exposes humans and the surrounding environment to suffer inert the snare of EDCs.

10.6   Policy Management and Recommendations

We need regulatory recommendation of EDCs, pointing on their identification, a policy geared to monitor and reduce exposure to better protect humans and the environmental health.

The first recommendation is on the identification of EDCs. This is the necessary basis for an effective action.

The Endocrine Disruptor Screening and Testing Advisory Committee [152] promotes evaluation of estrogen, androgen and thyroid receptor disruptions.

Despite these clear indications, in the USA, regulation requires testing for estrogen only for few materials in use.

To obtain better EDCs identification, we would probably need more sensitive tests to some EDCs and clarify more pathways for each estrogen, androgen and thyroid receptor.

For thyroid axis, identification disruptors are scarce, and, for many pathways, they are completely absent. Furthermore, we need to modify some test modalities for disruptor identification, which could give misleading results, for example, the changes in uterine weight test under high concentrations of estrogenic EDCs [153, 154]. At the same time, to this day, a whole series of capable tests for a broader number of nuclear receptors, and other receptor types, is available, which is also capable of assessing several mechanisms of action for EDCs [155]. This capable test should be inserted in regulatory requirement after validation. The Organisation for Economic Cooperation and Development guidance provides new assays to test more pathways than those required from EU and USA regulation.

As soon as possible, we need tests to identify EDCs involved in adipocyte development, steroidogenesis and all reproductive functions to begin from spermatogenesis and other female reproductive system pathologies or dysfunctions (e.g. endometriosis, polycystic ovary syndrome).

Human diseases should become the first-tier management to identify EDCs and their adverse impact [156]. The difficulty to recognize the effects induced by EDCs is found in the regulations that encourages just in vivo observation. This means that we need to use vertebrate animals or epidemiological evidence, but the regulatory authorities have proposed strong limitations to use mammals for regulatory testing and at the same time we do not have sufficient in vitro guidelines for this shortcoming. Likewise, the non-mammalian vertebrate and invertebrate models have not the possibility to overcome the shortcoming information.

We need more typical mammalian models (egg rodents) evidence to validate in vitro testing and have the possibility to examine complex effect on health.

All this limitation takes risk of EDCs identification. A determination of adverse effect should be sufficient for identification as an EDC and subsequent regulation.

Obviously, the search for already tested chemical safer substitutes should be supported and developed by different governments. Before chemicals get into the market to be used by the population, they should be tested in vitro and in vivo in order to give a guarantee not to alter the state of human health and the surrounding environment. Unfortunately, some substitutions used up to now for polybrominated diphenyl or BPA remain questionable and dangerous in other pathways. Another recommendation concerns the exposure to EDCs. Once identified, an EDC should be excluded from the possibility of human exposure. A risk-based regulatory approach currently prevails in which the effects are evaluated on the basis of exposure degree. This means that we need a specific monitoring for each single EDC conjugated to each environmental reality, economic development system lifestyle and profession.

It is essential that decision makers know how chemicals are being used.

Information campaigns could be organized for the general population or population particularly exposed to EDCs. An action of this dimension could change exposure levels by informing and raising awareness of how to avoid the possible and different contact with individual EDCs and defend health from the sneaky snare of EDCs.

10.7   Conclusion

EDC policies are justified on economic grounds and to further environmental justice.

The endocrine disruptor is constantly expanding its range of influence on the globe population. We are currently witnessing what's happening in the surrounding environment.

On the scientific level that works on the identification of chemical substances with altering characteristics of the functional systems of living organisms, it is necessary to expand the possibilities of directly investigating altered processes in vivo on animal models. The hazard-based approach, meant to guide and reach the regulations, should be preferred to the risk-based one.

The health risks are considerable as the social costs associated with poor health conditions are, in this sense, representative signs are emerging both in the number of women and men involved and in the developed dysfunctions.

In particular, since it is better measurable, great evidence of how much human activities are seriously impacting the quality of the planet, and, consequently, of human health, can be precisely deduced from the progressive semen quality decline, which probably represents the impact best. The reproductive system, in particular the male one, can be represented as a "Sentinel Organ System" due to its extreme sensitivity to environmental stress; recent evidence indicates human semen as an early and sensitive source of exposure, therefore a useful tool for measuring the presence and effects of chemical substances not only on the classic seminal parameters (number, motility, morphology and DNA sperm damage) but also on others that are now better studied in vivo with molecular biology techniques. Exposure assessment, in conjunction with information on the inherent toxicity of the chemical (i.e. the expected response at a given level of exposure), plays a key role in predicting the likelihood, nature and magnitude of adverse health effects. The use of reproductive biomarkers for early risk detection represents a possible new methodological approach where they could be exploited as early indicators of environmental toxicity and enhanced risk of chronic adverse effects not only for reproductive health. In particular, human semen as an early and reliable source of biomarkers, giving information on biologically active exposures, can be very useful for preventive health surveillance programs, especially in environmental risk areas. This approach appears very promising, above all, in young people (maximum fertile age:18–35 years), considering the possibility to reduce the chronic-degenerative diseases in future adults. In this context, many scientific findings regard the association between pollution and fertility problems. Therefore, the safeguard of germ cells is a new challenge to reduce the burden of epigenetically transmitted diseases.

References

1. Yilmaz B, Terecia H, Sandal S, Kelestimur F. Endocrine disrupting chemicals: exposure, effects on human health, mechanism of action, models for testing and strategies for prevention. Rev Endocr Metab Disord. 2020;21(1):127–47. https://doi.org/10.1007/s11154-019-09521-z.
2. ECHA. Endocrine disruptor assessment list of European Chemical Agency (ECHA). endocrine disruptor assessment list. 2023. https://echa.europa.eu/it/ed-assessment.
3. Colborn T, Vom Saal FS, Soto AM. Review developmental effects of endocrine-disrupting chemicals in wildlife and humans. Environ Health Perspect. 1993;101(5):378–84.
4. Gore AC, Chappell VA, Fenton SE, et al. EDC-2: the Endocrine Society's second scientific statement on endocrine-disrupting chemicals. Endocr Rev. 2015;36:e1–150.

5. Solecki R, Kortenkamp A, Bergman Å, Chahoud I, Degen GH, Dietrich D, Greim H, Håkansson H, Hass U, Husoy T, Jacobs M, Jobling S, Mantovani A, Marx-Stoelting P, Piersma A, Ritz V, Slama R, Stahlmann R, Martin van den Berg R, Zoeller T, Boobis AR. Scientific principles for the identification of endocrine-disrupting chemicals: a consensus statement. Arch Toxicol. 2017;91(2):1001–6. https://doi.org/10.1007/s00204-016-1866-9.
6. WHO. State of the science of endocrine disrupting chemicals. http://www.who.int/ceh/publications/endocrine/en/. Accessed 20 Nov 2020.
7. Pironti C, Ricciardi M, Proto A, Bianco PM, Montano L, Motta O. Endocrine-disrupting compounds: an overview on their occurrence in the aquatic environment and human exposure. Water. 2021;13(10):1347. https://doi.org/10.3390/w13101347.
8. Myers JP, Guillette LJ Jr, Palanza P, Parmigiani S, Swan SH, Vom Saal FS. The emerging science of endocrine disruption. In: Ragaini RC, editor. International Seminars on Planetary Emergencies, 30th Session. London: World Scientific Publishing; 2004. p. 105–12.
9. Heindel JJ, vom Saal FS, Blumberg B, Bovolin P, Calamandrei G, Ceresini G, Cohn BA, Fabbri E, Gioiosa L, Kassotis C, et al. Parma consensus statement on metabolic disruptors. Environ Health. 2015;14:5.
10. Trasande L, Zoeller RT, Hass U, Kortenkamp A, Grandjean P, Myers JP, DiGangi J, Bellanger M, Hauser R, Legler J, Skakkebaek NE, Heindel JJ. Estimating burden and disease costs of exposure to endocrine-disrupting chemicals in the European union. J Clin Endocrinol Metab. 2015;100(4):1245–55. https://doi.org/10.1210/jc.2014-4324.
11. Jagai JS, Messer LC, Rappazzo KM, Gray CL, Grabich SC, Lobdell DT. County-level cumulative environmental quality associated with cancer incidence. Cancer. 2017;123(15):2901–8. https://doi.org/10.1002/cncr.30709.
12. Siegel R, Naishadham D, Jemal A. Cancer statistics, 2013. CA Cancer J Clin. 2014;63:11–30.
13. National Heart, Lung, and Blood Institute. Fact book: fiscal year 2008. Bethesda, MD: National Institutes of Health; 2009.
14. Tomasetti C, Vogelstein B, Cancer etiology. Variation in cancer risk among tissues can be explained by the number of stem cell divisions. Science. 2015;347:78–81.
15. Wallace TA, Martin DN, Ambs S. Interactions among genes, tumor biology and the environment in cancer health disparities: examining the evidence on a national and global scale. Carcinogenesis. 2011;32:1107–21.
16. Wu S, Powers S, Zhu W, Hannun YA. Substantial contribution of extrinsic risk factors to cancer development. Nature. 2016;529:43–7.
17. Papalou O, Kandaraki EA, Papadakis G, Diamanti-Kandarakis E. Endocrine disrupting chemicals: an occult mediator of metabolic disease. Front Endocrinol. 2019;10:112. https://doi.org/10.3389/fendo.2019.00112. eCollection 2019.
18. Ogden CL, Carroll MD, Kit BK, Flegal KMJAMA. Prevalence of childhood and adult obesity in the United States, 2011-2012. JAMA. 2014;311(8):806–14.
19. International Diabetes Federation. IDF Diabetes Atlas. 8th ed; 2017. https://www.idf.org/e--library/epidemiology-research/diabetes-atlas.html
20. Lehnert T, Sonntag D, Konnopka A, Riedel-Heller S, Konig HH. Economic costs of overweight and obesity. Best Pract Res Clin Endocrinol Metab. 2013;27:105–15. https://doi.org/10.1016/j.beem.2013.01.002.
21. Huang ES, Laiteerapong N, Liu JY, John PM, Moffet HH, Karter AJ. Rates of complications and mortality in older patients with diabetes mellitus: the diabetes and aging study. JAMA Int Med. 2014;174:251–8. https://doi.org/10.1001/jamainternmed.2013.129569.
22. Anneet B, Raphael R. Endocrine disruptor chemicals. In: Endotext. South Dartmouth (MA): MDText.com, Inc; 2000.
23. American Diabetes. Economic costs of diabetes in the US. in 2017. Diabetes Care. 2018;41:917–28. https://doi.org/10.2337/dci18-0007.
24. Predieri B, Alves CAD, Iughetti L. New insights on the effects of endocrine-disrupting chemicals on children. J Pediatr. 2022;98(Suppl 1):S73–85. https://doi.org/10.1016/j.jped.2021.11.003.

25. Johnson RA, Harris RE, Wilke RA. Are pesticides really endocrine disruptors? WMJ. 2000;99(8):34–8.
26. Guillette LJ Jr. Organochlorine pesticides as endocrine disruptors in wildlife. Cent Eur J Public Health. 2000;8:34–544.
27. Guillette LJ, Gross TS, Masson GR, Matter JM, Percival HF, Woodward AR. Developmental abnormalities of the gonad and abnormal sex hormone concentrations in juvenile alligators from contaminated and control lakes in Florida. Environ Health Perspect. 1994;102(8):680–8.
28. Hayes TB, Collins A, Lee M, Mendoza M, Noriega N, Stuart AA, Vonk A. Hermaphroditic, demasculinized frogs after exposure to the herbicide atrazine at low ecologically relevant doses. Proc Natl Acad Sci U S A. 2002;99(8):5476–80.
29. Hayes TB, Anderson LL, Beasley VR, de Solla SR, Iguchi T, Ingraham H, Kestemont P, Kniewald J, Kniewald Z, Langlois VS, Luque EH, McCoy KA, Muñoz-de-Toro M, Oka T, Oliveira CA, Orton F, Ruby S, Suzawa M, Tavera-Mendoza LE, Trudeau VL, Victor-Costa AB, Willingham E. Demasculinization and feminization of male gonads by atrazine: consistent effects across vertebrate classes. J Steroid Biochem Mol Biol. 2011;127(1–2):64–73. https://doi.org/10.1016/j.jsbmb.2011.03.015. Epub 2011 Mar 23.
30. Grilo TF, Rosa R. Intersexuality in aquatic invertebrates: prevalence and causes. Sci Total Environ. 2017;592:714–28. https://doi.org/10.1016/j.scitotenv.2017.02.099. Epub 2017 Mar 18.
31. Baatrup E, Junge M. Antiandrogenic pesticides disrupt sexual characteristics in the adult male guppy Poecilia reticulate. Environ Health Perspect. 2001;109(10):1063–70.
32. Singleton DW, Khan SA. Xenoestrogen exposure and mechanisms of endocrine disruption. Front Biosci. 2003;1(8):S110–8.
33. Akingbemi BT, Hardy MP. Oestrogenic and antiandrogenic chemicals in the environment: effects on male reproductive health. Ann Med. 2001;33(6):391–403.
34. Sultan C, Balaguer P, Terouanne B, Georget V, Paris F, Jeandel C, Lumbroso S, Nicolas J. Environmental xenoestrogens, antiandrogens and disorders of male sexual differentiation. Mol Cell Endocrinol. 2001;178(1–2):99–105.
35. Hodges LC, Bergerson JS, Hunter DS, Walker CL. Estrogenic effects of organochlorine pesticides on uterine leiomyoma cells in vitro. Toxicol Sci. 2000;54(2):355–64.
36. Chrustek A, Hołyńska-Iwan I, Dziembowska I, Bogusiewicz J, Wróblewski M, Cwynar A, Olszewska-Słonina D. Current research on the safety of Pyrethroids used as insecticides. Medicina (Kaunas). 2018;54(4):E61. https://doi.org/10.3390/medicina54040061.
37. Garey J, Wolff MS. Estrogenic and antiprogestagenic activities of pyrethroid insecticides. Biochem Biophys Res Commun. 1998;251(3):855–9.
38. Bell EM, Hertz-Picciotto I, Beaumont JJ. A case-control study of pesticides and fetal death due to congenital anomalies. Epidemiology. 2001;12(2):148–56.
39. Quesada I, Fuentes E, Viso Leon MC, Soria B, Ripoll C, Nadal A. Low doses of the endocrine disruptor bisphenol-A and the native hormone 17β-estradiol rapidly activate transcription factor CREB. FASEB J. 2002;16:1671–3.
40. Takeuchi T, Tsutsumi O. Serum bisphenol a concentrations showed gender differences, possibly linked to androgen levels. Biochem Biophys Res Commun. 2002;291:76–8.
41. Swan SH, Elkin EP, Fenster L. The question of declining sperm density revisited: an analysis of 101 studies published 1934–1996. Enciron Health Perspect. 2000;108(10):961–6.
42. Andersson AM, Jørgensen N, Main KM, Toppari J, Meyts ER, Leffers H, Juul A, Jensen TK, Skakkebæk NE. Adverse trends in male reproductive health: We may have reached a crucial 'tipping point'. Int J Androl. 2008;31(2):74–80.
43. Mascarenhas MN, Flaxman SR, Boerma T, Vanderpoel S, Stevens GA. National, regional, and global trends in infertility prevalence since 1990: a systematic analysis of 277 health surveys. PLoS Med. 2012;9(12):e1001356. https://doi.org/10.1371/journal.pmed.1001356. Epub 2012 Dec 18.
44. Carlsen E, Giwercman A, Keiding N, Skakkebaek NE. Evidence for decreasing quality of semen during past 50 years. BMJ. 1992;305:609–13.

45. Travison TG, Araujo AB, O'Donnell AB, Kupelian V, John B, McKinlay JB. A population-level decline in serum testosterone levels in American men. J Clin Endocrinol Metab. 2007;92(1):196–202. https://doi.org/10.1210/jc.2006-1375. Epub 2006 Oct 24.
46. Atlantis E, Fahey P, Martin S, O'Loughlin P, Taylor AW, Adams RJ, Shi Z, Wittert G. Predictive value of serum testosterone for type 2 diabetes risk assessment in men. BMC Endocr Disord. 2016;16(1):26. https://doi.org/10.1186/s12902-016-0109-7.
47. Levine H, Jørgensen N, Andrade AM, Mendiola J, Weksler-Derri D, Mindlis I, Pinotti R, Swan SH. Temporal trends in sperm count: a systematic review and meta-regression analysis. Hum Reprod Update. 2017;23(6):646–59. https://doi.org/10.1093/humupd/dmx022.
48. Huang C, Li B, Xu K, Liu D, Hu J, Yang Y, Nie H, Fan L, Zhu W. Decline in semen quality among 30,636 young Chinese men from 2001 to 2015. Fertil Steril. 2017;107(1):83–88.e2. https://doi.org/10.1016/j.fertnstert.2016.09.035.
49. Montano L, Maugeri A, Volpe MG, Micali S, Mirone V, Mantovani A, Navarra M, Piscopo M. Mediterranean diet as a shield against male infertility and Cancer risk induced by environmental pollutants: a focus on flavonoids. Int J Mol Sci. 2022;23(3):1568. https://doi.org/10.3390/ijms23031568.
50. Sharpe RM. The "oestrogen hypothesis"- where do we stand now? Int J Androl. 2003;26:2–15.
51. Knez J. Endocrine-disrupting chemicals and male reproductive health. Reprod BioMed Online. 2013;26:440–8. https://doi.org/10.1016/j.rbmo.2013.02.005.
52. Del-Mazo J, Brieño-Enríquez MA, García-López J, López-Fernández LA, De-Felici M. Endocrine disruptors, gene deregulation and male germ cell tumors. Int J Dev Biol. 2013;57:225–39.
53. Sifakis S, Androutsopoulos VP, Tsatsakis AM, Spandidos DA. Human exposure to endocrine disrupting chemicals: effects on the male and female reproductive systems. Environ Toxicol Pharmacol. 2017;51:56–70. https://doi.org/10.1016/j.etap.2017.02.024.
54. Manikkam M, Haque MM, Guerrero-Bosagna C, Nilsson EE, Skinner MK. Pesticide methoxychlor promotes the epigenetic transgenerational inheritance of adult-onset disease through the female germline. PLoS One. 2014;9:e102091.
55. Manikkam M, Tracey R, Guerrero-Bosagna C, Skinner MK. Dioxin (TCDD) induces epigenetic transgenerational inheritance of adult onset disease and sperm epimutations. PLoS One. 2012;7:e46249.
56. Skinner MK, Anway MD. Seminiferous cord formation and germ-cell programming: epigenetic transgenerational actions of endocrine disruptors. Ann N Y Acad Sci. 2005;1061:18–32.
57. Anway MD, Cupp AS, Uzumcu M, Skinner MK. Epigenetic transgenerational actions of endocrine disruptors and male fertility. Science. 2005;308:1466–9.
58. Doyle TJ, Bowman JL, Windell VL, McLean DJ, Kim KH. Transgenerational effects of di-(2)-ethylhexyl phthalate on testicular germ cell associations and spermatogonial stem cells in mice. Biol Reprod. 2013;88:112.
59. Susiarjo M, Sasson I, Mesaros C, Bartolomei MS. Bisphenol A exposure disrupts genomic imprinting in the mouse. PLoS Genet. 2013;9:e1003401.
60. Goldsby JA, Wolstenholme JT, Rissman EF. Multi- and transgenerational consequences of bisphenol A on sexually dimorphic cell populations in mouse brain. Endocrinology. 2016;158:21–30.
61. Hao C, Gely-Pernot A, Kervarrec C, Boudjema M, Becker E, Khil P, et al. Exposure to the widely used herbicide atrazine results in deregulation of global tissue-specific RNA transcription in the adult testis and is associated with a global decrease of histone trimethylation in mice. Nucleic Acids Res. 2016;44:9784–802.
62. Horan TS, Marre A, Hassold T, Lawson C, Hunt PA. Germline and reproductive tract effects intensify in male mice with successive generations of estrogenic exposure. PLoS Genet. 2017;2(23):e1006885. https://doi.org/10.1371/journal.pgen.1006885.
63. Viluksela M, Pohjanvirta R. Multigenerational and transgenerational effects of dioxins. Int J Mol Sci. 2019;20(12):2947.

64. Barazani Y, Katz BF, Nagler HM, Stember DS. Lifestyle, environment, and male reproductive health. Urol Clin North Am. 2014;41:55–66.

65. Selevan SG, Borkovec L, Slott VL, Zudova Z, Rubes J, Evenson DP, Perreault SD. Semen quality and reproductive health of young Czech men exposed to seasonal air pollution. Environ Health Perspect. 2000;108:887–94.

66. Rubes J, Selevan SG, Evenson DP, Zudova D, Vozdova M, Zudova Z, Robbins WA, Perreault SD. Episodic air pollution is associated with increased DNA fragmentation in human sperm without other changes in semen quality. Hum Reprod. 2005;20:2776–83.

67. Guven A, Kayikci A, Cam K, Arbak P, Balbay O, Cam M. Alterations in semen parameters of toll collectors working at motorways: does diesel exposure induce detrimental effects on semen? Andrologia. 2008;40:346–51.

68. Deng Z, Chen F, Zhang M, Lan L, Qiao Z, Cui Y, An J, Wang N, Fan Z, Zhao X, Li X. Association between air pollution and sperm quality: a systematic review and meta-analysis. Environ Pollut. 2016;208:663–9.

69. Hammoud A, Carrell DT, Gibson M, Sanderson M, Parker-Jones K, Peterson CM. Decreased sperm motility is associated with air pollution in Salt Lake City. Fertil Steril. 2010;93(6):1875–9. https://doi.org/10.1016/j.fertnstert.2008.12.089.

70. Fathi Najafi T, Latifnejad Roudsari R, Namvar F, Ghavami Ghanbarabadi V, Hadizadeh Talasaz Z, Esmaeli M. Air pollution and quality of sperm: a meta-analysis. Iran Red Crescent Med J. 2015;17(4):e26930. https://doi.org/10.5812/ircmj.17(4)2015.26930.

71. Auger J, Kunstmann JM, Czyglik F, Jouannet P. Decline in semen quality among fertilemen in Paris during the past 20 years. N Engl J Med. 1995;332((9)):281–5.

72. Mendiola J, Jorgensen N, Andersson AM, Stahlhut RW, Liu F, Swan SH. Reproductive parameters in young men living in Rochester, New York. Fertil Steril. 2014;101:1064–71.

73. Le Moal J, Rolland M, Goria S, Wagner V, De Crouy-Chanel P, Rigou A, et al. Semen quality trends in French regions are consistent with a global change in environmental exposure. Reproduction. 2014;147:567.

74. Hauser R, Sokol R. Science linking environmental contaminant exposures with fertility and reproductive health impacts in the adult male. Fertil Steril. 2008;89:e59–e6.

75. Akre O, Cnattingius S, Bergstrom R, Kvist U, Trichopoulos D, Ekbom A. Human fertility does not decline: evidence from Sweden. Fertil Steril. 1999;71:1066–9.

76. Menchini-Fabris F, Rossi P, Palego P, Simi S, Turchi P. Declining sperm counts in Italy during the past 20 years. Andrologia. 1996;28:304–32.

77. Nordkap L, Joensen UN, Blomberg Jensen M, Jørgensen N. Regional differences and temporal trends in male reproductive health disorders: semen quality may be a sensitive marker of environmental exposures. Mol Cell Endocrinol. 2012;355(2):221–30.

78. Bergamo P, Volpe MG, Lorenzetti S, Mantovani A, Notari T, Cocca E, et al. Human semen as an early, sensitive biomarker of highly polluted living environment in healthy men: a pilot biomonitoring study on trace elements in blood and semen and their relationship with sperm quality and RedOx status. Reprod Toxicol. 2016;66:1–9.

79. Lettieri G, D'Agostino G, Mele E, Cardito C, Esposito R, Cimmino A, Giarra A, Trifuoggi M, Raimondo S, Notari T, Febbraio F, Montano L, Piscopo M. Discovery of the involvement in DNA oxidative damage of human sperm nuclear basic proteins of healthy young men living in polluted areas. Int J Mol Sci. 2020;21(12):E4198. https://doi.org/10.3390/ijms21124198.

80. Zhou N, Cui Z, Yang S, Han X, Chen G, Zhou Z, et al. Air pollution and decreased semen quality: a comparative study of Chongqing urban and rural areas. Environ Pollut. 2014;187:145–52.

81. Vecoli C, Montano L, Borghini A, Notari T, Guglielmino A, Mercuri A, Turchi S, Andreassi MG. Effects of highly polluted environment on sperm telomere length: a pilot study. Int J Mol Sci. 2017;18:1703. https://doi.org/10.3390/ijms18081703.

82. Bosco L, Notari T, Ruvolo G, Roccheri MC, Martino C, Chiappetta R, Carone D, Lo Bosce G, Carrillo L, Raimondo S, Guglielmino A, Montano L. Sperm DNA fragmentation: an early

and reliable marker of air pollution. Environ Toxicol Pharmacol. 2018;58:243–9. https://doi.org/10.1016/j.etap.2018.02.001.

83. Zani C, Zani D, Donato F, Marullo M, Viola GCV, Lorenzetti S, Montano L. Lifestyle, environmental exposures and male fertility in healthy young men in North Italy. Eur J Pub Health. 2019;29(Supplement_4):ckz186–459. https://doi.org/10.1093/eurpub/ckz186.459.

84. Montano L, Ceretti E, Donato F, Bergamo P, Zani C, Viola GCV, Notari T, Pappalardo S, Zani D, Ubaldi S, Bollati V, Consales C, Leter G, Trifuoggi M, Amoresano A, Lorenzetti S. Effects of a lifestyle change intervention on semen quality in healthy young men living in highly polluted areas in Italy: the FASt randomized controlled trial. Eur Urol Focus. 2021;8(1):351–9. https://doi.org/10.1016/j.euf.2021.01.017.

85. Pirastu R, Comba P, Conti S, Iavarone I, Fazzo L, Pasetto R, Zona A, Crocetti E, Ricci P. Sentieri-Epidemilogical study of residents in National Priority Contaminated Sites:mortality, cancer incidence and hospital discharges. Epidemiol Prev. 2014;38:25–33.

86. Tagliabue G, Borgini A, Tittarelli A, van Donkelaar A, Martin RV, Bertoldi M, Fabiano S, Maghini A, Codazzi T, Scaburri A, Favia I, Cau A, Barigelletti G, Tessandori R, Contiero P. Atmospheric fine particulate matter and breast cancer mortality: a population-based cohort study. BMJ Open. 2016;6(11):e012580. https://doi.org/10.1136/bmjopen-2016-012580. PMID: 2807627528–32.

87. Martuzzi M, Mitis F, Bianchi F, Minichilli F, Comba P, Fazzo L. Cancer mortality andcongenital anomalies in a region of Italy with intense environmental pressure due to waste. Occup Environ Med. 2009;66:725–32. https://doi.org/10.1136/oem.2008.044115.

88. Pasetto R, Zengarini N, Caranci N, De Santis M, Minichilli F, Santoro M, Pirastu R, Comba P. Environmental justice in the epidemiological surveillance system of residents in Italian National Priority Contaminated Sites (SENTIERI project). Epidemiol Prev. 2017;41(2):134–9.

89. Bay K, Asklund C, Skakkebaek N, Andersson AM. Testicular dysgenesis syndrome: possible role of endocrine disrupters. Best Pract Res Clin Endocrinol Metab. 2006;20(1):77–90.

90. Ségurel L, Wyman MJ, Przeworski M. Determinants of mutation rate variation in the human germline. Annu Rev Genomics Hum Genet. 2014;15:47–70.

91. Blumenstiel JP. Sperm competition can drive a male-biased mutation rate. J Theor Biol. 2007;249(3):624–32.

92. Aitken RJ, Gibb Z, Baker MA, Joel Drevet J, Parviz Gharagozloo P. Causes and consequences of oxidative stress in spermatozoa. Reprod Fertil Dev. 2016;28(2):1–10. https://doi.org/10.1071/RD15325.

93. Montano L, Bergamo P, Andreassi MG, Lorenzetti S. The role of human semen as an early and reliable tool of environmental impact assessment on human health. In: Spermatozoa—Facts and Perspectives. Rijeka: InTechOpen; 2018. https://doi.org/10.5772/intechopen.73231.

94. Choy JT, Eisenberg ML. Male infertility as a window to health. Fertil Steril. 2018;110(5):810–4. https://doi.org/10.1016/j.fertnstert.2018.08.015.

95. Barnhart KT. Introduction: fertility as a window to health. Fertil Steril. 2018;110(5):781–2. https://doi.org/10.1016/j.fertnstert.2018.08.031.

96. Pisarska MD. Fertility status and overall health. Semin Reprod Med. 2017;35(3):203–4. https://doi.org/10.1055/s-0037-1603728. Epub 2017 Jun 28.

97. Eisenberg ML, Li S, Brooks JD, Cullen MR, Baker LC. Increased risk of cancer in infertile men: analysis of U.S. claims data. J Urol. 2015;193(5):1596–601. https://doi.org/10.1016/j.juro.2014.11.080.

98. Hanson BM, Eisenberg ML, Hotaling JM. Male infertility: a biomarker of individual and familial cancer risk. Fertil Steril. 2018;109(1):6–19. https://doi.org/10.1016/j.fertnstert.2017.11.005.

99. Glazer CH, Bonde JP, Eisenberg ML, Giwercman A, Hærvig KK, Rimborg S, Vassard D, Pinborg A, Schmidt L, Bräuner EV. Male infertility and risk of nonmalignant chronic diseases: a systematic review of the epidemiological evidence. Semin Reprod Med. 2017;35(3):282–90. https://doi.org/10.1055/s-0037-1603568.

100. Brinton LA. Fertility status and cancer. Semin Reprod Med. 2017;35(3):291–7. https://doi.org/10.1055/s-0037-1603098.
101. Rogers MJ, Walsh TJ. Male infertility and risk of cancer. Semin Reprod Med. 2017;35:298–303.
102. Baker JA, Buck GM, Vena JE, Moysich KB. Fertility patterns prior to testicular cancer diagnosis. Cancer Causes Control. 2005;16:295–9.
103. Jorgensen N, Vierula M, Jacobsen R, Pukkala E, Perheentupa A, Virtanen HE, Skakkebaek NE, Toppari J. Recent adverse trends in semen quality and testis cancer incidence among Finnish men. Int J Androl. 2011;34:e37–48.
104. Rives N, Perdrix A, Hennebicq S, Saias-Magnan J, Melin MC, Berthaut I, Barthélémy C, Daudin M, Szerman E, Bresson JL, Brugnon F, Bujan L. The semen quality of 1158 men with testicular cancer at the time of cryopreservation: results of the French national CECOS network. J Androl. 2012;33:1394–401.
105. Eisenberg ML, Li S, Behr B, Pera RR, Cullen MR. Relationship between semen production and medical comorbidity. Fertil Steril. 2015;103:66–71.
106. Jensen TK, Jacobsen R, Christensen K, Nielsen NC, Bostofte E. Good semen quality and life expectancy: a cohort study of 43,277 men. Am J Epidemiol. 2009;170:559–6.
107. Montano L. Reproductive biomarkers as early indicators for assessing environmental health risk. In: Marfe G, Di Stefano C, editors. Toxic waste management and health risk. Bentham Science Publishers; 2020. https://doi.org/10.2174/97898114547451200101. eBook eISBN 978-981-14-5474-5. https://www.eurekaselect.com/185279/chapter.
108. Gennaro Lettieri G, Marra F, Moriello C, Prisco P, Notari T, Trifuoggi M, Giarra A, Bosco L, Montano L, Piscopo M. Molecular alterations in spermatozoa of a family case living in the land of fires—a first look at possible transgenerational effects of pollutants. Int J Mol Sci. 2020;21(18):6710.
109. Raimondo S, Gentile M, Esposito G, Gentile T, Ferrara I, Crescenzo C, Palmieri M, Cuomo F, De Filippo S, Lettieri G, Piscopo M, Montano L. Could Kallikrein-Related Serine Peptidase 3 Be an Early Biomarker of Environmental Exposure in Young Women? Int J Environ Res Public Health. 2021;18:8833. https://doi.org/10.3390/ijerph1816883.
110. Suzuki K. The developing world of DOHaD. Dev Orig Health Dis. 2018;9(3):266–9.
111. Bianco-Miotto T, Craig JM, Gasser YP, van Dijk SJ, Ozanne SE. Epigenetics and DOHaD: from basics to birth and beyond. J Dev Orig Health Dis. 2017;8(5):513–9.
112. Soubry A, Hoyo C, Jirtle RL, Murphy SK. A paternal environmental legacy: evidence for epigenetic inheritance through the male germ line. BioEssays. 2014;36(4):359–71. https://doi.org/10.1002/bies.201300113. Epub 2014 Jan 16.
113. Soubry A. POHaD: why we should study future fathers. Environ Epigenet. 2018;4(2):dvy007.
114. Soubry A. Epigenetics as a driver of developmental origins of health and disease: did we forget the fathers? BioEssays 2018; 40(1).
115. Curley JP, Mashoodh R, Champagne FA. Epigenetics and the origins of paternal effects. Horm Behav. 2011;59(3):306–14.
116. Braun K, Champagne FA. Paternal influences on offspring development: behavioural and epigenetic pathways. Neuroendocrinology. 2014;26(10):697–706.
117. Zhao ZH, Schatten H, Sun QY. Environmentally induced paternal epigenetic inheritance and its effects on offspring health. Reprod Dev Med. 2017;1(2):89–99.
118. Xavier MJ, Roman SD, Aitken RJ, Nixon B. Transgenerational inheritance: how impacts to the epigenetic and genetic information of parents affect offspring health. Hum Reprod Update. 2019;25:dmz017.
119. Bergman A, Brandt I, Brouwer A, Harrison P, Holmes P, Humfrey C, Keiding N, Randall G, Sharpe R and Skakkebaek N. European workshop on the impact of endocrine disrupters on human health and wildlife. 2–4 December 1996. Report of Proceedings, Weybridge, UK, 1997.
120. European Commission. Communication from the Commission to the Council and the European Parliament—Community strategy for endocrine disrupters—A range of substances suspected

of interfering with the hormone systems of humans and wildlife. 1999. COM/99/0706 final. https://eur-lex.europa.eu/legal-content/EN/ALL/?uri=CELEX:51999DC0706.

121. European Parliament, Council of the European Union. Decision No 1386/ 2013/EU of the European Parliament and of the Council of 20 November 2013 on a General Union Environment Action Programme to 2020 Living well, within the limits of our planet. 2014. http://data.europa.eu/eli/dec/2013/1386/oj.

122. Regulation (EC) No 1907/2006 of the European Parliament and of the Council of 18 December2006 concerning the Registration, Evaluation, Authorisation and Restriction of Chemicals (REACH), establishing a European Chemicals Agency, amending Directive 1999/45/EC and repealing Council Regulation (EEC) No 793/93 and Commission Regulation (EC) No 1488/94 as well as Council Directive 76/769/EEC and Commission Directives 91/155/EEC, 93/67/EEC,93/105/EC and 2000/21/EC.

123. Regulation (EC) No 1107/2009 of the European Parliament and of the Council of 21 October 2009 concerning the placing of plant protection products on the market and repealing Council Directives 79/117/EEC and 91/414/EEC.

124. Mantovani A, Fucic A. Challenges in endocrine disruptor toxicology and risk Assessment. London: The Royal Society of Chemistry; 2021. https://doi.org/10.1039/9781839160738.

125. Commission Regulation (EU) 2018/605 of 19 April 2018 setting out scientific criteria for the determination of endocrine-disrupting and amending Annex II to Regulation (EC) 1107/2009. 2018. http://data.europa.eu/eli/reg/2018/605/oj.

126. Regulation (EU) No 528/2012 of the European Parliament and of the Council of 22 May 2012 concerning the making available on the market and use of biocidal products. https://echa.europa.eu/regulations/biocidal-products-regulation/understanding-bpr.

127. European Commission, Commission Delegated Regulation (EU) 2017/2100 of 4 September 2017 setting out scientific criteria for the determination of endocrine-disrupting properties pursuant to Regulation (EU) No 528/2012 of the European Parliament and Council. 2018. http://data.europa.eu/eli/reg_del/2017/2100/oj.

128. Regulation (EU) 2017/745 of the European Parliament and of the Council of 5 April 2017 on medical devices, amending Directive 2001/83/EC, Regulation (EC) No 178/2002 and Regulation (EC) No 1223/2009 and repealing Council Directives 90/385/EEC and 93/42/EEC (Text with EEA relevance). http://data.europa.eu/eli/reg/2017/745/2020-04-24.

129. Regulation (EU) 2017/746 of the European Parliament and of the Council of 5 April 2017 On in vitro diagnostic medical devices and repealing directive 98/79/EC and commission decision 2010/227/EU (text with EEA relevance). ELI: http://data.europa.eu/eli/reg/2017/746/2022-01-28.

130. Commission Regulation (EU) No 10/2011 of 14 January 2011 on plastic materials and articles intended to come into contact with food). Bundesinstitut für Risikobewertung (BfR), Database BfR Recommendations on Food Contact Materials. https://bfr.ble.de/kse/faces/DBEmpfehlung_en.jsp.

131. UN Environment Programme. Strategic approach to international chemicals management. Endocrine-disrupting chemicals. 2020. http://www.saicm.org/Implementation/EmergingPolicyIssues/EndocrineDisruptingChemicals/tabid/5476/language/en-US/Default.aspx. Accessed 31 Mar 2020.

132. UN Environment Programme. Scientific knowledge of endocrine disrupting chemicals. 2019. https://www.unenvironment.org/explore-topics/chemicals-waste/what-we-do/emerging-issues/scientific-knowledge-endocrine-disrupting. Accessed 30 Mar 2020.

133. Schmidt CW. TSCA 2.0: a new era in chemical risk management. Environ Health Perspect. 2016;124:A182–6.

134. Trasande L, Massey RI, DiGangi J, Geiser K, Olanipekun AI, Gallagher L. How developing nations can protect children from hazardous chemical exposures while sustaining economic growth. Health Aff (Millwood). 2011;30:2400–9.

135. Bellanger M, Demeneix B, Grandjean P, Zoeller RT, Trasande L. Neurobehavioral deficits, diseases, and associated costs of exposure to endocrine-disrupting chemicals in the European Union. J Clin Endocrinol Metab. 2015;100:1256–66.
136. Hauser R, Skakkebaek NE, Hass U, et al. Male reproductive disorders, diseases, and costs of exposure to endocrine-disrupting chemicals in the European Union. J Clin Endocrinol Metab. 2015;100:1267–77.
137. Legler J, Fletcher T, Govarts E, et al. Obesity, diabetes, and associated costs of exposure to endocrine-disrupting chemicals in the European Union. J Clin Endocrinol Metab. 2015;100:1278–88.
138. Attina TM, Malits J, Naidu M, Trasande L. Racial/ethnic disparities in disease burden and costs related to exposure to endocrine-disrupting chemicals in the United States: an exploratory analysis. J Clin Epidemiol. 2019;108:34–43.
139. Forouzanfar MH, Alexander L, Anderson HR, et al. Global, regional, and national comparative risk assessment of 79 behavioural, environmental and occupational, and metabolic risks or clusters of risks in 188 countries, 1990–2013: a systematic analysis for the global burden of disease study 2013. Lancet. 2015;386:2287–323.
140. Grosse SD, Teutsch SM, Haddix AC. Lessons from cost-effectiveness research for United States public health policy. Annu Rev Public Health. 2007;28:365–91.
141. Salkever DS. Assessing the IQ-earnings link in environmental lead impacts on children: have hazard effects been overstated? Environ Res. 2014;131:219–30.
142. Trasande L, Zoeller RT, Hass U, et al. Estimating burden and disease costs of exposure to endocrine-disrupting chemicals in the European Union. J Clin Endocrinol Metab. 2015;100:1245–55.
143. Attina TM, Hauser R, Sathyanarayana S, et al. Exposure to endocrine-disrupting chemicals in the USA: a population-based disease burden and cost analysis. Lancet Diabetes Endocrinol. 2016;4:996–1003.
144. Trasande L, Zoeller RT, Hass U, et al. Burden of disease and costs of exposure to endocrine disrupting chemicals in the European Union: an updated analysis. Andrology. 2016;4:565–72.
145. Diamanti-Kandarakis E, Bourguignon JP, Giudice LC, et al. Endocrine-disrupting chemicals: an Endocrine Society scientific statement. Endocr Rev. 2009;30:293–342.
146. Zoeller RT, Bergman A, Becher G, et al. A path forward in the debate over health impacts of endocrine disrupting chemicals. Environ Health. 2014;13:118.
147. Vandenberg LN, Colborn T, Hayes TB, et al. Regulatory decisions on endocrine disrupting chemicals should be based on the principles of endocrinology. Reprod Toxicol. 2013;38:1–15.
148. Shaffer RM, Sellers SP, Baker MG, et al. Improving and expanding estimates of the global burden of disease due to environmental health risk factors. Environ Health Perspect. 2019;127:105001.
149. Pumarega J, Gasull M, Lee DH, López T, Porta M. Number of persistent organic pollutants detected at high concentrations in blood samples of the United States population. PLoS One. 2016;11:e0160432.
150. Solecki R, Kortenkamp A, Bergman Å, et al. Scientific principles for the identification of endocrine-disrupting chemicals: a consensus statement. Arch Toxicol. 2017;91:1001–6.
151. Danish Centre on Endocrine Disrupters. Report on Interpretation of knowledge on endocrine disrupting substances (EDs) – what is the risk? 2019. http://www.cend.dk/files/ED_Risk_report-final-2019.pdf.
152. US Environmental Protection Agency. Endocrine Disruptor Screening and Testing Advisory Committee (EDSTAC) final report. Washington, DC: US Environmental Protection Agency; 2021.
153. Markey CM, Michaelson CL, Veson EC, Sonnenschein C, Soto AM. The mouse uterotrophic assay: a reevaluation of its validity in assessing the estrogenicity of bisphenol A. Environ Health Perspect. 2001;109:55–60.

154. Ceccatelli R, Faass O, Schlumpf M, Lichtensteiger W. Gene expression and estrogen sensitivity in rat uterus after developmental exposure to the polybrominated diphenylether PBDE 99 and PCB. Toxicology. 2006;220:104–16.
155. La Merrill MA, Vandenberg LN, Smith MT, et al. Consensus on the key characteristics of endocrine-disrupting chemicals as a basis for hazard identification. Nat Rev Endocrinol. 2020;16:45–57.
156. Schug TT, Abagyan R, Blumberg B, et al. Designing endocrine disruption out of the next generation of chemicals. Green Chem. 2013;15:181–98.

Chapter 11
Talking with Patients and the Public About Endocrine-Disrupting Chemicals

Linda C. Giudice

11.1 Scope of the Problem

Over the past 75 years there has been a remarkable rise in noncommunicable diseases globally [1] that threaten reproductive health and capacity directly and indirectly. Genetic mutations at the population level in this timeframe are unlikely, and environmental chemical exposures have been considered as potential contributors due to marked increases in industrial chemical production occurring in parallel with the noncommunicable disease trends [2]. A recent study reports about 350,000 chemicals and mixtures registered globally, although most are unregulated in their production, disposal, and recycling with wide differences across countries and regions [3]. Many of these chemicals are EDCs (i.e., chemicals or mixtures of chemicals that affect any stage of human development [4]). EDCs are part of environmental chemical contaminants that are present in all populations studied [5], and it has been said that "babies are born pre-polluted" [6]. Exposures occur by ingestion, inhalation, trans-dermally, and trans-placentally, with greatest vulnerability pre-conceptually, during in utero development, neonatally, and in the adolescent period [7, 8]. EDCs are distributed widely, contaminating our air, water, soil, and food, and there is now strong experimental and epidemiologic evidence that EDCs harm reproductive health and outcomes—specifically gametogenesis, embryogenesis, reproductive tract development, fertility, pregnancy outcomes, and child neurodevelopment, which may be preventable in some cases by adopting mitigating strategies [8–12].

L. C. Giudice (✉)
University of California, San Francisco, San Francisco, CA, USA
e-mail: Linda.Giudice@ucsf.edu

© The Author(s) 2023
R. Marci (ed.), *Environment Impact on Reproductive Health*,
https://doi.org/10.1007/978-3-031-36494-5_11

11.2   Empowering Healthcare Professionals to Talk with Patients and the Public About EDCs

Despite a voluminous amount of experimental evidence from animal models and in vitro cellular studies that provide the mechanistic underpinnings of these observations, as well as human epidemiologic and wildlife field data, EDC effects on reproduction are still not widely appreciated by the public and by healthcare professionals. In 2014, the American College of Obstetricians and Gynecologists (ACOG) published the 2012 survey of its ~50,000 members on the topic of prenatal environmental exposures [13]. Of the 2514 respondents, 78% said they believed they could reduce patient exposures to environmental health hazards through patient counseling. However, 50% reported rarely taking an environmental health history, <20% routinely inquired about exposures commonly found in pregnant women in the United States, and <7% had had any training on the topic. Barriers to counseling included lack of knowledge and uncertainty about the evidence, concerns that patients may not be able to reduce harmful exposures, and fears about provoking patient anxiety [13]. An important message from this survey was that physicians believed they could play a key role in preventing environmental exposures that harm patient health and reproductive outcomes, although they did not feel empowered to do so. Physician education, training, having reliable sources and following evidence-based guidelines, and having access to communication tools were identified as unmet needs that could enhance the impact of physicians talking with patients about environmental exposures and mitigating strategies.

In the intervening years (2012–present), there has been a plethora of educational opportunities for physicians and other healthcare professionals and researchers regarding EDCs and reproductive health. Systematic and narrative reviews, opinions, and white papers on this topic have been issued by leading global organizations (e.g., UNEP-WHO [14]), and professional societies and their committees, e.g., the Royal College of Obstetricians and Gyneacologists [15]; the American College of Obstetricians and Gynecologists [16, 17]; The Endocrine Society [9, 11]; the International Federation of Gynaecology and Obstetrics (FIGO) [7, 18]; and Project TENDR [19], who have acknowledged the need for educating their constituents with information passed to patients and the public at large. These organizations, which are reliable and respected sources, have convened experts and collaborated with patient advocate groups to examine the evidence and provide educational guidance in the form of webinars, scholarly works, systematic reviews, and calls to action about the effects of EDCs on reproductive health. Some are developing educational modules to train the trainers; for example, the Philippine Obstetrical and Gynecological Society (https://pogsinc.org/) Subcommittee on Reproductive and Developmental Environmental Health (RDEH) developed a "Training of Trainers on RDEH" post-graduate course for their Annual meeting in September 2021 that was attended by over 100 participants. Moreover, some institutions are developing courses for undergraduates and medical and graduate students about EDCs and other toxic environmental exposures, laying the ground for a more informed and empowered work force in the future.

11.3 Taking An Environmental Exposure History

In July 2021, ACOG published an update to its Committee Opinion of 2013 [16] that demonstrated remarkable advances in the field and tools for empowering clinicians. Therein, it underscored the importance for obstetrical healthcare providers to be knowledgeable regarding EDCs and presented approaches for exposure risk assessment and risk reduction strategies, as well as clinical counseling [17]. ACOG suggested key elements of an environmental exposure history, reported associations with reproductive health outcomes, and recommended counseling approaches for patients to minimize exposure to toxic environmental agents—most of which are EDCs. Elements in the environmental history included: lead exposures, home environment and lifestyle (e.g., cigarette smoking, flame retardants in foam furniture), cleaning products, pesticides, personal care products, diet, nutrition, produce, food preparation and storage containers, fast foods, and occupational exposures. While the Committee Opinion was focused on obstetric care clinicians, the recommendations and conclusions apply equally well to those contemplating pregnancy and, also, anyone undergoing fertility evaluation and treatment.

The University of California San Francisco (UCSF) Program on Reproductive Health and the Environment (PRHE) hosts an extensive website for clinicians, researchers, patients, and families. Specifically relevant to clinicians are the links to environmental history forms at http://prhe.ucsf.edu/prhe/clinical_resources.html, replete with references and links to additional clinically relevant information. Also key in environmental history taking are several items that complement the ACOG recommendations:

1. Assessing patient risk including toxicant, dose, frequency, duration and route of exposure, timing to vulnerable developmental windows, and comorbidities.
2. Inquiring about hazardous occupations and hobbies.
3. Underscoring behaviors that can reveal possible mitigation strategies during history taking that can help minimize exposures.

Thus, environmental exposure history assessment has expanded the practitioner's access to risk assessment for individual patients, and some practices have incorporated "smart sets" in electronic medical record as checklists and for efficiency of information gathering.

11.4   Patient (and Clinician) Resources

Several groups have also put together valuable resources for clinicians to provide to patients and their families or to be accessed directly by the public. Some examples are [8]:

- Patient- and family-friendly information is available on the UCSF PRHE website as down loadable brochures ("Toxic Matters," "Food Matters," "Work Matters," "Pesticides Matter"), in English and Spanish (http://prhe.ucsf.edu/prhe/families. html) (Fig. 11.1a). These can be printed and distributed in waiting rooms, doctors' offices, or included in new patient information packages and are downloadable, free of charge, from the website.
- The Environmental Working Group and Skin Deep website has commentaries and updates on current environmental issues and consumer information (http:// www.ewg.org/ They host a searchable database of personal care products detailed and ranked by toxicity of ingredients (http://www.weg.org/skindeep/) that is available free of charge.
- SafetyNest is a personalized health education platform for prenatal care. It has an app that enables clinicians and pregnant women with a toolkit to reduce diseases and adverse events liked to toxic chemical exposures (http://www. nysafetynest.org/).
- The International Federation of Gynaecology and Obstetrics (FIGO) is a global professional organization that has collaborated with UCSF PRHE and nonprofit organizations, e.g., Health and the Environment (HEAL), the National Resources Defense Council (NRDC) (https://www.nrdc.org/), and the FREIA Project (http://freiaproject.eu/wp/subscribe/) to produce scholarly, evidence-based

Fig. 11.1 Patient educational materials. (**a**) Patient brochures about environment and reproductive health freely downloadable from https://prhe.ucsf.edu. (**b**) Patient infographic on toxic chemicals and pregnancy available from https://www.figo.org

information about EDCs and mostly reproductive outcomes. Recently, FIGO, HEAL, and PRHE have produced patient infographics regarding EDCs and pregnancy in several languages (https://www.figo.org/reproductive-and-developmental-environmental-health-committee). An example is shown in Fig. 11.1b that can be of value for practical steps patients and the public can take to minimize EDC risk.

11.5 Resources for Healthcare Professionals, Trainees, and Others

While the resources above are valuable for patients and the public, they are equally valuable for healthcare professionals to increase awareness among themselves and their colleagues, as well as for patients and the public. Some additional resources include:

- The Endocrine Disruptor Exchange (TEDX (http://www.endomcrinedisruption. org/). TEDX has a database of ~10,000 chemicals with endocrine disruption potential and also has an interactive timeline of critical windows of EDC exposures during development of several organ systems based on animal and human studies (http://www.endocrinedisruption.org/prenatal-origins-of-endocrine-disruption/critical-windows-of-development/timeline-test/). This is a valuable resource to clinicians, trainees and others to understand these critical periods of vulnerability to EDC effects.
- The Collaborative on Health and the Environment (CHE) is a U.S.-based nongovernmental organization whose mission is to engage scientific dialogue on environmental impacts on human health (http://www.healthandenvironment. org/), with the ultimate goal of disease and disability prevention. Their website contains linkages between chemical contaminants and ~ 200 human diseases and disorders that can be searched by disease or chemical (http://www. healthandenvironment.org/tddb/).
- The UCSF Program for Reproductive Health and the Environment (PRHE) hosts a website (http://ucsf.prhe.edu) replete with references of EDC exposures and effects on reproductive health, as well as updates on health policies mainly in the United States, but also around the globe.
- The Endocrine Society has published two major scientific statements about EDCs [9, 11] which give abundant, evidence-based information about the chemicals and their effects on all endocrine systems derived from experimental animal models and human epidemiologic studies. Moreover, the Society hosts an up-to-date website about EDCs and human health, of use to researchers, clinicians, and patients (https://www.hormone.org/your-health-and-hormones/endocrine-disrupting-chemicals-edcs).
- The International Federation of Gynaecology and Obstetrics (FIGO) through its Committee on Reproductive Health and the Environment (name changed in 2021

to the Committee on Climate Change and Toxic Environmental Exposures) periodically issues white papers and publishes in peer-reviewed manuscripts on the effects of EDCs and other environmental toxicants on reproductive health (http://www.FIGO.org). It issued a major treatise on EDCs and reproductive health in 2015 [7] and has a recent publication on climate change with citations about the interplay of climate change, air pollution, and EDCs [18] that is valuable for healthcare professions to know.

- The American College of Obstetricians and Gynecologists has just updated its Committee Opinion on "Reducing Prenatal Exposures to Environmental Chemicals" available at:

    https://www.acog.org/clinical/clinical-guidance/committee-opinion/articles/2021/07/reducing-prenatal-exposure-to-toxic-environmental-agents. This is a valuable resource for clinicians in practice and those in training.

11.6 Talking with Patients and the Public About EDCs and Risk Assessment and Mitigation

Healthcare providers have multiple options for discussing EDCs and reproductive health with their patients—in the office, providing resources to be accessed online or to be read in hard copy at home, talking at community gatherings, religious events, health fairs, schools, parent meetings, and other venues. Physician voices are respected and carry much weight, and the ability to prevent harm is huge, along with increased patient and public health safety [20]. Using guidelines and recommendations from professional organizations can facilitate healthcare professionals' knowledge base and comfort level in addressing these issues, which are so important to minimize risk.

11.7 How Healthcare Providers Can Advocate for Change

Some professional societies have assembled white papers, systematic reviews, and other documents to prepare healthcare professionals to advocate for laws that transcend individual, personal behavior in mitigating risks and commit governments to prioritize environmental impacts on health as part of their political and economic agendas. Healthcare providers and scientists have powerful voices in legislative bodies and forums to advocate for changes in health policies to protect the public health from EDCs and other environmental threats. Primers have been developed, e.g., by the Endocrine Society in collaboration with the IPEN, a leading global network of 700 nongovernmental organizations working in more than 100 developing countries and countries with economies in transition [21]. These efforts are essential in parallel to individual efforts to minimize harm for this and future generations.

11.8 What Will It Take for People to Pay Attention to EDCs and Human Health?

The question arises what it will take to get people's attention (individuals and governments) to prioritize environmental health. A recent commentary suggests that there are three key data elements to quantify the health burden of specific industrial chemicals [22]. These include the relationship between exposures and health outcomes, the prevalence of the exposures, and monetary cost per case of illness [22]. The analytical approaches and quality and accuracy of the data are key for all three areas. Education and dialogue are critical components of raising awareness of the issues and priority setting.

11.9 Summary

Physicians and other healthcare professionals are uniquely positioned to talk with patients about EDCs and reproductive health, and it is essential that patients and the public are aware of EDCs and their effects on reproductive health and children's health and ways to mitigate these impacts. Also, voices of clinicians and scientists can be powerful in advocating for environmental health policy changes at the local and national levels to protect the public health. Moreover, individuals and groups can also make a huge difference by letting leaders know what health issues are important to them, what the importance of reproductive environmental health is to this and future generations, and why it should be prioritized in local and national health agendas. By so doing, patients, the general public, and professional and political leaders can and should work together toward equitable and impactful solutions for this and future generations.

Acknowledgments The author acknowledges the efforts of individuals and leaders across the globe who have contributed to increasing knowledge and awareness about environmental effects on the reproductive health of this and future generations and strategies to mitigate them.

Financial Support and Sponsorship The Robert B. Jaffe, MD, Professorship in the Reproductive Sciences in the Department of Obstetrics, Gynecology and Reproductive Sciences at the University of California, San Francisco, enabled the author's effort for this contribution.

Conflicts of Interest The author is Chair of the FIGO Committee on Climate Change and Toxic Environmental Exposures, a voluntary position, and has no financial conflicts of interest related to this manuscript.

References

1. WHO. 2021. https://www.who.int/health-topics/noncommunicable-diseases#tab=tab_1. Accessed 17 Oct 2021.
2. Kumar M, Sarma DK, Shubham S, Kumawat M, Verma V, Prakash A, Tiwari R. Environmental endocrine-disrupting chemical exposure: role in non-communicable diseases. Front Public Health. 2020;8:553850. https://doi.org/10.3389/fpubh.2020.55385.
3. Wang Z, Walker GW, Muir DCG, Nagatani-Toshida K. Toward a global understanding of chemical pollution: a first comprehensive analysis of national and regional chemical inventories. Environ Sci Technol. 2020;54:2575–84.
4. Zoeller RT, Brown TR, Doan LL, Gore AC, Skakkebaek NE, Soto AM, Woodruff TJ, Vom Saal FS. Endocrine-disrupting chemicals and public health protection: a statement of principles from the Endocrine Society. Endocrinology. 2012;153:4097–110.
5. Woodruff TJ, Zota AR, Schwartz JM. Environmental chemicals in pregnant women in the United States: NHANES 2003–2004. Environ Health Perspect. 2011;119:878–85.
6. US DHHS. President's cancer panel 2008–2009 annual report 2010. 2010. https://deainfon-cinihgov/advisory/pcp/annualreports/pcp08-09rpt/pcp_report_08-09_508.pdf. Accessed 17 Oct 2021.
7. Di Renzo GC, Conry JA, Blake J, DeFrancesco MS, DeNicola N, Martin JN, McCue KA, Richmond D, Shah A, Sutton P, Woodruff TJ, van der Poel SZ, Giudice LC. International Federation of Gynecology and Obstetrics opinion on reproductive health impacts of exposure to toxic environmental chemicals. Int J Gynaecol Obstet. 2015;131:219–25.
8. Giudice LC, Zlatnik MG, Segal T, Woodruff TJ. Environmental factors and reproduction. In: Strauss JF, Barbieri R, Dokras A, Williams CJ, Williams Z, editors. Yen & Jaffe's Reproductive Endocrinology: Physiology, Pathophysiology, and Clinical Management. 9th ed. Amsterdam: Elsevier; 2021a.
9. Diamanti-Kandarakis E, Bourguignon JP, Giudice LC, Hauser R, Prins GS, Soto AM, Zoeller RT, Gore AC. Endocrine-disrupting chemicals: an Endocrine Society scientific statement. Endocr Rev. 2009;30:293–342.
10. Giudice LC. Environmental impact on reproductive health and risk mitigating strategies. Curr Opin Obstet Gynecol. 2021;33(4):343–9.
11. Gore AC, Chappell VA, Fenton SE, Flaws JA, Nadal A, Prins GS, Toppari J, Zoeller RT. EDC-2: the Endocrine Society's second scientific statement on endocrine-disrupting chemicals. Endocr Rev. 2015;36(6):E1–E150. https://doi.org/10.1210/er.2015-1010.
12. Segal TR, Giudice LC. Before the beginning: environmental exposures and reproductive and obstetrical outcomes. Fertil Steril. 2019;112:613–21.
13. Stotland NE, Sutton P, Trowbridge J, Atchley DS, Conry J, Trasande L, Gerbert B, Charlesworth A, Woodruff TJ. Counseling patients on preventing prenatal environmental exposures–a mixed-methods study of obstetricians. PLoS One. 2014;9(6):e98771. https://doi.org/10.1371/journal.pone.0098771.
14. Bergman A, Heindel JJ, Jobling S, et al. State of the science of endocrine disrupting chemicals 2012. An assessment of the state of the science of endocrine disruptors prepared by a group of experts for the United Nations Environment Programme and World Health Organization, Geneva. 2013. www.who.int/ceh/publications/endocrine/en/.
15. Royal College of Obstetricians and Gynaecologists. Guideline: chemical exposures during pregnancy (Scientific impact paper 37). 2013. https://rcog.org.uk/guidance/browse-all-guidance/scientific-impact-papers/chemical-exposures-during-pregnancy-dealing-with-potential-but-unproven-risks-to-child-health-scientific-impact-paper-no-37/. Accessed 17 Oct 2021.
16. ACOG. American College of Obstetricians and Gynecologists Committee on Health Care for Underserved Women, American Society for Reproductive Medicine Practice Committee, The University of California San Francisco & Program on Reproductive Health and the

Environment: Exposure to toxic environmental agents. Committee Opinion No. 575 Companion Piece, Washington, DC. 2013.

17. ACOG. Committee opinion, committee on obstetric practice, DeNicola N, Borders AE, Singla V, Woodruff TJ. No. 832, Washington DC. 2021.

18. Giudice LC, Llamas-Clark EF, DeNicola N, Pandipati S, Zlatnik MG, DCD D, Woodruff TJ, Conry JA, FIGO Committee on Climate Change and Toxic Environmental Exposures. Climate change, women's health, and the role of obstetricians and gynecologists in leadership. Int J Gynaecol Obstet. 2021b;155(3):345–56.

19. Bennett D, Bellinger DC, Birnbaum LS, et al. Project TENDR: targeting environmental neuro-developmental risks the TENDR consensus statement. Environ Health Perspect. 2016;124(7):A118–22.

20. Sweeney E. The role of healthcare professionals in environmental health and fertility decision-making. J Environ Occup Health policy. 2017;27:28–50.

21. Gore AC, Crews D, Doan LL, La Merrill M, Patisaul H, Zota A. Introduction to endocrine disrupting chemicals. A guide for public interest organizations and policy-makers. 2014. https://ipen.org/sites/default/files/documents/ipen-intro-edc-v1_9a-en-web.pdf. Accessed 17 Oct 2021.

22. Woodruff TJ. Making it real–the environmental burden of disease. What does it take to make people pay attention to the environment and health? J Clin Endocrinol Metab. 2015;100:1241–4.